普通高等教育“十一五”国家级规划教材

全国高等学校计算机教育研究会“十四五”规划教材

21世纪大学本科计算机专业系列教材

丛书主编 李晓明

Java程序设计

（第3版）

朱庆生 古 平 主编

刘 骥 葛 亮 杨瑞龙 编著

清华大学出版社

北 京

内容简介

本书充分融入作者在承担“Java程序设计”国家精品课程和国家精品资源共享课程建设中的成果，遵循“兴趣为先、任务驱动、学以致用”的教学理念，每章均采用“Why-What-How”的渐进风格编写，并有大量精选案例分析。

全书以最新版JDK 17为基础，内容分为4篇。第一篇为基础篇(第1～6章)，介绍Java的基础语法，包括基础类型与结构、数组、类与对象、内部类、Lambda表达式、Java高级特性(泛型、反射、注解)、Java系统类(集合、Stream)等。第二篇为提高篇(第7～10章)，从实用性的角度，重点介绍Java的异常处理、输入输出处理、多线程、图形用户界面设计等。第三篇为网络篇(第11～14章)，重点面向网络应用，介绍URL与Socket通信技术、Web编程技术、JDBC技术、Web站点构建等。第四篇为实例篇(第15章)，分别以Servlet＋JSP技术和Spring Boot开发框架为例，展示其在“学生信息管理系统”中的应用开发过程和技巧。

本书定位于Java语言的初学者，适合作为高等学校计算机类专业“Java程序设计”课程的教材，也可供Java自学人员、软件开发人员参考使用。

图书在版编目(CIP)数据

Java程序设计/朱庆生，古平主编；刘骥，葛亮，杨瑞龙编著．—3版．—北京：清华大学出版社，2023.7
(2024.12重印)
21世纪大学本科计算机专业系列教材
ISBN 978-7-302-63816-2

Ⅰ.①J… Ⅱ.①朱… ②古… ③刘… ④葛… ⑤杨… Ⅲ.①JAVA语言－程序设计－高等学校－教材
Ⅳ.①TP312.8

中国国家版本馆CIP数据核字(2023)第106011号

责任编辑：张瑞庆
封面设计：常雪影
责任校对：郝美丽
责任印制：刘海龙

出版发行：清华大学出版社
网　　址：https://www.tup.com.cn，https://www.wqxuetang.com
地　　址：北京清华大学学研大厦A座　　**邮　　编**：100084
社 总 机：010-83470000　　**邮　　购**：010-62786544
投稿与读者服务：010-62776969，c-service@tup.tsinghua.edu.cn
质量反馈：010-62772015，zhiliang@tup.tsinghua.edu.cn
课件下载：https://www.tup.com.cn，010-83470236
印 装 者：三河市铭诚印务有限公司
经　　销：全国新华书店
开　　本：185mm×260mm　　**印　　张**：25　　**字　　数**：644千字
版　　次：2011年5月第1版　2023年8月第3版　　**印　　次**：2024年12月第3次印刷
定　　价：72.80元

产品编号：099407-01

前言

FOREWORD

本书第 2 版于 2017 年由清华大学出版社出版，得到了广大高校师生和读者的好评，也收到了很多好的建议和意见，在此表示衷心的感谢。

习近平总书记在党的二十大报告中指出，“教育、科技、人才是全面建设社会主义现代化国家的基础性、战略性支撑”“深入实施科教兴国战略，全方位加强基础学科人才培养，部署国家关键领域急需高层次人才培养专项，加快卓越工程师培养，加快造就更多拔尖创新人才”。为响应党和国家号召，培养、造就一批创新能力强，适应经济、社会发展需要的创新技术人才，同时保持与 Java 知识体系的同步迭代更新，有必要对原教材进行改版，一方面对知识体系进一步扩展，纳入更多 Java 8～Java 17 中的新功能、新特性；另一方面增强案例的启发性和实用性。本次改版具体修改内容如下：

(1) 在基础篇与提高篇中，依据 Java 17 的最新功能及特性，对相关类及知识体系进行了全面更新，补充了一系列新的编程方式和程序框架。

例如，第 4 章新增了模块（module），使得代码组织更安全。第 5 章新增了接口 Comparator 和 Comparable，以及接口中的默认方法和静态方法等，方便了对象的比较排序和接口的扩展；Lambda 表达式和函数式接口作为 Java 8 提供的一种新特性，使得 Java 能进行“函数式编程”，从而写出更简洁、灵活的代码。第 6 章进一步丰富了集合框架，新增了接口 Queue 和 Stream，Stream 使用类似 SQL 语句查询数据的方式提供对 Java 集合运算和表达的高阶抽象，让程序员可以写出高效率、干净的代码。第 7 章结合实际工程项目场景将异常信息记录到日志文件中。第 8 章在传统流处理方法的基础上引入了新的 NIO 组件，支持面向缓冲区的、基于通道的 I/O 操作，从而实现更加高效的文件读写。第 9 章引入了守护线程，新增了基于 Timer、Callable 以及线程池的多种线程创建方式，优化了线程的管理调度和安全性。

(2) 在网络篇和实例篇中，在更新 JSP 与 Servlet 技术的同时，引入了新的 Web 开发框架 Spring Boot，支持快速 Web 项目构建。

例如，第 12 章更新 Servlet API 到最新版 4.0，使用注解方式指定 Servlet 与 URL 之间的映射关系。第 14 章引入了数据库连接池技术，提高应用程序访问数据库的效率。第 15 章通过两个案例展示了不同技术方案（Servlet＋JSP 技术和 Spring Boot 框架）下同一项目的实现过程差异，让读者体会到如何基于 Spring Boot、以优雅的“零配置”方式快速构建一个 Web 项目。

(3) 根据补充的知识点，对相关案例进行了同步更新，撰写思路仍然沿袭了本书一贯的风格，强调案例的层次性和实用性。

参与本书第3版修订的有朱庆生、古平、刘骥、葛亮、杨瑞龙等，感谢老师们的辛勤付出，也感谢广大读者在本次修订过程中给予的意见和建议，感谢清华大学出版社的大力支持。

由于Java技术涵盖面广并且发展迅速，作者水平有限，书中难免有不足之处，诚请广大读者批评指正，以便作者改进、完善。

作　者

2023年1月

目 录

CONTENTS

第一篇 基 础 篇

第二篇 提 高 篇

第三篇　网　络　篇

第四篇 实 例 篇

第一篇　基础篇

第1章 Java概述

Java是目前最流行的编程语言之一,它以强大的移植能力、多线程处理和网络处理能力成为研究人员和开发人员瞩目的焦点。本章将开启Java的学习之旅,除了解为什么要使用Java,以及如何使用Java外,读者还需要掌握以下内容:

(1) Java语言的特点。

(2) Java开发运行环境的配置。

(3) Java程序的种类。

(4) 简单Java程序的开发。

1.1 认识Java语言

Java语言是20世纪90年代由Sun Microsystems公司(简称Sun公司)开发的革命性的编程语言,被美国著名的专业杂志PC Magazine评为1995年十大优秀科技产品之一。之所以称Java为革命性的编程语言,是因为传统的软件往往与具体的实现环境有关,一旦环境有所变化,就需要对软件做一番改动,既费时又耗力;而用Java语言编写的软件能在执行码的层次上兼容,只要计算机提供了Java虚拟机环境,用Java编写的软件就能在其上运行。

Java技术的通用性、高效性、平台移植性和安全性,使之成为网络计算的理想技术。从笔记本电脑到数据中心,从游戏控制台到科学超级计算机,从手机到互联网,Java无处不在。它在各个重要的行业部门得到了广泛的应用,而且出现在各种各样的设备和网络中。

迄今为止,Java平台是全球最大的、最具活力的开发团队之一。凭借其卓越的通用性、高效性和移植性,Java对开发者具有不可估量的价值,使他们可以:①在一个平台上编写软件,然后即可在几乎所有其他平台上运行;②创建可在Web浏览器和Web服务器中运行的程序;③开发适用于在线论坛、存储、投票、HTML格式处理以及其他用途的服务器端应用程序;④将采用Java语言的应用程序或服务组合在一起,形成高度定制的应用程序或服务;⑤为移动电话、远程处理器、低成本的消费产品以及其他任何具有数字核心的设备编写强大而高效的应用程序。

Java已经成为最流行的编程语言之一。

1.1.1 Java 语言的特点

Java 语言是一种高级编程语言,具有简单、结构中立、面向对象、可移植、分布式、高性能、多线程、健壮、动态、安全等特点。

在用 Java 语言进行程序开发时,首先以纯文本的方式编写所有的 Java 源程序,并保存成以.java 为后缀名的文件;然后将这些源程序用 javac 编译器编译成以.class 为后缀名的字节代码文件;字节代码不是被本地处理器执行的代码,而是能够被 Java 虚拟机(Java Virtual Machine,JVM)执行的代码。最后用 Java 运行工具在 Java 虚拟机中执行 Java 应用程序。Java 程序的开发过程如图 1-1 所示。

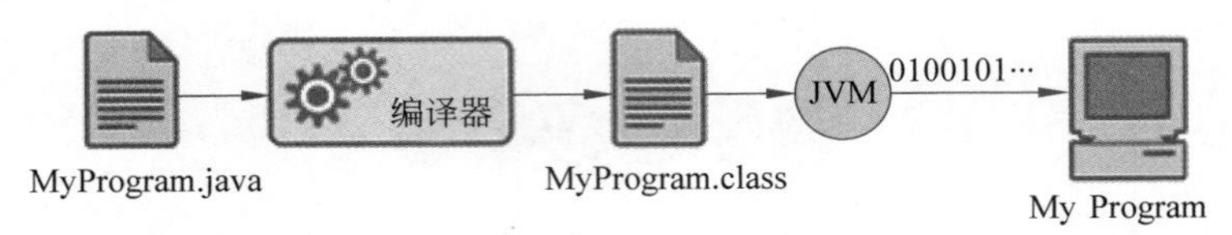

图 1-1 Java 程序的开发过程

由于 Java 虚拟机可以运行在不同的操作系统之上,因此同一个字节代码文件可以在 Windows、Solaris OS、Linux、macOS 等操作系统上运行,如图 1-2 所示。

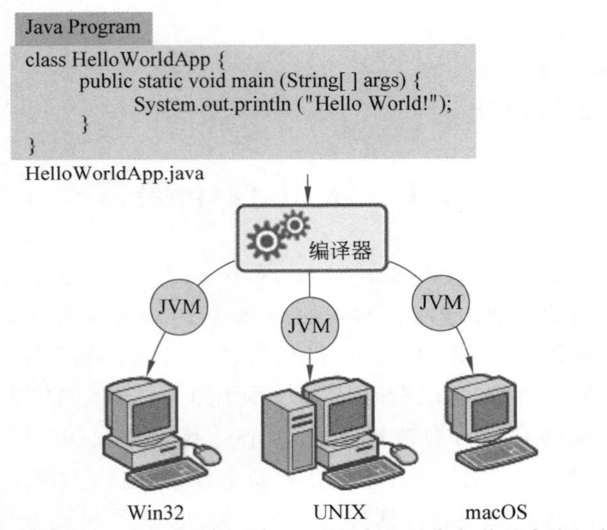

图 1-2 通过 Java 虚拟机,同一个 Java 程序可以在各种平台上运行

1.1.2 Java 平台的体系结构

Java 不仅是编程语言,还是一个强大的软件平台。Java 技术为程序员的软件开发提供了多种支持,主要包括:

(1) **各种开发工具**。如编译器、解释器、文档生成器和文件打包工具等。一位初级程序员主要使用的工具是编译器 javac、解释器 java、文档生成器 javadoc 和打包工具 jpackage。

(2) **应用程序编程接口**(Application Programming Interface,API)。应用程序编程接口包含了对软件常用功能的支持,涉及基本数据对象、用户界面、网络与安全、XML 生成、数据库访问等众多方面。这些 API 主要可以分为 3 类:①Java 核心 API,是由 Sun 制定的基本 API,任何 Java

平台都必须提供；②Java 标准扩展 API(javax)，是由 Sun 制定的扩充 API，Java 平台可以选择性地提供或加载；③厂商或者组织所提供的 API，由各家公司或组织所提供。

Java API 的制定过程是公开的，有许多业界技术领先的公司共同参与，所以相当完善而优异。与 Java 标准相关的资料都可以在 https://docs.oracle.com/en/java/javase/index.html 获得。

目前，Java 的体系结构已经变得相当庞大。Sun 公司在 1998 年将 Java 平台划分成 J2EE、J2SE、J2ME 共 3 个版本（现已改称为 Java EE、Java SE、Java ME），针对不同的市场目标和设备进行了定位。Java EE(Java Platform, Enterprise Edition)的主要目的是为企业计算提供一个应用服务器的运行和开发平台。Java EE 本身是一个开放的标准，任何软件厂商都可以推出自己的符合 Java EE 标准的产品，使用户可以有多种选择。IBM、Oracle、BEA、HP 等多家公司已经推出了自己的产品，例如 BEA 公司的 WebLogic、IBM 公司的 WebSphere 等。Java SE(Java Platform, Standard Edition)的主要目的是为台式机和工作站提供一个开发和运行平台。Java SE 是 Java 技术的基础，在学习 Java 的过程中，将首先学习 Java SE。Java ME(Java Platform, Micro Edition)主要面向电子消费产品，目的是为电子消费产品提供一个 Java 的运行平台，使得 Java 程序可以在手机、PDA、机顶盒等产品上运行。由于技术的发展 Java ME 已经很少使用，但近几年来占市场主流的智能手机操作系统——Android，使用 Java 作为开发语言。因此，Java 也是移动开发领域最常用的语言。

1.2 Java 运行环境与开发环境

Java 提供了一个免费的 Java 开发工具集(Java Development Kits,JDK)，编程人员可以利用这些工具来开发或者调试 Java 程序。通常以 JDK 的版本来定义 Java 的版本。JDK 1.0 版于 1996 年 1 月公开，从 1.0 版本到 1.6 版本，基本上每两年发布一个主要版本。2009 年，Sun 公司被 Oracle 公司收购，之后在 2011 年发布了 1.7 版本，2014 年发布了 1.8 版本，即 JDK 8。JDK 8 是现在企业中使用范围最广的一个版本，其更新会至少持续到 2030 年 12 月，当前 JDK 8 的最新版是 8u341。从 2018 年开始，每 6 个月就会发布一个 Java 版本，以便更快地引入新特性，其中 LTS 版是长期支持版，维护周期更长。目前的最新版是 JDK 18.0.2.1 和 JDK 17.0.4.1，其中 JDK 17 是 LTS 版，其更新会至少持续到 2024 年 9 月。

JDK 简单易学，是开发和运行 Java 程序的基本环境。但 JDK 不是一个集成开发环境，不方便进行复杂的 Java 软件开发，也不利于团体协作开发。目前主流的 Java 开发方法都是通过 Eclipse、NetBeans、JBuilder 等集成开发环境进行的。本书将采用 Eclipse 进行 Java 程序的开发和调试，有关 Eclipse 的安装和基本使用方法参见 1.2.3 节。

1.2.1 Java 运行环境

如果只想运行已有的 Java 程序，可以只安装 Java 运行环境(Java Runtime Environment, JRE)，JRE 由 Java 虚拟机、Java 的核心类以及一些支持文件组成。可以在网站 http://www.java.com/上免费下载 JRE(见图 1-3)，然后根据提示安装即可。

1.2.2 安装 JDK

为了开发 Java 应用程序，需要安装 JDK。在安装 JDK 的同时将安装 JRE。在网站 https://www.oracle.com/java/technologies/downloads/上免费下载 JDK。例如，根据提示可

图 1-3　下载 JRE

以将支持 64 位 Windows 操作系统的 JDK 安装程序 jdk-17_windows-x64_bin.exe 下载到本地硬盘，然后根据提示进行安装。成功安装后，JDK 目录下有 bin、lib 等子目录，如图 1-4 所示。其中 bin 目录存放了 javac、java 等命令程序，lib 目录存放了 Java 的类库文件。

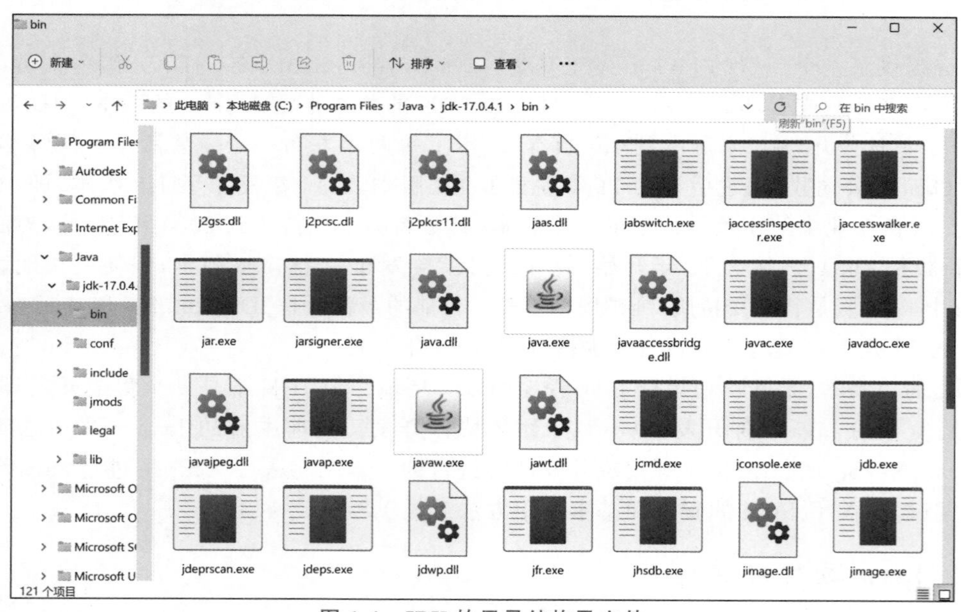

图 1-4　JDK 的目录结构及文件

1.2.3　安装 Eclipse 开发工具

Eclipse 是一款开放源代码、跨平台、基于 Java 的可扩展集成开发环境（IDE）。它最初由 IBM 公司开发，在 2001 年 11 月贡献给开源社区，现在它由非营利软件供应商联盟 Eclipse 基金会（Eclipse Foundation）管理。目前 Eclipse 已经发布了 4.24 版。时至今日，Eclipse 已经成为最主要的 Java 开发工具。但 Eclipse 的用途并不局限于 Java 领域。Eclipse 提供了一种称为插件(Plug-in)的扩展机制，允许软件开发人员构建与 Eclipse 环境无缝集成的插件。通过插件可以让 Eclipse 具有更多的功能。这使得 Eclipse 不仅可以作为 Java 的开发工具，还可以成为 C/C++、Ruby、Python 等语言的开发工具。

要使用 Eclipse 首先需要安装 JDK 环境(见 1.2.2 节)。然后从 Eclipse 官方网站(http://www.eclipse.org/downloads/packages/)选择 Eclipse IDE for Java Developers 和对应的操作系统进行下载，如图 1-5 所示。本书选择下载 64 位 Windows 版本，下载的文件为 eclipse-java-2022-06-R-win32-x86_64.zip。下载完成后，解压下载的文件到任意目录即可完成安装。

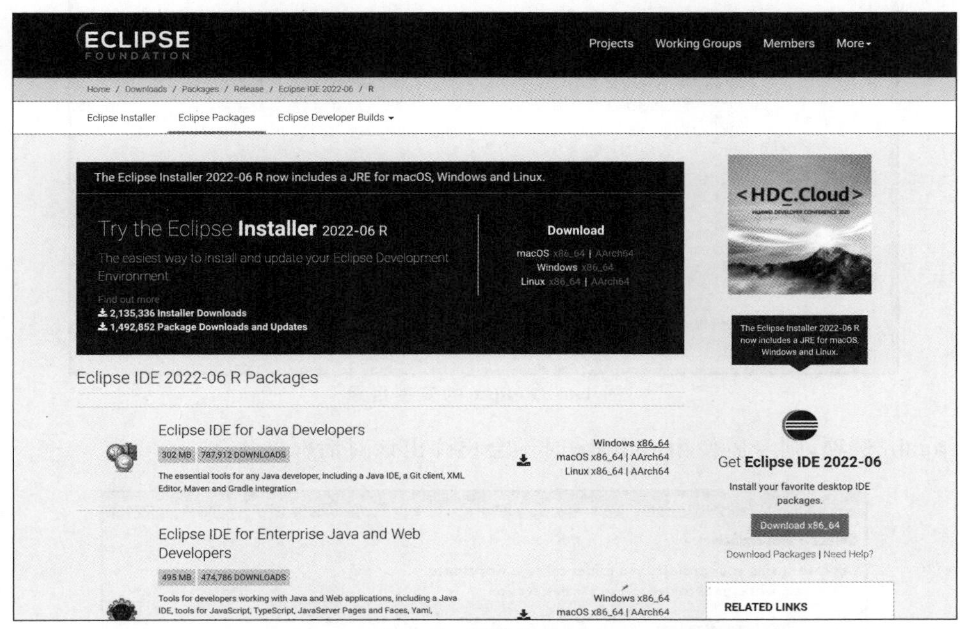

图 1-5　下载 Eclipse

1.3　Java 程序举例

常见的 Java 程序主要有两类：Application(应用程序)和 Servlet(服务器端小程序)。应用程序在计算机中单独运行，而 Servlet 是运行在服务器端的小程序，它可以处理客户端传来的请求(request)，然后将处理结果以响应(response)的方式传回给客户端。本节将介绍简单的 Java 应用程序，关于服务器端小程序的内容将在第 12 章中介绍。本节中的例子将在 Eclipse 中进行开发。

1.3.1　用 Eclipse 编写第一个 Java Application

【例 1.1】 用 Eclipse 编写 Java Application，要求在命令行窗口中显示“Hello，world!”。

1. 运行 Eclipse

首先在 Eclipse 的安装目录双击 eclipse.exe，以运行 Eclipse，如图 1-6 所示。

首次运行 Eclipse 会出现如图 1-7 所示的界面。该界面提示用户为 Eclipse 选择一个工作空间(workspace)。所谓工作空间，是 Eclipse 存放源代码的目录。本书选择 C:\workspace 作为工作空间，今后创建的 Java 源程序就存放在该目录。勾选 Use this as the default and do

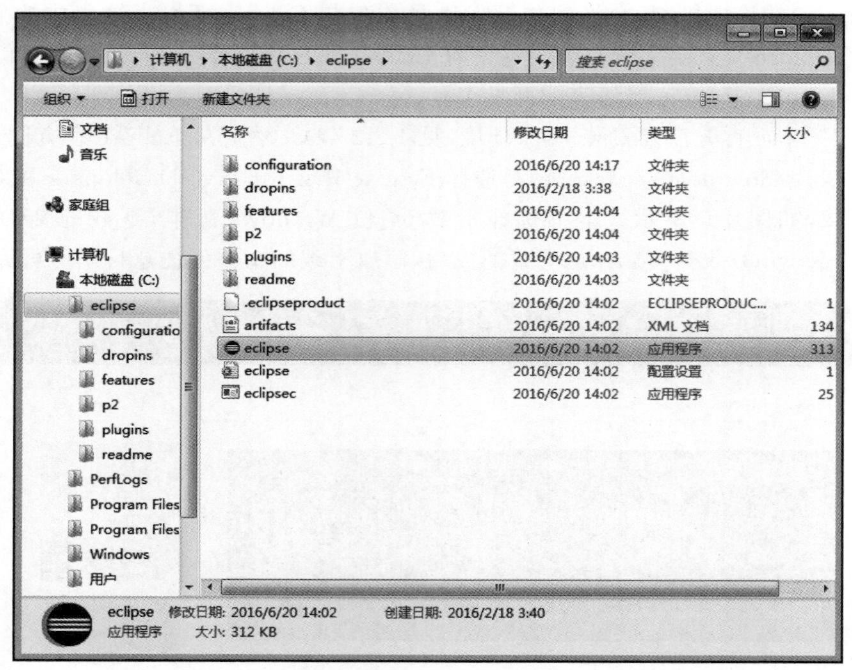

图 1-6　Eclipse 的安装目录

not ask again 选项,则今后使用 Eclipse 时不会再弹出该对话框。

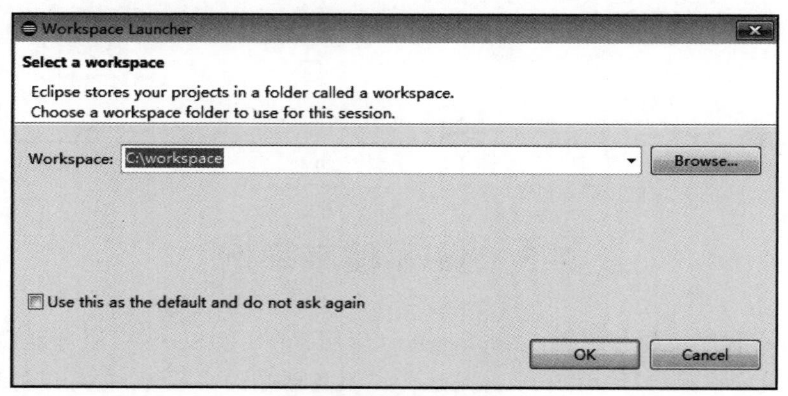

图 1-7　选择工作空间

Eclipse 运行后的界面如图 1-8 所示。界面左侧列出了当前工作空间里已有的 Java 工程。界面中部是 Java 源代码的编辑窗口。界面右侧是函数大纲,列出了当前编辑的 Java 代码的所有函数。界面下方是错误等信息提示。

2. 新建 Java 工程

若要在 Eclipse 中编写 Java 代码,必须首先新建一个 Java 工程(Java Project)。选择菜单项 File,然后选择 New 选项,最后选择 Java Project 选项,就会弹出新建 Java 工程的对话框,如图 1-9 所示。在新建 Java 工程对话框中输入 Java 工程的名称(如 code01),单击 Finish 按钮,就会在工作空间中新建一个 Java 工程,如图 1-10 所示。

3. 新建 Java 类

在 Eclipse 中选中 code01 下的 src 目录(src 目录用于存放 Java 程序的源代码),右击,选择

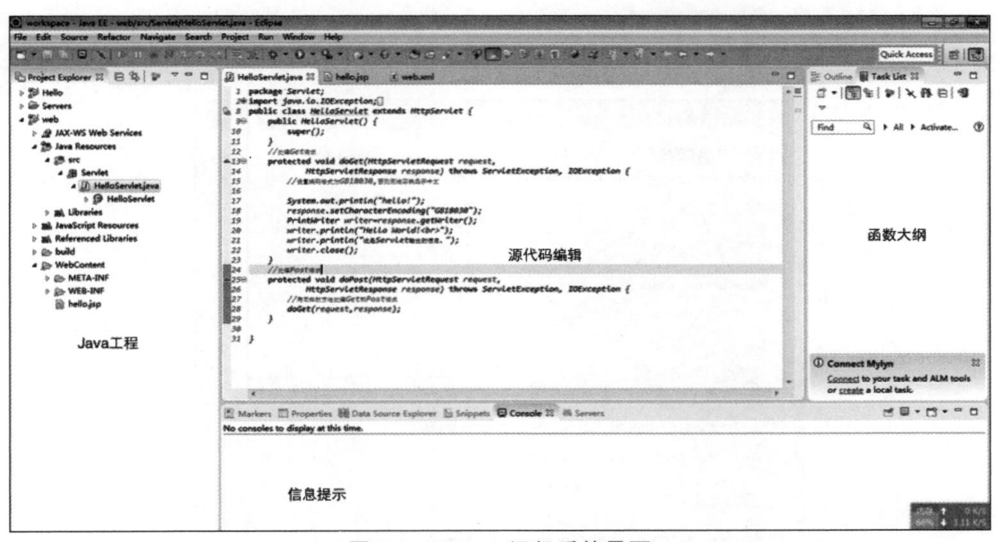

图 1-8 Eclipse 运行后的界面

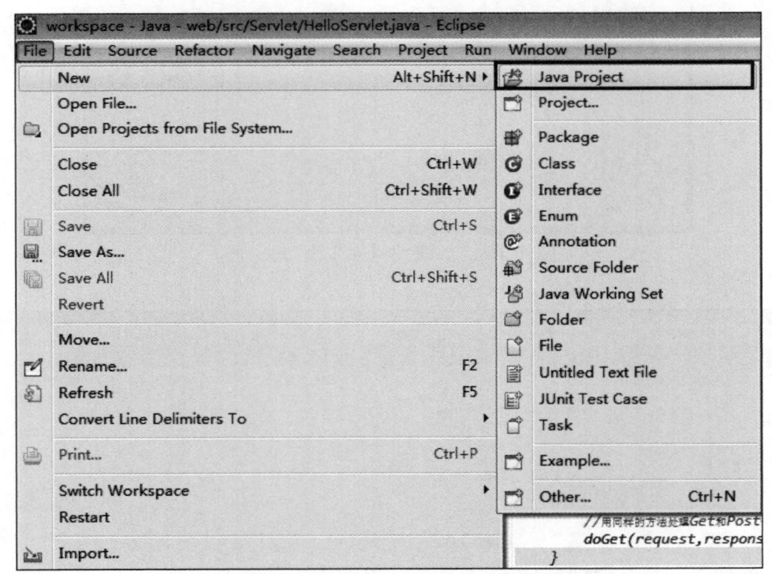

图 1-9 通过菜单新建 Java 工程

New,然后选择 Class,新建一个 Java 类(如图 1-11 所示),弹出如图 1-12 所示的对话框。在图 1-12 对话框中输入新建 Java 类的包名、类名等信息,单击 Finish 按钮就可以完成 Java 类的新建。

4. 运行 Java 程序

在新建的 HelloWorld 类中编辑如图 1-13 所示的代码。选中 HelloWorldApp 类,选择工具栏上的图标(如图 1-14 所示),选择 Run as,然后选择 Java Application,即可运行 HelloWorldApp 程序。程序运行的结果会显示在界面的下方,如图 1-15 所示。

分析:上例中的程序代码展示了 Java 应用程序的主要结构,主要包含 3 种组成:HelloWorldApp 类的定义、main()方法和注释。

(1) HelloWorldApp 类的定义:用 class 关键字来声明一个新的类,其类名为 HelloWorldApp,它是一个公共类(用 public 修饰)。整个类的定义由大括号“{ }”括起来。

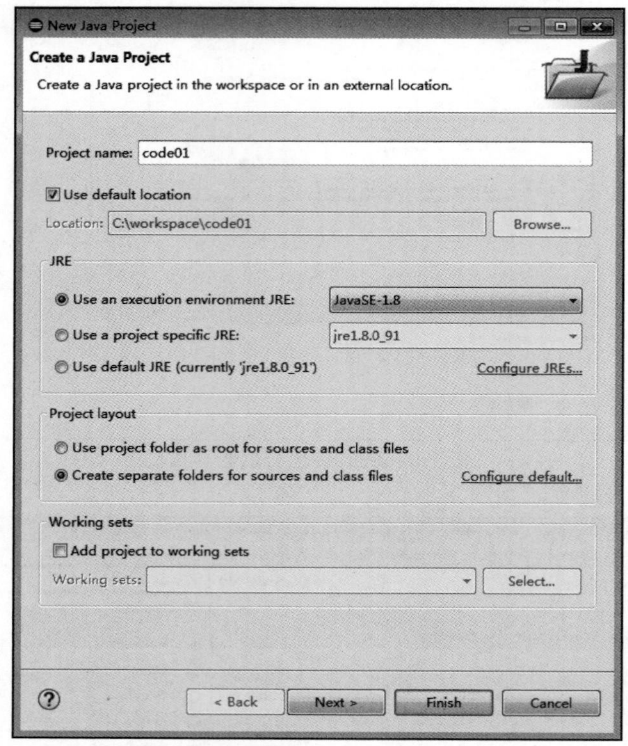

图 1-10　新建“Java 工程”对话框

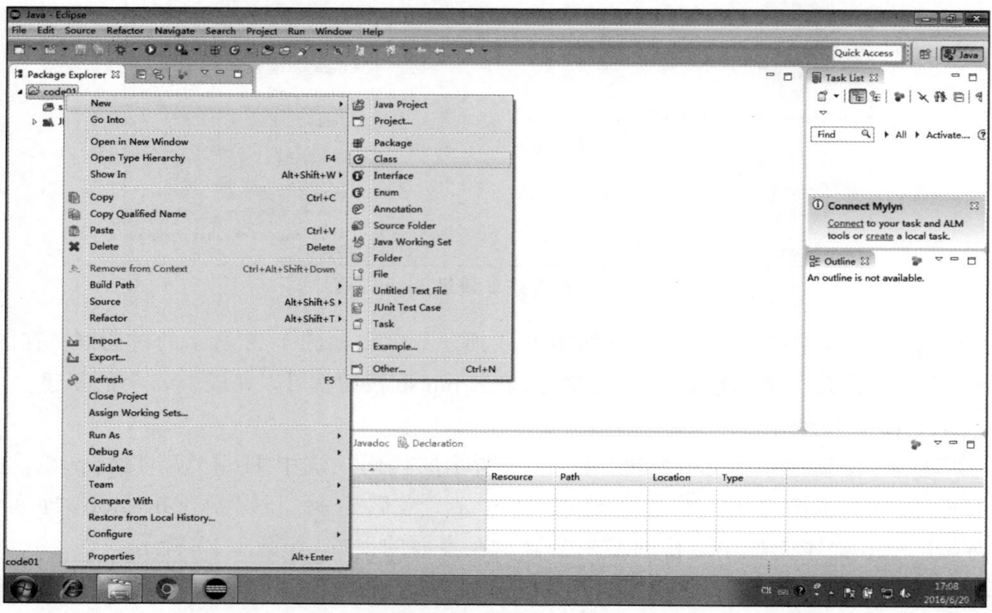

图 1-11　新建 Java 类

图 1-12 “新建 Java 类”对话框

图 1-13 编辑 HelloWorldApp

图 1-14 运行 Java 程序

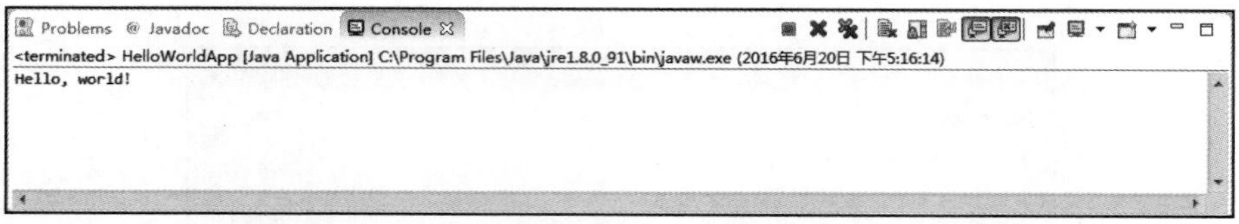

图 1-15　HelloWorldApp 的运行结果

(2) main()方法：该方法的声明形式是

```
public static void main(String[] args)
```

其中访问权限 public 指明所有的类都可以使用这一方法；static 指明该方法是一个类方法，它可以通过类名直接调用；void 则指明 main()方法不返回任何值；String[] args 是 main()方法接收的参数，其类型是字符串数组，它使得运行时系统可以向 Java 应用程序传递参数。Java 应用程序以 main()方法作为入口来执行程序，因此对于 Java 应用程序来说，main()方法是必需的。在 Java 程序中，可以定义多个类，每个类中可以定义多个方法，但是最多只有一个公共类，main()方法也只能有一个。在 main()方法的实现(大括号内)中，只有一条语句

```
System.out.println("Hello, world!");
```

它的作用是将字符串输出到标准输出(命令行窗口)中。

(3) //后的内容为注释。注释在程序编译时将被忽略掉，但它可以帮助程序员更好地理解程序代码。

提示：Java 源程序是区分大小写的。公共类必须放在与其同名的文件中。

1.3.2　用 Eclipse 编写带命令行参数的应用程序

【例 1.2】 用 Eclipse 编写一个带命令行参数的应用程序。

首先在 Eclipse 中编写如下所示的应用程序。

```
package code0103;
public class CommandLine {
    public static void main(String[] args) {
        System.out.println("打印命令行参数");
        for (int i = 0; i < args.length; i++) {
            System.out.println(args[i]);
        }
    }
}
```

运行上述程序的方法与普通应用程序略有区别。首先如图 1-16(a)所示选择 Run Configurations。在图 1-16(b)所示界面中选择 Java Application，右击，在菜单中选择 New，新建一个名为 CommandLine 的 Java 应用程序运行配置。修改 CommandLine 的配置，如图 1-16(c)所示。在“Arguments”选项卡中，选中 Program arguments 并在其中输入命令行参数。单击 Run 按钮就可以运行该程序，程序运行的结果如图 1-16(d)所示。

分析：main()方法的 args 参数就是命令行参数，它使得在运行程序时可以传递数据。命令行参数的类型是字符串数组(String[])。通过 Eclipse 运行时，需要配置运行参数，添加待传入的命令行参数值。

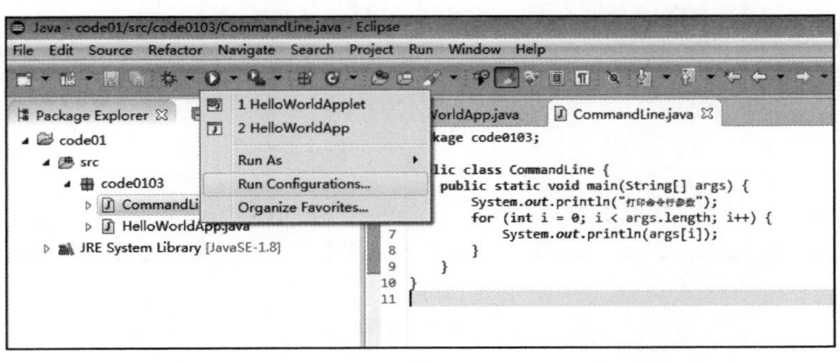

(a)

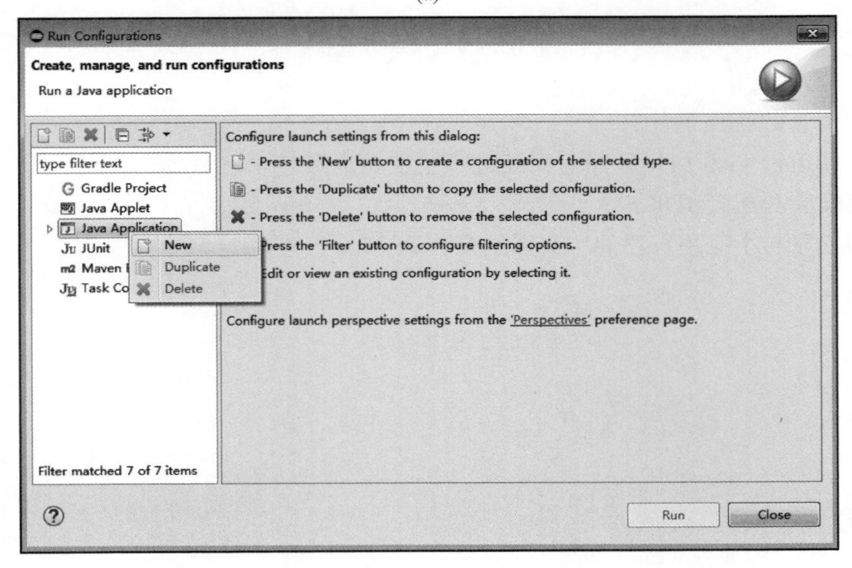

(b)

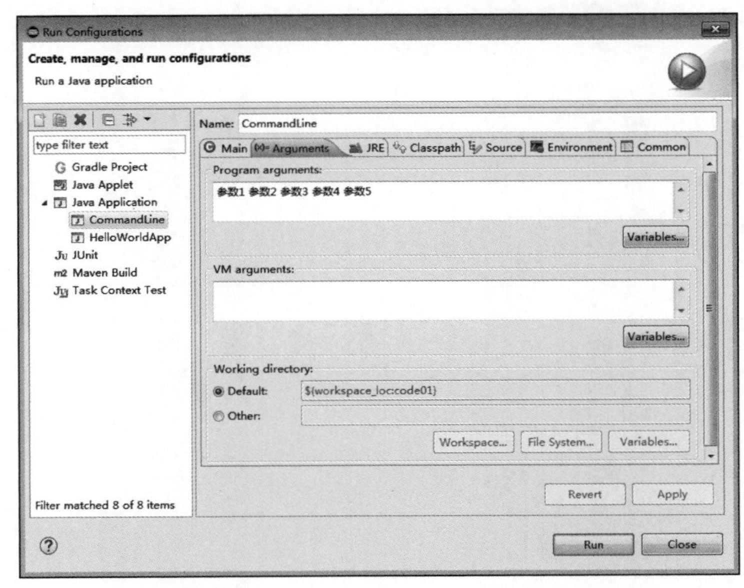

(c)

图 1-16　运行带命令行参数的应用程序

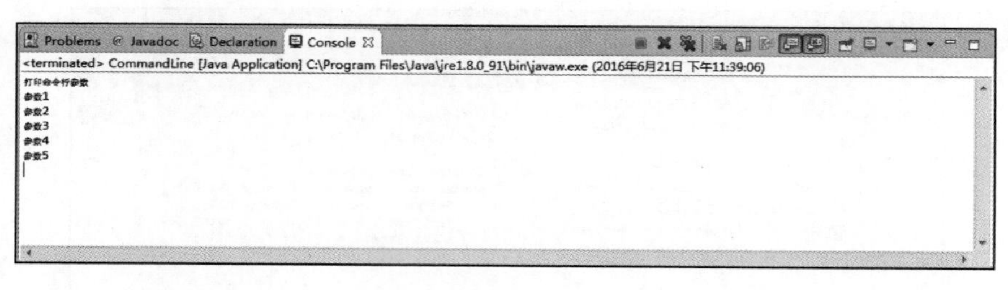

(d)

图 1-16 (续)

1. Java 语言的特点是什么?
2. 开发和运行 Java 应用程序需要经过哪些主要步骤?
3. 如何安装和配置 JDK?
4. 应用程序与小应用程序的主要区别是什么?

第 2 章 Java 程序设计基础

本章学习目标

一个 Java 程序是由一个或多个类组成的，类是由最基本的标识符、变量、运算符、语句和方法等组成的。本章将介绍 Java 编程中这些最基本元素的定义和使用，以及它们所构成的简单程序。通过本章学习，重点掌握以下内容：

(1) 变量的定义和使用。

(2) 基本数据类型和运算符的使用。

(3) 数据类型的转换。

(4) 程序流程控制语句的使用。

2.1 标识符和关键字

2.1.1 如何定义标识符

就像现实世界中的任何事物都有自己的名字，在程序中可以为各个元素进行命名，这种命名的记号就是标识符(Identifier)。在 Java 中正确地定义标识符需要遵守一些规定。

在 Java 中标识符一般是以字母、下画线(_)、美元符号($)等开始的一个字符序列，后面可以跟字母、下画线、美元符号、数字等字符，不能包含运算符和一些特殊字符，如#、*等。

Java 语言使用 Unicode 字符集，在虚拟机中用 16 位二进制表示一个字符，并且在 0～255 编码区与 ASCII 字符集是兼容的。为了使所有的字符都可以用 16 位二进制表示，Unicode 定义了 UTF-16 编码格式，Java 语言中的 char(字符)类型就是用它来表示的。

在 Java 中，标识符是大小写字母敏感的，没有最大长度的限制，不能和关键字相同。下面是合法的标识符：

```
Good,_exam,$fine
```

下面是非法标识符：

```
9Ten,book*,java#,super
```

非法标识符的原因：第一个标识符以数字开始；第二个标识符包含运算符*；第三个包含特殊符号#，第四个是关键字。

提示：也可以使用汉字作为标识符。例如，int 汉字＝5；其中，“汉字”就是一个合法的标识符。特别注意不要使用 ASCII 码中的特殊字符和空格。

2.1.2 关键字

关键字,又称保留字,是具有特殊含义的字符序列。Java 不允许对关键字赋予别的含义。所有的关键字都是小写字母,如果被大写就不是关键字了。

(1) 定义数据类型的关键字:

```
byte  short  int  long  float  double  char  boolean  var
```

(2) 流程控制的关键字:

```
if  else  switch  case  default  do  while  for  break  continue  yield
```

(3) 方法、类型、变量的修饰关键字:

```
private  public  protected  final  static  abstract  synchronized  volatile
native  strictfp
```

(4) 异常处理关键字:

```
try  catch  finally  throw  throws  assert
```

(5) 对象相关关键字:

```
new  extends  implements  class  instanceof  this  super  enum
```

(6) 字面常量关键字:

```
false  true  null
```

(7) 方法相关关键字:

```
return  void
```

(8) 包、模块相关关键字:

```
package  import  module  exports  open  requires  uses  provides
```

提示:与 C 语言不同,Java 语言中无 sizeof 关键字,所有基本数据类型的长度是确定的,不依赖执行环境,所以不需要此关键字。

2.1.3 注释

在源程序文件中添加注释可以增加程序的可读性,编译器不会对注释的内容进行编译。Java 允许有以下 3 种形式的注释。

(1) 单行注释:以//开头,至该行结尾,其格式为

```
//注释内容…
```

(2) 多行注释:以/ * 开始,遇到 * /结束,其格式为

```
/ * 注释文本
    …
* /
```

(3) 文档注释:以/**开头,遇到 * /结束,其他行以 * 开始,注释中可以使用以@开始的

标记(Tag),注明其后面文本的含义。例如,@author 后面的文本是"作者"。Javadoc 命令可以提取这种注释产生专门的 API 文档。JDK 的 API 文档就是用此种方式从其源代码产生的。其格式为

```
/**注释文本
 * @author 张三
 * @version 1.0…
 * /
```

2.2 常量和变量

2.2.1 变量

变量是 Java 程序中的基本存储单元,它的值可以修改。变量的定义包括类型、变量名和值 3 个部分。变量名必须是标识符,变量的类型决定了变量的数据性质、范围、存储在内存中所占的字节数以及可以进行的合法操作。其定义格式为

```
[修饰符] <类型名>  <变量名> [=<初值>][,<变量名>[=<初值>]…];
```

其中,<>表示一个占位符,需要替换成具体的类型名称和变量标识符。变量名的长度没有限制。[ ]中的内容是可选项。例如:

```
int i;
public int j=5, k=4;
```

第一个语句声明了 int 类型的变量 i,声明后,系统将给变量分配内存空间。

第二行语句,在同一行用逗号分隔,连续声明了两个 public int 型的变量 j 和 k,并且赋了初值。

变量的作用域指明了该变量能够被访问到的有效范围。声明一个变量的同时也就指明了它的作用域。按作用域划分,变量可分为局部变量、类成员变量和方法参数。

局部变量是在方法内部或代码块中声明的变量,它的作用域为它所在的代码块,一般以大括号"{}"为边界。

类成员变量,它的作用域是整个类;方法参数的作用域是方法体。

在一个作用域中,变量名应该是唯一的。在一个作用域中,如果有多个同名的变量可以访问,则按照"邻近"原则,在当前域中定义的变量隐藏其他同名的变量。

【例 2.1】 变量的使用。

```
package code0202;
public class HelloWorldApp {
    int j=5;                                        //此处 j 为类成员变量
    public static void main(String[] args) {        //args 为方法 main 的参数
        double j=10;                                //此处 j 为方法 main 的局部变量
        System.out.println("Hello, world!");
        System.out.println(j);                      //输出 double 类型的变量 j
    }                                               //double j 作用域结束
}                                                   //int  j 作用域结束
```

程序运行结果如下:

```
Hello, world!
10
```


分析：第一个变量 j 定义在所有方法(包括 main)之外，因此它是一个类的成员变量；在 main 方法中定义的第二个同名变量 j，其作用域是方法体的{ }之间，根据“邻近”原则，它将隐藏其他同名变量，因此输出的 j 值应该是局部变量的 j 值。

2.2.2 常量

在变量类型的前面加上修饰符 final 关键字，该变量被限定为常量。例如：

```
final int MAX=100;
```

定义了一个 int 型的常量 MAX，第一次赋了初值后就不能修改它的值了。一般地，Java 约定常量标识符全部用大写字母，如果有多个单词组成，则每个单词均为大写字母，并用下画线连接。例如：

```
final  int  MAX_COUNT=15;
```

思考：请将 Java 语言定义常量的方式和 C/C++ 语言进行比较。

2.3 基本数据类型

Java 语言是一种强类型的语言，要求每个变量、每个表达式都必须有确定的数据类型。Java 编译器对所有的表达式和参数都要进行类型相容性的检查，以保证类型是兼容的。任何类型的不匹配都是错误的，编译器完成编译以前，错误必须被改正。

Java 语言的数据类型可以分为两类：基本数据类型(Primitive Type)和引用类型(Reference Type)。详细的分类见图 2-1。

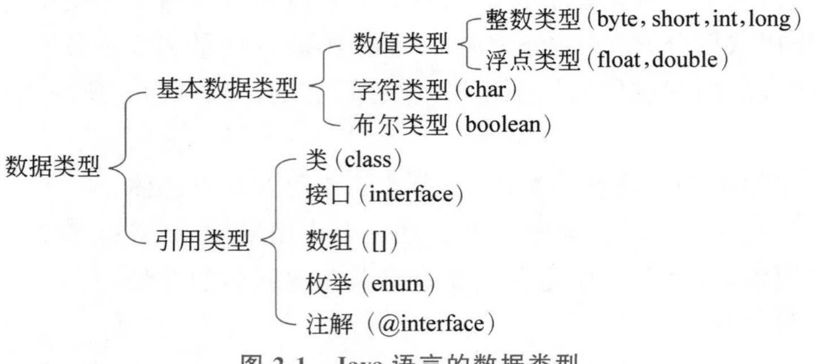

图 2-1 Java 语言的数据类型

Java 的基本数据类型都有固定的长度，不随运行平台的变化而变化；引用类型都是用类或对象实现的。

提示：Java 语言不支持 C/C++ 语言中的指针类型、结构类型(struct)、联合类型(union)。

2.3.1 布尔类型

布尔类型数据类型用关键字 boolean 表示，只有 true 和 false 两个值，它们不对应于任何整数值，经常在流程控制语句中使用。例如：

```
boolean b=false, t=true;
```

2.3.2 字符类型

1. 字符常量

字符常量是用单引号括起来的一个字符，例如'A'。用双引号括起来的是字符串，例如"chongqing university"。一个字符也可以用一个16位的Unicode码表示。

Java提供了转义字符，以反斜杠(\)开头，见表2-1。

表2-1 Java中的转义字符

转义字符	描述
\ddd	1～3位八进制数所表示的字符(ddd)
\uxxxx	1～4位十六进制数所表示的字符(xxxx)
\'	单引号字符
\"	双引号字符
\\	反斜杠
\r	回车
\n	换行
\f	走纸换页
\t	横向跳格
\b	退格

例如，'A'的Unicode码是\u0061，那么在Java程序中'\u0061'就表示字符'A'。'中'的Unicode码是\u4e2d，那么在Java程序中'\u4e2d'就表示字符'中'。

2. 字符变量

字符类型变量用char表示，在JVM中用16位Unicode码表示一个char值，范围为0～65535。字符型变量定义格式如下：

```
char c, c1='a';
```

其中定义了两个字符型变量c和c1，且c1的初始值为'a'。

Java语言的字符型数据不同于整数，但是可以和整数在一起运算，从字符型向整数型发生自动类型转换，从整数向字符型转换时需要强制类型转换。例如：

```
int i=22122;
char nine='9';
int j='a';                                  //由字符向整数，自动类型转换
char c=(char)(i+nine+j);
```

在第4条语句中，字符型和整数在一起运算，字符型变量首先转换为整型，这样i+one+j的运算结果为22276；再强制转换成字符型，字符变量c的值为中文字符'圄'。

2.3.3 整数类型

Java 语言中的整数有以下 4 种进制表示形式。

(1) 十进制：用 0～9 的数表示，首位不能为 0。例如，124，－100。

(2) 八进制：以 0 开头，后跟多个 0～7 的数字。例如，0134。

(3) 十六进制：以 0x 或 0X 开头，后跟多个 0～9 的数字或 A～F 字母的大小写形式。a～f 或 A～F 分别表示 10～15。例如，0x23FE 等于十进制数 9214。

(4) 二进制：以 0b 或 0B 开头，后跟多个 0～1 的数字，数字之间还可以使用下画线分隔。例如，int b=0b1010_0101_0101;等于十进制数 2645。

Java 语言定义了 4 种整数类型，见表 2-2。

表 2-2 整数类型

数据类型	所占位数	数的范围
byte	8	$-2^{7}\sim(2^{7}-1)$
short	16	$-2^{15}\sim(2^{15}-1)$
int	32	$-2^{31}\sim(2^{31}-1)$
long	64	$-2^{63}\sim(2^{63}-1)$

一个整数数字隐含为 int 型，在表示 long 型常量时需要在数字后面加上后缀 L 或 l。例如，3L 表示一个 long 型的常量，而不是 int 型常量。

整数变量定义如下：

```
byte   b=5;
int    i=300;
long   j=300;                    //把一个 int 型值赋给 long 型变量 j
long   j2=300L;                  //把一个 long 型常量赋值给 long 型变量 j2
int    i=4L;                     //错误,不能把 long 型值赋给 int 型变量
```

提示：在定义变量时，所赋值不能超过数据类型的表示范围。例如，变量 b 的值不能超过 127。

2.3.4 浮点类型

浮点数表示数学中的实数，即有整数和小数部分。浮点数有以下两种表现形式。

(1) 标准记数法：由整数部分、小数点和小数部分组成。例如，2.0、345.789。

(2) 科学记数法：由十进制数、小数点、小数和指数构成，指数部分由字母 E 或 e 跟上正负号的整数表示。例如，345.789 可以表示成 3.45789E+2。

Java 语言有两种浮点数类型(简称浮点类型或浮点型)：单精度浮点数(float)和双精度浮点数(double)，见表 2-3。

表 2-3 浮点类型

数 据 类 型	所 占 位 数	数的绝对值范围
float(单精度浮点数)	32	1.4E－45～3.4E＋38
double(双精度浮点数)	64	4.9E－324～1.8E＋308

一个浮点数默认为 double 型。在一个浮点数后加字母 F 或 f，则表示为 float 型。常量值 3.45 的类型是 double，3.45F 的类型是 float。

浮点数的定义如下：

```
double  d;                       //定义一个 double 型变量 d
double  d1=3.4;                  //可以在定义的时候赋予初始值
```

```
double   d2=3.4d;              //在定义 double 变量时，可以加后缀 D 或 d，也可以不加
float    f=3.4F;               //在定义 float 型变量时，需要在数值后面加 F 或 f
float    f1=3.4;               //常量值 3.4 为 double，编译时会发生类型不匹配的错误
```

2.3.5 各类型数据间的相互转换

各种数据类型的数据可以在一起进行混合运算。运算时，不同类型的数据先转换为相同类型的数据，再进行运算。数据类型之间的转换分为自动类型转换和强制类型转换。

1. 自动类型转换

自动类型转换是从低级到高级发生自动类型转换，也就是从表示范围小的类型向表示范围大的类型发生自动类型转换。不同数据类型的转换如下：

低————————————————————→高

byte，short，char ——→int ——→long ——→float ——→double

提示：byte、short 和 char 在一起运算时，首先转换为 int 类型再进行运算。

【例 2.2】 分析下面程序中的错误。

```
byte     b1=5;
short    s1=6;
short    s2 ;
s2=b1+s1;
```

分析：byte 类型和 short 类型的数据进行运算时首先都转换为 int 类型。b1＋s1 的结果类型为 int，在第 4 行中就会发生赋值类型不匹配的编译错误，int 类型的值不能赋给 short 类型的变量 s2。

2. 强制类型转换

由高级向低级转换数据类型时，需要强制类型转换。即在变量前面，把需要转换的“目标类型”放到圆括号“()”里面。例如：

```
int     i=65;
char    c;
c=(char)i;        //把 int 型变量转换成 char 型，需要强制类型转换
```

2.4 运 算 符

运算符负责对数据进行计算和处理。按运算符需要操作数据的个数，可将其分为一元运算符、二元运算符和三元运算符。例如，＋＋(自增运算)运算符为一元运算符，* 运算符为二元运算符，条件运算符为三元运算符。如果按操作数的类型分，又可将其分为算术运算符、赋值运算符、自增/自减运算符、条件运算符、位运算符、关系运算符、逻辑运算符等。通过运算符将各操作数连接成表达式，例如，a＋6、b＊7－9。

运算符一般由一个或多个符号构成，例如，＋、＞＝、＜＜＝。少数运算符有两种含义，应根据上下文理解，如－3 中的－是作为一元运算符(负号)使用。而 a－6 中的－是作为二元运算符(减号)使用。

运算符有优先级，()的优先级最高，＝的优先级最低，当一个表达式中有多个运算符时，先计算优先级较高的，再计算优先级较低的。运算符也有结合性。

2.4.1 算术运算符

算术运算符主要用于整型或浮点型数据的运算。算术运算符如表 2-4 所示。

表 2-4 算术运算符

	运算符	用法	含义	结合性
二元运算符	＋	op1＋op2	加法	左
	－	op1－op2	减法	左
	*	op1 * op2	乘法	左
	/	op1/op2	除法	左
	%	op1%op2	模运算(求余)	左
一元运算符	＋	＋op1	正数	右
	－	－op1	负数	右
	＋＋	＋＋op1,op1＋＋	自增	右,左
	－－	－－op1, op1－－	自减	右,左

1. 算术运算符的运算特点

(1) 对于二元运算符,运算结果的数据类型一般为两个操作数中表达范围较大的类型。例如,一个整数和浮点数运算的结果为浮点数。

(2) 对于一元运算符,运算结果的类型与操作数的类型相同。

(3) 自增、自减运算符有前缀和后缀两种形式,若是前缀形式(即＋＋、－－符号出现在变量的左侧)时,对变量实施的运算是“先运算后使用”; 若是后缀形式(即＋＋、－－符号出现在变量的右侧)时,对变量实施的运算是“先使用后运算”。

2. 算术运算符的注意事项

(1) 在 Java 语言中,"%"(求模运算符)的操作数可为浮点数。例如,52.3%10＝2.3。

(2) Java 语言对"＋"运算进行了扩展,可连接字符串。例如,"ab"＋"efd"得"abefd"。

(3) 做"＋"运算时,如果一个操作数是字符串,其他操作数自动转换成字符串。例如,String s; s＝"s: "＋6 * 5; //结果是 s＝"s:30"。

(4) byte、short、char 等类型进行混合运算时,会先自动转换为 int 类型再运算。

【例 2.3】 算术运算符举例。

```
package code0203;
public class TestOperator{
    public static void main(String[] args) {
        int  i=29;
        int  j=3;
        float  a=23.5f;
        double  b=4.0 ;
        System.out.println("i+a="+(i+a));        //整数与浮点数相加
        System.out.println("i * j="+(i * j));    //两个整数相乘
        System.out.println("i/j="+(i/j));        //对于整数,运算结果为整数
        System.out.println("i%j="+(i%j));        //求余数
        System.out.println("a * b="+(a * b));    //两个浮点数相乘
```

```
        System.out.println("a/b="+(a/b));      //对于浮点数,运算结果为浮点数
        System.out.println("a%b="+(a%b));      //浮点数求余,结果为浮点数
        System.out.println("i++="+(i++));      //先使用,后自增
        System.out.println("++i="+(++i));      //先自增,后使用
    }
}
```

程序运行结果：

```
i+a=52.5
i*j=87
i/j=9
i%j=2
a*b=94.0
a/b=5.875
a%b=3.5
i++=29
++i=31
```

分析：在打印输出语句中，＋除了可以进行整型或浮点数的加法运算外，还可以进行字符串的连接，可以把字符串和数值类型在一起进行运算，最后转换为字符串。

2.4.2 赋值运算符

赋值运算符是二元运算符，左边的操作数必须是变量，右边的操作数为表达式，左右两边的类型如果一致，则直接将右边的值赋给左边的变量；如果不一致，则将表达式的值转换为左边变量的类型，再赋值。与其他运算符相比，赋值运算符的优先级最低，且具有右结合特性。

1. 基本赋值运算符

赋值运算符的作用是使变量获得值，其基本使用格式如下：

```
<变量名>=<表达式>
```

其中，＝是赋值运算符，＜变量名＞获得计算出的表达式的值。例如：

```
int i,j;        i=10;        j=i+20;
```

2. 扩展赋值运算符

在赋值运算符“＝”前面加上其他运算符即构成扩展赋值运算符。例如：

```
a+=5;                        等价于 a=a+5;
a*=b+5;                      等价于 a=a*(b+5);因为赋值运算符的优先级最低
```

在Java语言中，大部分的运算符都可以加到＝的前面构成扩展赋值运算符，常见的扩展赋值运算符见表2-5。

表 2-5 扩展赋值运算符

运 算 符	示例(num 为变量)	含 义
+=	num += 3	num = num + 3
-=	num -= 3	num = num - 3
*=	num *= 3	num = num * 3

续表

运　算　符	示例(num 为变量)	含　义
/=	num /= 3	num = num / 3
%=	num %= 3	num = num % 3

3. 赋值相容

如果变量的类型与表达式的类型相同,就可以赋值,这称为类型相同;如果两者类型不同,并且变量类型比表达式类型长时,系统会自动将表达式的结果转换为较长的类型。例如,int 转换为 long,也可以赋值,这称为赋值相容(assignment compatible)。例如:

```
long  value=98L;                    //类型相同
long  value2=4;                     //int 向 long 自动转换,赋值相容
```

如果变量类型比表达式类型短,则赋值不兼容,编译时产生"可能存在的精度丢失"的错误。例如:

```
int  i=99L;                         //不能把 long 数据赋值给 int 型变量
```

赋值不兼容时,使用强制类型转换。例如:

```
int i=(int)123123123123L;         //强制类型转换
```

强制类型转换可能会发生精度丢失。

2.4.3 条件运算符

条件运算符由? 和:组成,其格式如下:

```
(boolean_expr) ?true_statement : false_statement;
```

其含义为:若 boolean_expr 为真,则执行语句 true_statement;若 boolean_expr 为假,则执行语句 false_statement。两条语句需要相同(相容)类型的计算结果。

例如:

```
int  result =  sum==0 ? 100 : 2*num;
```

如果 sum 等于 0,则 result 赋值为 100;否则赋值为 2 * num。

条件运算符可以替代简单的 if-else 语句。

2.4.4 位运算符

Java 语言提供位操作运算符。所有的数在 Java 虚拟机中都会转换为补码二进制表示。

例如:

```
整数 1 表示成补码二进制为 00000000 00000000 00000000 00000001(4 字节)
整数-1 表示成补码二进制为 11111111 11111111 11111111 11111111(4 字节)
```

位运算不是对整个数进行运算,而是对该数的二进制位上的 0 或 1 进行运算。位运算符见表 2-6。

表 2-6　位运算符

<table>
<tr><th>运算符</th><th>示　例</th><th>含　义</th></tr>
<tr><td>&</td><td>op1 & op2</td><td>使 op1 和 op2 按位相与</td></tr>
<tr><td>|</td><td>op1 | op2</td><td>使 op1 和 op2 按位相或</td></tr>
<tr><td>～</td><td>～op</td><td>对 op 按位取反</td></tr>
<tr><td>^</td><td>op1 ^ op2</td><td>使 op1 和 op2 按位异或</td></tr>
<tr><td>＜＜</td><td>op1 ＜＜ op2</td><td>使 op1 左移 op2 位，右补 0</td></tr>
<tr><td>＞＞</td><td>op1 ＞＞ op2</td><td>使 op1 右移 op2 位，左边补充符号位</td></tr>
<tr><td>＞＞＞</td><td>op1 ＞＞＞ op2</td><td>使 op1 无符号右移 op2 位，左边始终补添 0</td></tr>
</table>

位运算符的运算表见表 2-7。

表 2-7　位运算符的运算表

<table>
<tr><th>A</th><th>B</th><th>A&B</th><th>A|B</th><th>A^B</th><th>～A</th><th>～B</th></tr>
<tr><td>0</td><td>1</td><td>0</td><td>1</td><td>1</td><td>1</td><td>0</td></tr>
<tr><td>1</td><td>1</td><td>1</td><td>1</td><td>0</td><td>0</td><td>0</td></tr>
<tr><td>0</td><td>0</td><td>0</td><td>0</td><td>0</td><td>1</td><td>1</td></tr>
</table>

注意事项：

(1) 除 ～ 为右结合外，其余均为左结合。

(2) 操作数的类型一般为整型或字符型。

(3) 若两个数据的长度不同，例如 a&b，a 为 byte 型，b 为 int 型，系统首先会将 a 的左侧 24 位填满，若 a 为正则填满 0，若 a 为负则填满 1，即进行“符号扩充”。

下面通过示例详解位运算符。

(1) 按位与运算符 &。

例如，5&9，其结果等于 1。

&	十进制数	二进制数
操作数 1	5	00000000 00000000 00000000 00000101
操作数 2	9	00000000 00000000 00000000 00001001
运算结果	1	00000000 00000000 00000000 00000001

除了末位，其他位都清 0。

(2) 按位或运算符|。

例如，5|9，其结果等于 13。

(3) 按位异或运算符 ^。

例如，5^9，其结果等于 12。

(4) 按位取反运算符～。

例如，～5，其结果等于－6。

(5) 左移运算符＜＜。

例如，5＜＜2，其结果等于 20，左移一位相当于乘 2。

(6) 右移运算符>>。

例如,16>>2,其结果等于 4,右移一位相当于除以 2。

(7) 无符号右移运算符>>>。

在移动时,高位填 0,最低位被舍弃。

例如,−1>>>1;结果为 2147483647。

思考:

(1) 5>>32 的运算结果是多少? 答案是 5。

(2) 5>>34 的运算结果是多少? 答案是 1。

(3) 5L>>64 的运算结果是多少? 答案是 5L。

分析:在移位运算中,对于 int 类型,第二个操作数先对 32 取模,余数是实际移动的位数。对于 long 型,第二个操作数先对 64 取模,余数是实际移动的位数。

2.4.5 关系运算符

关系运算符用来比较两个值的大小,结果返回布尔值 true 或 false。关系运算符有 6 种,如表 2-8 所示。

表 2-8 关系运算符

运 算 符	示 例	含 义
==	op1 == op2	比较两个数据是否相等
!=	op1 != op2	比较两个数据是否不等
<	op1 < op2	比较一个数是否小于另一个数
>	op1 > op2	比较一个数是否大于另一个数
<=	op1 <= op2	比较一个数是否小于或等于另一个数
>=	op1 >= op2	比较一个数是否大于或等于另一个数

注意事项:

(1) 注意==不要写成=。

(2) 在 Java 中,任何类型的数据,都可以通过==或!=来比较是否相等。

(3) 关系运算的结果是 true 或 false,而不是 1 或 0。与数字没有对应关系。

(4) 关系运算符的优先级高于逻辑运算符的优先级。

2.4.6 逻辑运算符

逻辑运算,又称布尔运算符,它只能处理布尔类型的数据,所得结果也是布尔值。逻辑运算符主要有 3 种:逻辑与“&&”,逻辑或“||”、逻辑非“!”。逻辑运算规则见表 2-9。

表 2-9 逻辑运算规则

x	y	x&&y	x||y	!x	!y
true	true	true	true	false	false
true	false	false	true	false	true
false	false	false	false	true	true

逻辑运算符支持“短路运算”(short-circuit)。所谓短路运算，是指从左向右依次计算，如果在前面的计算中已经可以得出整个表达式的计算结果，则后面的剩余部分就不计算了。例如，int i=5，对于表达式(i>8)&&(i<4)的计算，当 i>8 时计算结果为 false，不论后面的计算结果如何，整个表达式的计算结果都为 false。后面剩余的部分就无须计算了。

2.5 表达式与计算的优先级

表达式是由运算符和操作数组成的符号序列，根据运算符的优先级和结合性执行计算，并得到某个值。

在表达式中使用的变量必须已经被初始化。表达式的运算结果类型就是表达式的类型。把一个表达式赋值给某个变量，需要进行类型检查，如果两边类型不同，则需要进行强制或自动类型转换。

表达式进行运算时，要按运算符的优先级从高到低进行，同级的运算符则按从左到右的方向进行。表 2-10 列出了 Java 运算符的优先级。

表 2-10 Java 运算符的优先级

<table>
<tr><td rowspan="15">高
↓
低</td><td>1</td><td>. [] () ++ -- ! ~ instanceof</td></tr>
<tr><td>2</td><td>new (类型)</td></tr>
<tr><td>3</td><td>* / %</td></tr>
<tr><td>4</td><td>+ -</td></tr>
<tr><td>5</td><td>>> >>> <<</td></tr>
<tr><td>6</td><td>< > <= >=</td></tr>
<tr><td>7</td><td>== !=</td></tr>
<tr><td>8</td><td>&</td></tr>
<tr><td>9</td><td>^</td></tr>
<tr><td>10</td><td>|</td></tr>
<tr><td>11</td><td>&&</td></tr>
<tr><td>12</td><td>||</td></tr>
<tr><td>13</td><td>? :</td></tr>
<tr><td>14</td><td>= += -= *= /= %= ∧=</td></tr>
<tr><td>15</td><td>&= != <<= >>= >>>=</td></tr>
</table>

通过对表 2-10 的分析，可以得出以下结论：

(1) 赋值运算符的优先级最低，因为赋值运算符要使用表达式的值。

(2) 关系运算符的优先级比逻辑运算符的优先级高。

(3) 一元运算符的优先级也比较高。

(4) 算术运算符比关系运算符和二元逻辑运算符的优先级高。

在表达式中，为了使表达式的计算过程和结构更清晰，可以用括号“()”标明运算次序，括

号中的表达式首先被计算。例如:

```
x<y&&x>10||y>0&x<50|y>50;
```

可以加上括号使表达式的结构更清晰:

```
((x<y)&&(x>10))||(((y>0)&(x<50))|(y>50));
```

2.6 程序控制语句

Java 程序控制语句分为 3 类:选择(分支)、循环和跳转。根据条件表达式的计算结果,选择语句可使程序选择不同的执行路径;循环语句使程序能够重复执行一个或多个语句;跳转语句允许程序以非线性的方式执行。

2.6.1 if 语句

if 语句是 Java 中的条件分支语句。它能将程序的执行路径分为两条。if 语句的完整格式如下:

```
if (condition)
    statement1;
else
    statement2;
```

其中,if 或 else 后面可以是单个语句(statement),也可以是程序块(block)。条件 condition 是任何返回布尔值的表达式,else 子句是可选的。

if 语句的执行过程:如果条件为真(true),就执行 if 后面的语句(statement1);否则,执行 else 后面的语句(statement2)。任何时候两条语句都不可能同时执行。

如果一个 if 语句中包含另外一个 if 语句,则为嵌套 if 语句。例如:

```
if(i == 10) {                       //第一个 if
  if (j < 20)                       //第二个 if
    a = b;
  if (k > 100)                      //第三个 if
    c = d;
  else a = c;                       //与上面最近的 if 对应,即第三个 if
} else                              //与第一个 if 对应
{
  a = d;
}
```

另外一种嵌套形式:if-else-if 阶梯,相当于 else 后面嵌套了 if-else 语句。其语法如下:

```
if(condition1)
    statement;
else if(condition2)
    statement;
else if(condition3)
    statement;
...
else
    statement;
```

条件表达式从上到下被求值。一旦找到为真的条件,就执行与它关联的语句,该阶梯的其他部分就被忽略了。如果所有的条件都不为真,则执行最后的else语句。

【例2.4】 通过使用if-else if阶梯来确定某月是什么季节。

```
package code0204;
public class IfElseDemo {
    public static void main(String args[]) {
        int month = 9;
        String season;
        if (month == 12 || month == 1 || month == 2)
            season = "冬季";
        else if (month == 3 || month == 4 || month == 5)
            season = "春季";
        else if (month == 6 || month == 7 || month == 8)
            season = "夏季";
        else if (month == 9 || month == 10 || month == 11)
            season = "秋季";
        else
            season = "错误的数据!";
        System.out.println(month+"月属于" + season + ".");    //用+连接不同类型的数据
    }
}
```

程序运行结果:

```
9月属于秋季.
```

2.6.2 switch语句与表达式

switch语句是Java的多路分支语句,根据表达式的值使程序执行不同语句序列。switch语句的通用形式如下:

```
switch (expression) {
case value1:
  ... //语句序列
  break;
case value2:
  ... //语句序列
  break;
case valueN:
  ... //语句序列
  break;
default:
  ... //缺省语句序列
}
```

表达式expression的计算结果必须为byte、short、int、char、String、enum类型,每个case语句后的值value必须是与expression类型兼容的常量。重复的case值是不允许的。

switch语句的执行过程:表达式的值首先与每个case语句中的常量进行比较。如果发现了一个与之相匹配的,则执行该case语句后的代码,不再匹配后面的case语句;如果没有一个case常量与表达式的值相匹配,则执行default语句。当然,default语句是可选的。如果没有相匹配的case语句,也没有default语句,则什么也不执行。

case 语句只是起到一个标号作用,用来查找匹配的入口,并从此处开始执行其后的语句序列。switch 语句的执行过程和 default 语句的位置没有关系,不会因为把 default 语句放在 switch 的开始处而执行 default 语句。

在 case 语句序列中的 break 语句将使程序执行流从整个 switch 语句退出,程序将从整个 switch 语句后的第一行代码开始继续执行。如果没有遇到 break 语句,则 switch 语句将一直执行到结束。

【例 2.5】 switch 语句使用(请与例 2.4 比较)。

```
package code0206;
public class SwitchBreakDemo {
    public static void main(String[] args) {
        int month = 6;
        String season;
        switch (month) {
        case 12:
        case 1:
        case 2:
            season = "冬季";
            break;
        case 3:
        case 4:
        case 5:
            season = "春季";
            break;
        case 6:
        case 7:
        case 8:
            season = "夏季";
            break;
        case 9:
        case 10:
        case 11:
            season = "秋季";
            break;
        default:
            season = "错误月份";
        }
        System.out.println(month+ "月属于" + season + ".");
    }
}
```

程序运行结果:

```
6月属于夏季.
```

分析:正如该程序所演示的那样,如果没有 break 语句,则程序将继续执行下面的每一个 case 语句,直到遇到 break 语句(或 switch 语句的末尾)。

switch 语句也可以嵌套。即将一个 switch 语句作为另一个 switch 语句序列的一部分。

switch 表达式是自 JDK14 引入的新功能,可以通过 yield 语句返回一个值,多个 case 语句可以缩写为一行,但是必须有 default 语句,使用运算符->代替了冒号,不需要使用 break 跳出 switch 表达式。下面重写例 2.5。

【例 2.6】 switch 表达式的使用(请与例 2.5 比较)。

```
package code0206;
public class SwitchExpressionDemo {
    public static void main(String[] args) {
        int month = 8;
        String season;
        season = switch (month) {
            case 12, 1, 2 -> {
                yield "冬季";
            }
            case 3, 4, 5 -> {
                yield "春季";
            }
            case 6, 7, 8 -> {
                yield "夏季";
            }
            case 9, 10, 11 -> {
                yield "秋季";
            }
            default -> {
                yield "错误月份";
            }
        };
        System.out.println(month + "月属于" + season + ".");
    }
}
```

程序运行结果:

```
8月属于夏季.
```

分析:多个 case 标号合并成了一行,用 yield 语句替代了 break 语句,并为 switch 表达式返回值,不会再执行其他 case 语句。变量 season 得到 switch 表达式的计算值。用运算符->连接一个语句、语句块或表达式。如果运算符->后面是一个表达式,则可以省略 yield。

2.6.3 while 与 do-while 语句

while 语句是 Java 最基本的循环语句。当它的条件表达式是 true 时,while 语句重复执行循环体,循环体可以是一个语句或者语句块。它的通用格式如下:

```
while (condition) {
    ... //循环体
}
```

【例 2.7】 使用 while 计算大于 100 且小于 200 的数的和。

```
package code0206;
public class SampleWhile
{
    public static void main(String[] args)
    {
        int sum = 0, i = 100;
        while (i < 200) {
```

```
            i++;
            sum += i;
        }
        System.out.println("the sum is " + sum);
    }
}
```

程序运行结果：

```
the sum is 15050
```

while 循环(或 Java 的其他任何循环)的循环体可以为空。例如：

```
int i=100, j=200;
while (++i < --j)
    ; //没有循环体
```

如果 while 循环一开始的条件表达式就是 false,那么循环体就根本不执行。有时需要先执行循环体,再判断条件表达式,可以使用 do-while 循环：它的循环体至少一次,因为它的条件表达式在循环的结尾。do-while 循环的通用格式如下：

```
do {
    ...  //循环体
} while (condition);
```

do-while 循环总是先执行循环体,然后再计算条件表达式。如果表达式为 true,则循环继续;否则,循环结束。

【例 2.8】 使用 do-while 计算大于 100 且小于 200 的数的和(重写例 2.7)。

```
package code0206;
public class SampleDowhile
{
    public static void main(String[] args)
    {
        int sum = 0, i = 100;
        do {
            i++;
            sum += i;
        } while (i < 200);
        System.out.println("the sum is " + sum);
    }
}
```

程序运行结果：

```
the sum is 15050
```

2.6.4 for 语句

for 循环是一个功能强大且形式灵活的结构。for 循环的通用格式如下：

```
for (initialization; condition; iteration) {
    ...  //循环体
}
```

for 循环的执行过程如下：

① 当循环启动时，先执行其初始化部分。仅被执行一次。通常，这是设置循环控制变量初值的一个表达式，作为控制循环的计数器。

② 计算条件 condition 的值。

③ 如果条件表达式为 true，则执行循环体；如果为 false，则循环终止。

④ 执行循环体的迭代(iteration)部分，这部分通常是增加或减少循环控制变量的一个表达式。该步结束后，转到②执行条件判断。

控制 for 循环的变量经常只是用于该循环，而不用在程序的其他地方。在这种情况下，可以在循环的初始化部分中声明局部变量。

【例 2.9】 使用 for 计算大于 100 且小于 200 的数的和(重写例 2.8)。

```
package code0206;
public class SampleFor
{
    public static void main(String[] args)
    {
        int sum = 0;
        for (int i = 101; i <= 200; i++) {
            sum += i;
        }
        System.out.println("the sum is " + sum);
    }
}
```

程序运行结果：

```
the sum is 15050
```

Java 允许两个或两个以上的变量控制 for 循环，可以在初始化部分和迭代部分声明和使用多个变量，每个变量之间用逗号分开。请看下面的程序：

```
int a, b;
for (a = 1, b = 4; a < b; a++, b--) {
    System.out.println("a = " + a);
    System.out.println("b = " + b);
}
```

for 循环的初始化和迭代部分可以为空。如果 for 循环的 3 个部分全为空，就是一个无限循环(死循环)。例如：

```
for(; ;) {
  ...  //循环体
}
```

这个循环将始终运行。在循环条件不容易直接确定时，可以使用死循环。在满足特定条件时，可以使用 break 语句来终止循环。

为了更方便地遍历数组或集合中的所有元素，Java 提供了 for-each 语句。下面通过例子来说明。

【例 2.10】 使用 for-each，求大于 100 且小于 200 的整数的和。

```
package code0206;
```

```
public class ForEachDemo {
    public static void main(String[] args) {
        int sum = 0;
        int a[] = new int[] {1,2,3,4,5,6,7,8,9};    //初始化数组,参见第 3 章
        //for-each 语句的使用
        for (int e : a)
            sum = sum + e;
        System.out.println("the sum is " + sum);
    }
}
```

该程序中,:表示 in 的意思,for(int e: a)就是 for each int e in a,即“对于数组 a 中的每个整数 e”。通过定义一个整数变量 e 表示数组中的每个元素。for-each 循环看上去比一般的 for 循环漂亮得多,不需要使用下标或者循环变量。

2.6.5 break 语句

break 语句有 3 种作用:①在 switch 语句中,break 语句用来终止一个语句块;②break 语句用来退出一个循环;③break 后面加语句标签。下面对后两种用法进行说明。

循环中遇到 break 语句时,循环被终止,执行循环后面的语句。

【例 2.11】 用 break 语句退出循环。

```
package code0206;
class BreakLoop {
  public static void main(String args[]) {
    for(int i=0;   ; i++) {                          //条件判断为空,表示死循环
      if(i == 5) break;                              //如果 i 等于 5,则终止循环
        System.out.println("i: " + i);
    }
    System.out.println("Loop complete.");
  }
}
```

程序运行结果:

```
i: 0
i: 1
i: 2
i: 3
i: 4
Loop complete.
```

分析:尽管 for 循环被设计为死循环,但是当 i 等于 5 时,break 语句终止了循环。

break 语句能用于任何 Java 循环中,包括 while、do-while 循环。在一系列嵌套循环中使用 break 语句时,它仅仅终止最里面的循环。

提示:关于 break 语句,一个循环中可以有一个以上的 break 语句;switch 语句中的 break 语句仅仅影响该 switch 语句,而不会影响其外部的任何循环;如果 switch 内嵌有循环语句,则循环语句内的 break 语句跳出循环,而不是终止 switch 语句。

break 语句加标签可以用于从嵌套很深的循环中跳出,其定义格式如下:

```
break label;
```

标签 label 是标识代码块的标签。label 可以是任何合法有效的 Java 标识符后跟一个冒号。当这种形式的 break 执行时,执行流程跳出指定的代码块。被加标签的代码块必须包围此 break 语句,但不需要直接包围 break。可以使用一个加标签的 break 语句跳出一系列的嵌套语句块。

例如,下面的程序示例了 3 个嵌套循环,每一个循环都有它自己的标签。break 语句跳出了标签 second 的代码块。

【例 2.12】 带标签的 break 的使用。

```
package code0206;
class BreakLabel
{
    public static void main(String args[])
    {
        boolean t = true;
        first: {
            second: {
                third: {
                    System.out.println("Before the break.");
                    if (t)
                        break second;
                    System.out.println("This won't execute");
                }
                System.out.println("This won't execute");
            }
            System.out.println("This is after second block.");
        }
    }
}
```

程序运行结果:

```
Before the break.
This is after second block.
```

2.6.6 continue 语句

有时需要提前结束本次循环,忽略这次循环剩余的循环体的语句,从而进行下一次循环。这时要用到 continue 语句。

在 while 和 do-while 循环中,continue 语句直接跳转到循环的条件表达式,继续循环过程。continue 后面的语句被忽略。

下例使用 continue 语句,使每行打印 5 个数字。

【例 2.13】 continue 的使用。

```
package code0206;
class SampleContinue
{
    public static void main(String args[])
    {
        for (int i = 1; i < 20; i++) {
            System.out.print(i + " ");
              if (i % 5 != 0)
```

```
                continue;
            System.out.println();
        }
    }
}
```

程序运行结果:

```
1 2 3 4 5
6 7 8 9 10
11 12 13 14 15
16 17 18 19
```

分析:该程序使用%运算符来检验变量 i 是否能被 5 整除。如果不能被 5 整除,则循环继续执行而不输出一个新行;如果能被 5 整除,则输出一个新行。

类似 break 语句,continue 语句也可以指定一个标签来继续包围它的循环。下面的示例运用 continue 语句打印三角形乘法口诀表的结果,请自行完成完整的口诀表。

【例 2.14】 带标签的 continue。

```
package code0206;
class ContinueLabel
{
    public static void main(String args[])
    {
        outer: for (int i = 1; i < 10; i++) {
            for (int j = 1; j < 10; j++) {
                if (j > i) {
                    System.out.println();
                    continue outer;
                }
                System.out.print(" " + (i * j));
            }
        }
        System.out.println();
    }
}
```

程序运行结果:

```
1
2 4
3 6 9
4 8 12 16
5 10 15 20 25
6 12 18 24 30 36
7 14 21 28 35 42 49
8 16 24 32 40 48 56 64
9 18 27 36 45 54 63 72 81
```

2.6.7 return 语句

return 语句用于从一个方法返回,或者返回一个值。使程序执行流程返回到调用此方法的地方。

下面示例中，Java 运行时系统调用 main()方法，因此 return 语句结束 main()方法，使程序执行返回到 Java 运行时系统。

【例 2.15】 带标签的 continue。

```
package code0206;
class ReturnTest
{
    public static void main(String args[])
    {
        boolean t = true;
        System.out.println("Before the return.");
        if (t)
            return;
        System.out.println("This won't execute.");
    }
}
```

程序运行结果：

```
Before the return.
```

分析：最后的 println 语句没有被执行。一旦 return 语句被执行，程序执行流程将转到它的调用者。

使用 return 语句返回一个值，其格式如下：

```
return 返回值;
```

如果 return 语句未出现在方法中，则执行方法的最后一条语句后将自动返回到主调方法。

2.6.8 实用案例：计算斐波那契（Fibonacci）数列

输出斐波那契数列的前 40 项。该数列的前两项都是 1，从第 3 项开始每一项都是前两项之和。

```
package code0206;
public class Fibonacci {
    public static void main(String[] args) {
        long f1,f2,f3,n=40;
        f1=f2=1;
        System.out.print(f1+" "+f2+" ");
        for(int i=3;i<=n;i++)          //i 从 3 开始,前面已经存在两项
        {
            f3=f1+f2;                  //第 3 项是前两项之和
            f1=f2;                     //更新前两项 f1
            f2=f3;                     //更新 f2
            System.out.print(f3+" ");
            if(i%10==0)                //输出 10 项后换行
            {
                System.out.println();
            }
        }
    }
}
```

程序运行结果:

```
1 1 2 3 5 8 13 21 34 55
89 144 233 377 610 987 1597 2584 4181 6765
10946 17711 28657 46368 75025 121393 196418 317811 514229 832040
1346269 2178309 3524578 5702887 9227465 14930352 24157817 39088169 63245986 102334155
```

2.7 实训任务

【任务描述】

百钱买百鸡:假定公鸡每只 5 元,母鸡每只 3 元,小鸡 1 元钱 3 只。现在有 100 元钱,要求买 100 只鸡,每种鸡至少买一只。请编程列出所有可能的购鸡方案。

【任务分析】

这是典型的约束满足问题,使用穷举法求解,即从可能的解空间(3 种鸡的个数)中找到满足要求(鸡的总数等于 100,总价等于 100)的解。基本的解法是使用循环语句遍历整个解空间,判断当前找到的解能否满足约束要求,如果满足则输出;然后继续穷举,直至搜索完毕所有的解。

假定 i 为公鸡数(i<=20),j 为母鸡数(j<=33),k 为小鸡数(k<=300),如果条件(5 * i+3 * j+k/3==100)和(i+j+k==100)同时成立,则意味着找到一个满足约束的解。继续遍历下一个解,可以将 i 或 j 加 1,或将 k 加 3(加 1 或加 2 可能出现无法除尽的情况)。

【任务解决】

完整示例代码如下:

```java
package code0207;
public class chick {
    public static void main(String[] args) {
        int count = 0;
        for (int i = 1; i <= 100/5; i++) {                          //公鸡数量
            for (int j = 1; j < 100/3; j++) {                       //母鸡数量
                for (int k = 0; k <= 100 * 3; k = k + 3) {          //小鸡数量
                    if ((5 * i + 3 * j + k / 3 == 100) && (i + j + k == 100)) {
                        System.out.println("公鸡:" + i + " 母鸡:" + j + " 小鸡:" + k);
                        count++;
                    }
                }
            }
        }
        System.out.println("共有" + count + "种买法");
    }
}
```

程序运行结果:

```
公鸡:4 母鸡:18 小鸡:78
公鸡:8 母鸡:11 小鸡:81
公鸡:12 母鸡:4 小鸡:84
共有 3 种买法
```

习题与思考

1. 试分析基本数据类型和引用数据类型的基本特点。

2. 请使用异或运算符^实现两个整数的交换。

3. 编写程序，显示如下螺旋方阵：

1	2	3	4
12	13	14	5
11	16	15	6
10	9	8	7

4. 下列哪个是合法的标识符？

A. a=b　　B. _Hello　　C. 2nd　　D. Chong qing

5. 下列哪些是合法的标识符？

A. new　　B. class　　C. int　　D. const2

6. 如果定义有变量 double d1，d2=4.0；则下列哪种说法是正确的？

A. 变量 d1 和 d2 均初始化为 4.0

B. 变量 d1 没有初始化，d2 初始化为 4.0

C. 变量 d1 和 d2 均未初始化

D. 变量 d2 没有初始化，d1 初始化为 4.0

7. 所有的变量在使用前都必须进行初始化，这种说法对否？

8. 数据类型 byte 的取值范围是多少？

A. 0～65 535　　B. －128～127

C. －32 768～32 767　　D. －256～255

9. 下列哪些是不能通过编译的语句？

A. int i = 32；　　B. float f = 45.0；

C. double d = 45.0；　　D. char a='c'；

10. 编写一个程序，求 1!＋2!＋…＋10!。

11. 编程验证哥德巴赫猜想：任何大于 6 的偶数可以表示为两素数之和。例如，10=3＋7。

第3章 数组

本章学习目标

数组是由类型相同的元素顺序组成的一个集合，每个数组都有一个唯一的名称，即数组名。通过数组名和下标可以方便地访问数组中的每个成员。本章主要讲述Java中一维和二维数组的创建和使用方法。通过本章的学习，应该重点掌握以下主要内容：

(1) 一维数组的定义和产生方法。

(2) 一维数组的初始化方法。

(3) 一维数组的引用方法。

(4) 二维数组的创建和引用方法。

3.1 数组使用初探

Java程序中通常使用变量来存放各种类型的数据，如将10个数分别存放在10个变量中，但这会非常麻烦，因为需要的变量名实在太多了。这时就可以考虑用一个数组变量来存放它们，并通过一个下标来访问存入数组中的每个成员。

【例3.1】 求10个学生某门课程的平均成绩。

分析：如果用变量s1，s2，…，s10来保存10个学生的成绩，显然不合适(变量太多且随学生人数递增)，而如果使用数组s来保存这些成绩，则问题可以处理如下：

```
package code0301;
public class AvgGrade {
    public static void main(String args[]) {
        int total = 0;
        int s[] = { 75, 69, 80, 85, 93, 97, 79, 77, 68, 90 };      //定义数组变量 s 并赋值
        double avg = 0;
        for (int i = 0; i < 10; i++) {
            total = total + s[i];                                   //访问数组成员 s[i]
        }
        avg = total / 10.0;                                         //求平均分 avg
        System.out.println("The average score is :" + avg);
    }
}
```

程序运行结果：

```
The average score is :81.3
```

该问题的求解涉及两个关键点：

(1) 定义数组变量 s,将所有值{75, 69, 80, 85, 93, 97, 79, 77, 68, 90}存入其中。

(2) 通过数组变量 s 加下标 i 的形式访问数组中每个成员：total = total + s[i]。

3.2 一维数组

与简单变量一样,数组必须先定义后使用。定义数组时,除了要给定数组的名称、数组成员类型,还要为其分配内存空间并进行初始化。

3.2.1 定义数组

一维数组的定义格式如下：

```
数据类型 数组名[ ];
```

其中,数据类型是指数组成员的类型,可以为基本数据类型,也可以为引用数据类型。[ ]为数组标记。在定义数组时,不允许在[ ]内指定数组元素的个数。例如：

```
int a[ ],  boolean temp[ ],  String s[ ];
```

表示数组 a 的数据类型为 int,temp 的数据类型为 boolean,s 的数据类型为类 String。

提示：定义数组时,另外一种可选形式为直接将方括号放在变量类型的后边。例如：

```
int[ ] a,  boolean[ ] temp,  String[ ] s
```

3.2.2 生成数组

数组定义只是建立了一种数组的引用,还必须使用关键字 new 为其分配内存空间,否则它是无法被访问的。基本形式如下：

```
数组变量名= new 数据类型[数组长度]
```

例如：

```
char s[];  s=new char[5];
```

的作用是为数组 s 分配 5×2B 大小的空间。也可以在定义数组的同时为之分配内存空间,例如：

```
int temp[]=new int[10];
```

3.2.3 初始化数组

当使用关键字 new 生成数组时,数组中的每个成员都会被自动初始化,初始值依据数组的类型而定。例如：

```
数值型的初始值:0;              字符型的初始值:'\0';
布尔型的初始值:false;          类对象的初始值:null;
```

也可以通过主动赋值的方式对数组进行初始化。例如

```
int s[ ]=new int[3];        s[0]=1; s[1]=2; s[2]=3
```

还可以将数组的定义、内存空间分配、初始化工作放在一条语句中,格式如下:

```
数据类型 数组名[ ]={值 1,…,值 n}
```

这种数组初始化形式的特点是无须给出数组的长度,系统会根据初始值的数量自动确定数组的长度,并依次将每个数组成员放在数组中。例如:

```
int s[ ]={1,2,3};
```

使得数组长度为 3,且 s[0]=1;s[1]=2;s[2]=3。

数组不能整体赋值,例如,下列语句是错误的:

```
char s[]=new char[5];        s[]={'a','b','c','d','e'};    //错误
```

提示:必须给出数组长度,而且数组一旦创建,就不允许再增加它的空间。

3.2.4 访问数组

数组成员的访问是通过数组名和下标来实现的,Java 中的下标是从 0 开始的,依次递增 1,如果数组 temp 的长度为 10,则数组下标为 0~9。这些成员分别表示为

```
temp[0],temp[1],…,temp[9]
```

对数组中任意成员的引用形式为:数组名[下标]。

例如,temp[4]表示引用数组 temp 的第 5 个元素。数组元素的用法和平常使用的普通变量没有什么差异,也可以被赋值或参与其类型允许的各种运算等。例如:

```
int i=temp[0]%2;
temp[1]=temp[0] * 5;
System.out.println("value="+temp[2]);
```

使用数组时要注意数组是否越界,每个数组一个属性 length,可以通过数组名.length 来获取数组的长度,即数组元素的个数。

3.2.5 实用案例 3.1:求一维数组的最大值及位置

为给一维数组进行随机赋值,本例采用 Math.random()方法产生随机数。

【例 3.2】 求一维数组的最大值及位置。

```
package Code0302;
public class MaxNum {
    public static void main(String[] args) {
        final int ARRAY_SIZE = 10;
        int a[] = new int[ARRAY_SIZE];
        int max = 0;
        int index = 0;                              //存储最大元素的位置
        for (int i = 0; i < a.length; i++) {        //本例中 a.length=10
            a[i] = (int) (Math.random() * 10);      //产生随机数,并对数组成员赋值
            System.out.print("  " + a[i]);
        }
```

```
        System.out.println();
        max = a[0];
        for (int j = 1; j < ARRAY_SIZE; j++) {
            if (a[j] > max)                              //判断当前位置的数是否大于最大值 max
            {
                max = a[j];
                index = j;                               //替换当前的最大值,记住对应位置
            }
        }
        System.out.println("A[" + index + "] has maximum value " + a[index]);

    }
}
```

程序可能的输出结果：

```
2  0  0  1  9  5  1  4  2  7
A[4] has maximum value 9
```

分析：第一个 for 循环依次将随机数读入下标为 i 的数组中，完成初始化操作；第二个 for 循环依次取出每个数组元素 a[j]的值，并判断其是否为当前最大值。

提示：如果例 3.2 中第二个 for 循环改为 for(int j＝1;j＜＝ARRAY_SIZE;j＋＋)，则出现数组下标越界的情况，因为 a[10]并不存在。

3.3 二维数组

到目前为止使用的都是一维数组，但有时应用中需要使用多维数组才能充分地存储数据。例如，(x,y)坐标系统中，如果需要存储坐标 x 和坐标 y 来标识一个点，则需要使用二维数组，其中一维用于存储坐标 x，另一维用于存储坐标 y。

将数组的下标数称为数组的维数，维数大于 1 的数组称为多维数组。因为多维数组的操作比较相似，本节重点介绍二维数组。

3.3.1 定义二维数组

二维数组的定义与一维数组相似：

```
数据类型  数组名[ ][ ];
```

两个方括号表示这是一个二维数组。当然，它也支持另一种形式：

```
数据类型[ ][ ] 数组名;
```

与一维数组相似，二维数组的创建可以先定义数组，再分配内存空间，也可以同时进行这两项工作。例如：

```
int temp[][];  temp=new int[2][3];
int temp[][]=new int[2][3];
```

这样即创建了一张二维结构的数组，其中行数为 2，列数为 3。也可将其看作一个长度为 2 的一维数组，每个数组成员又是一个长度为 3 的一维数组。这些数组成员按行排列为

```
temp[0][0], temp[0][1], temp[0][2]
temp[1][0], temp[1][1], temp[1][2]
```

同样，可以在定义数组的同时对其进行初始化。例如：

```
int temp[][]={{1,2},{3,4},{5,6}}
```

系统自动根据初始值的情况，为数组指定大小长度，并依次将值填入相应的单元中，例如初始化后 temp[0][1]＝2，temp[2][0]＝5。

注意：等式右边大括号内嵌套的大括号不能省略，它代表数组的一行及其组成。

3.3.2 二维数组元素的引用

引用二维数组元素的方式如下：

```
数组名[下标 1][下标 2]
```

下标仍然是从 0 开始逐渐递增的，同样在引用时要注意数组的越界问题。二维数组也有 length 属性，可以求每一维数组的长度。例如：

```
int temp[][]=new int[3][5];
System.out.println(temp.length);          //求二维数组的长度实际是求它的行数 3
System.out.println(temp[0].length);       //每个数组成员又是一个一维数组，其长度为 5
```

【例 3.3】 求二维数组中各元素的和。

```
package code0303;
public class SumAll{
    public static void main(String args[]){
    int total=0;
    int arr[][]=new int[3][4];
    for(int i=0;i<arr.length;i++){                //初始化并显示二维数组
            for(int j=0;j<arr[i].length;j++){
                arr[i][j]=i+j;
                System.out.print(" " + arr[i][j]);
            }
            System.out.println();
        }
        for(int i=0;i<arr.length;i++)             //求和
            for(int j=0;j<arr[i].length;j++){
                total = total +arr[i][j];
            }
        System.out.println(" The Sum is:"+total);
    }
}
```

程序运行结果：

```
0 1 2 3
1 2 3 4
2 3 4 5
The Sum is:30
```

3.3.3 实用案例 3.2：求两个矩阵的乘积

根据矩阵的乘法规则，两个矩阵相乘将产生一个新的矩阵，如 a[4,3]×b[3,2]将产生一个 r[4,2]的新矩阵,其中的某个元素 r[i][j]=a[i][0]×b[0][j]+a[i][1]×b[1][j]+a[i][2]×b[2][j]。例如:

$$x=\begin{pmatrix}1 & 6\\3 & 8\end{pmatrix}\times\begin{pmatrix}2 & 2\\9 & 7\end{pmatrix}$$

其计算过程为

$$1\times2+6\times9=56$$
$$1\times2+6\times7=44$$
$$3\times2+8\times9=78$$
$$3\times2+8\times7=62$$

因此结果为$\begin{pmatrix}56 & 44\\78 & 62\end{pmatrix}$。

【例 3.4】 求两个矩阵的乘积。

```
public class MatrixMultiply {
    public void multiply(int[][] a,int[][] b) {
        int [][] r = new int[4][2];            //r 用于存放运算结果
        int tmp =0;
        for (int k =0; k < r[0].length; k++) {
            //双重循环,遍历 a 矩阵
            for (int i = 0; i< a.length; i++ ) {
                tmp = 0;
                for (int j = 0; j < a[0].length; j++) {
                    tmp += a[i][j] * b[j][k] ;
                }
                r[i][k]=tmp;
            }
        }
         for (int i = 0; i < r.length; i++) {
            for (int j = 0; j < r[0].length; j++) {
                System.out.print(r[i][j] + "\t");
            }
            System.out.println();

        }
    }
    public static void main(String[] args) {
        int[][] a = new int[][] {
                    { 1, 2, 3 },
                    { 4, 5, 6 },
                    { 7, 8, 9 },
                    { 11, 12, 13} };
        int[][] b = new int[][]{
                    {1,2},
                    {3,4},
                    {5,6} } ;
        MatrixMultiply ma=new MatrixMultiply();
        ma.multiply(a,b);
    }
}
```

3.4 Arrays 类

为了更方便地操作数组,Java 在包 java.util 定义一个叫 Arrays 的类,该类包含了如下几个用 static 修饰的静态方法。

(1) static type[] copyOf(type[] original,int length):将 original 数组复制为一个新数组,其中 length 为新数组的长度。

(2) static int binarySearch(type[] a,type key):使用二分搜索法在类型为 type 的数组中搜索类型为 type 的指定值 key。

(3) static boolean equals(type[] a,type[] b):比较两个类型为 type 的数组是否相等。

(4) static void fill(type[] a, type val):用一个指定的值 val 填充类型为 type 的数组 a。

(5) static void fill(type[] a, int fromIndex, int toIndex, type val):与前一个方法类似,但填充时仅仅针对下标为 fromIndex 到 toIndex−1 的数组元素赋值为 val。

(6) static void sort(type[] a):对类型为 type 的数组排序。

上述每个方法的具体描述,请参考 JDK 文档。

【例 3.5】 Arrays 类的基本使用。

```
package code0304;
import java.util.Arrays;
public class ArraysDemo {
    public static void main(String[] args) {
        Integer array[] = new Integer[9];
        for (int i = 1; i < 10; i++)
            array[i - 1] = (int) (Math.random() * 100);
        //显示,排序数组
        System.out.print("原内容: ");
        display(array);
        Arrays.sort(array);
        System.out.print("排序后: ");
        display(array);
        //将值-1 分配给数组 array 中下标从 0~3-1 位置上的元素
        Arrays.fill(array, 0, 3, -1);
        System.out.print("执行 fill()后: ");
        display(array);
        //搜索 39
        System.out.print("值 39 的位置 ");
        int index = Arrays.binarySearch(array, 39);
        System.out.println(index);
    }
    static void display(Integer array[]) {
        for (int i = 0; i < array.length; i++)
            System.out.print(array[i] + " ");
        System.out.println("");
    }
}
```

程序运行结果:

```
原内容: 90 48 81 14 3 35 95 4 97
排序后: 3 4 14 35 48 81 90 95 97
```

```
执行 fill()后: -1 -1 -1 35 48 81 90 95 97
值 39 的位置 -5
```

提示：Java 8 对 Arrays 类的功能进行了增强。如增加了一些新的方法，可以利用多 CPU 的并行处理能力提高数组排序、赋值等操作性能；parallelSort(type[] a)方法与之前介绍的 sort()方法相似，但可以在多 CPU 上并行实现。

实用案例 3.3 对数组按中文名称排序

【例 3.6】 对数组按中文名称排序。

```
package code0304;
import java.text.Collator;
import java.util.Arrays;
import java.util.Comparator;
public class SortByChinese {
    public static void main(String[] args) {
        String[] arrStrings = {"计算机","长江", "通信","数学"};
        Arrays.sort(arrStrings);
        for (int i = 0; i < arrStrings.length; i++)
            System.out.println(arrStrings[i]);
        System.out.println("-----------------------------");
        //Collator 类用来执行区分语言环境的字符串比较,这里选择用 CHINA
        Comparator comparator = Collator.getInstance(java.util.Locale.CHINA);
        //根据指定比较器产生的顺序对指定对象数组进行排序
        Arrays.sort(arrStrings, comparator);
        for (int i = 0; i < arrStrings.length; i++)
            System.out.println(arrStrings[i]);
    }
}
```

程序运行结果：

```
数学
计算机
通信
长江
-----------------------------
长江
计算机
数学
通信
```

Arrays 类中的 sort()方法缺省，那么按照数组中数值的大小或字母顺序进行排序，但这种处理方式对中文是无效的，如上例运行结果。为此，我们使用了类 Arrays 中另一种形式的 sort()方法：sort(T[] a, Comparator<? super T> c)，它可以根据指定比较器 Comparator 产生的顺序对对象数组进行排序。为获取 Comparator 对象，可以通过方法 Collator.getInstance()实现，其中参数 java.util.Locale.CHINA 表示按中文语言进行排序。

3.5 数组实训任务

【任务描述】

编写一个模拟的 Java 发牌程序,要求将 2 副牌(即 108 张牌)发给 4 个人,并留 8 张底牌,最后输出底牌和每个人手中的牌。

【任务分析】

(1) 首先确定扑克牌在计算机中的表达方式,因为计算机中无法表达扑克牌中的花色,因此最好为每张牌设定一个编号,拟定的编号规则如下:

红桃按照从小到大依次为 1~13。

方块按照从小到大依次为 14~26。

黑桃按照从小到大依次为 27~39。

梅花按照从小到大依次为 40~52。

小王为 53,大王为 54。

(2) 由于可发的牌和每个玩家手中的牌都需要记录,且其由多个有序数据组成,因此需要设计两个数组,以存放 108 张牌和每个玩家手中的牌。

(3) 整个程序由以下 4 部分构成:

① 按照以上编号规则初始化一个包含 108 个数字的数组。

② 每次随机从该数组中抽取一个数字,分配给保存玩家数据的数组。

③ 循环输出每个玩家手中的牌。

④ 最后输出底牌。

【任务解决】

【例 3.7】 模拟发牌。

```
package code0305;
import java.util.*;
public class CardPlay{
    public static void main(String[] args){
        int[] total = new int[108];                       //存储 108 张牌的数组
        int[][] player = new int[4][25];                  //存储 4 个玩家的牌
        int leftNum = 108;                                //当前剩余牌的数量
        int ranNumber;
        Random random = new Random();                     //生成 Random 对象,用以生成随机数
        for(int i = 0;i < total.length;i++){              //初始化一维数组
            total[i] = (i + 1) % 54;
            if(total[i] == 0){                            //处理大小王编号
                total[i] = 54;
            }
        }
        //循环发牌
        for(int i = 0;i < 25;i++){//为每个人发牌
            for(int j = 0;j < player.length;j++){     /生成随机下标
                ranNumber = random.nextInt(leftNum); //random.nextInt 方法生成随机数
                player[j][i] = total[ranNumber];     //发牌
                total[ranNumber] = total[leftNum - 1];      //删除已经发过的牌
                leftNum--;
```

```
            }
        }
        //循环输出玩家手中的牌
        for(int i = 0;i < player.length;i++){        //通过两层循环遍历二维数组
            System.out.print("玩家" + i+ "的牌:");
            for(int j = 0;j < player[i].length;j++){
                System.out.print(" " + player[i][j]);
            }
            System.out.println();
        }
        //底牌
        System.out.print("底牌:");
        for(int i = 0;i < 8;i++){
            System.out.print(" " + total[i]);
        }
        System.out.println();
    }
}
```

程序运行结果：

```
玩家 0 的牌: 52 15 12 53 32 24 48 36 24 11 54 27 14 22 3 8 10 40 43 46 40 45 37 41 22
玩家 1 的牌: 49 25 5 47 20 35 45 51 29 17 10 26 20 51 3 1 29 43 31 9 33 30 32 4 53
玩家 2 的牌: 42 7 13 38 34 50 12 6 39 49 33 23 14 31 7 19 39 48 16 18 35 5 41 9 11
玩家 3 的牌: 44 2 30 26 28 46 44 19 21 23 25 21 15 13 1 28 34 52 4 8 6 47 17 2 27
底牌: 50 42 18 37 16 38 36 54
```

习题与思考

1. 为了定义 3 个整型数组 a1、a2、a3，下面声明正确的语句是哪一个？
   A. int Array[] a1,a2;　int a3[]={1,2,3,4,5};
   B. int [] a1,a2;　int a3[]={1,2,3,4,5};
   C. int a1,a2[];　int a3={1,2,3,4,5};
   D. int [] a1,a2;　int a3=(1,2,3,4,5);
2. 编程实现两个同行列数矩阵的加法运算。
3. 将给定数组 int a[]={78,23,56,34,12,45,67,89}按从小到大顺序进行排序并输出。
4. 给出下列程序段运行的结果：

```
public static void main(String args[])
{
    int array[]={1,2,3,4,5};
    printarray(array);
    could_modify(array);
    printarray(array);
}
```

```
static void could_modify(int a[]){
    for(int i=0;i<a.length;i++)
        a[i] *=i;
}
static void printarray(int a[])
{
    for(int i=0;i<a.length();i++)
    System.out.println(a[i]);
}
```

5. 数组创建后，可否通过设置变量 length 的大小来增减数组的大小？

第4章 类和对象设计

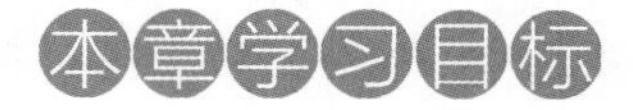

Java程序一般由一个主类和支持这个主类的其他相关类构成，前面几章已经初步接触和使用到Java类库中的一些基本类(如String等)，本章将建立属于自己的类，并通过其构建实用的应用程序。本章主要内容包括：

(1) 类的定义。

(2) 对象的创建和使用。

(3) 构造方法定义和使用。

(4) 类的成员变量与成员方法。

(5) 方法的重载。

(6) 权限修饰符。

4.1 面向对象基础

要设计一个计算机程序，一种方法是根据问题解决的步骤，将程序组织为一系列的线性代码或函数，然后依次调用或访问它们即可，如C或Pascal程序设计语言。

面向对象程序设计(Object Oriented Programming，OOP)语言则是从另一角度看待计算机程序，其核心是引入了对象的概念，使之可以逼真地模拟现实世界中的任何事物，从而达到与人类的思维习惯一致。

现实世界中存在很多不同的对象，如人、车、家用电器等，每个对象都有自己独特的属性和行为，可以提供不同的功能和服务，一方面对象之间各自独立，另一方面它们又相互联系，通过协作的方式完成复杂的任务。例如，“扬声器”可以播放中频和高频声音；“喇叭”可以播放低音，“高频头”可以接收无线电广播，“CD播放器”可以读取光盘音频数据。这些对象似乎彼此独立，但一旦将其连接、驱动，则可以构成一个复杂的音响系统，播放各种音乐。

OOP充分借鉴这一思想，将现实世界中的事物及其关系映射到问题空间中，设计多个对象并使其各司其职，分别完成一组相关的任务。如果一个对象依赖于一个不在其控制范围内的方法，就需要访问包含该方法的其他对象，即请求其他对象提供该方法或服务，利用OOP的术语称作一个对象向另一个对象发送消息。

如“CD播放”任务的实现，需要“CD播放器”和“扬声器”两个部件的同时参与才能完成，因此，可以对应设计两个对象，分别实现“读取数据”和“发声”的功能，并将其组合起来，以完成上述功能。

其实，这是 OOP 中一种典型的设计思想——抽象：将现实世界中的实体理解为由属性数据和对这些属性实施行为的统一体，即对象。属性表示对象的性质，行为则定义了对象可以提供的外部服务。例如，将“CD 播放器”视为对象，则它具有播放速度、重量、形状等属性，对外可提供读数据、写数据等操作，而且这些操作将部分影响或改变“CD 播放器”的属性。OOP 中其他 3 个典型特征是封装、继承和多态。

（1）封装：是一种将操作和操作所涉及的数据捆绑在一起，使其免受外界干扰和误用的机制。

例如，将录音机视为一个对象，如图 4-1 所示。它的属性包括运转方向、运转速度，提供的行为包括播放、后退、录音、前进。我们使其围绕在属性数据周围。这样所有属性均无法直接访问，因为它们被上述方法封装保护起来，使用者只能通过提供的方法改变或访问属性的值，就像我们只能通过录音机的按键来使用它，而不是直接改变其速度、方向一样。

在 Java 中，最基本的封装单元是类，一个类定义为具有相似特征（属性和行为）对象的一种抽象。

（2）继承：是指一个新的类继承原有类的基本特征，并可增加新的特性。原有的类称为父类或基类，新的类称为子类或派生类。在子类中，不仅包含父类的所有数据和方法，还可增加新的属性和方法。继承具有传递性。如果 B 继承自 A，C 继承自 B，则 C 间接继承自 A。这样就形成了一个层次关系。人之间的继承关系如图 4-2 所示。

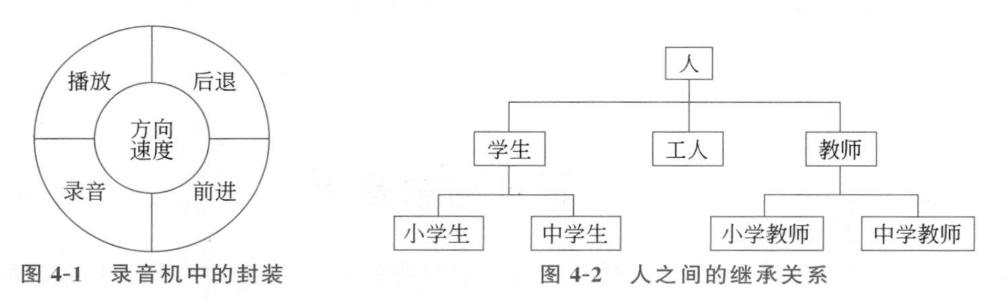

图 4-1　录音机中的封装　　图 4-2　人之间的继承关系

根据类的继承机制，在父类中只定义各子类共同需要的属性和方法。类派生时，可以增加新的属性和方法。因此，父类的基本特征可被所有子类对象共享，提高了类的重复利用率。

（3）多态：是指类中同一名称的行为（方法）可以有多种不同的功能，或者相同的接口有多种实现方法，具体知识参见第 5 章。

4.2　类和对象初探

类是 Java 中最重要的复合数据类型，是组成 Java 程序的基本元素，它封装了一类对象的属性和改变这些属性的方法，是这一类对象的原型。与简单数据类型的使用不同，用户必须先定义类，并生成该类的实例，然后才能通过该实例访问其成员变量和方法。

例如，圆可以有很多种，但所有的圆均由圆心位置和半径决定，其基本行为至少包括计算面积和周长等。通过抽象，可以定义类 Circle。

【例 4.1】 定义类 Circle,并输出任意圆的面积和周长。

```
package code0402;
class Circle {
    float x, y;                                          //定义圆心 x,y 坐标
    float radius;                                        //定义成员变量(半径)
    double getArea() {                                   //定义计算面积的方法
        return radius * radius * Math.PI;
    }
    double getCircumference() {                          //定义计算周长的方法
        return 2 * radius * Math.PI;
    }
    public static void main(String args[]) {
        Circle c = new Circle();                         //创建类 Circle 的对象 c
        c.radius = 3;
        c.x = 0;
        c.y = 0;                                         //为对象 c 中变量赋值
        System.out.println("Area=" + c.getArea());       //访问成员方法
        System.out.println("Circumference=" + c.getCircumference());
    }
}
```

程序运行结果:

```
Area=28.27
Circumference=18.84
```

分析:整个程序可分为以下 3 部分:

(1) 通过关键字 class 定义类 Circle,包括圆心、半径属性和周长、面积计算方法。

(2) 在主函数 main 中通过语句 Circle c=new Circle 生成类 Circle 的对象 c。

(3) 通过点运算符“.”访问对象 c 的成员变量和方法。

从对象的整个生命周期来看,它包含了类的定义、对象实例化、对象初始化、对象使用和对象清除等若干阶段,下面逐一对这些阶段的工作进行分析。

4.3 定义类

类通过关键字 class 进行标识,基本形式如下:

```
[类修饰符] class 类名 [extends 父类名] [implements 接口名]
{
    //类体,包括定义类的成员变量和方法
}
```

类名由用户指定,可以是任意合法的标识符。

类体是定义在大括号中的部分,是整个类的核心,可以分为类的成员变量和方法两部分。

类修饰符为可选项,它决定了类在程序运行过程中以何种方式被处理。定义类时,可以接受缺省的修饰符,或其他类型修饰符(如 abstract、final 等)。

“extends 父类名”也为可选项,表示所定义的类继承自其他父类,这时该类自动获得父类中所有可能的属性和方法,并可添加父类没有的其他成员。

“implements 接口名”也为可选项,它表示所定义的类需要通过实现某个接口完成,接口

实际也是一种特殊的类,它所定义的方法一般为空,需要在派生的类中实现该方法。类的继承和实现将在第 5 章介绍。

4.3.1 定义成员变量

成员变量的定义格式如下:

```
[修饰符]类型  成员变量名列表;
```

成员变量的类型可以为任意数据类型,包括:简单类型(如整形、浮点型),数组,类(如String)等。在一个类中,成员变量名是唯一的。类的成员变量定义在所有的方法体之外,因此它的作用域是整个类,即从变量定义开始一直到标识类体结束的}处。

类中的所有方法均可以直接访问成员变量。

成员变量的修饰符(如 public、protected、private、final、static 等)为可选项,主要起到对变量进行访问控制和限定的作用,具体将在 4.6 节和 5.4 节讨论。

【例 4.2】 定义学生类,其中含有学号、姓名、出生日期等属性,出生日期又含有年、月、日等属性。

```
class Student{
    int no;
    String name;
    Birthday day;                   //day 的类型为类 Birthday
}
class Birthday{                     //定义类 Birthday
    int year;
    int month;
    int date;
}
```

4.3.2 定义成员方法

方法表示对象所具有的行为,其基本定义格式如下:

```
[修饰符] 返回值类型 方法名([参数列表])
{
    //方法体
}
```

返回值类型可以是任意基本类型或类,如果方法不返回任何值,它必须声明为 void (空)。参数列表由零个或多个参数构成,参数之间用逗号分隔,每个参数由一个数据类型和一个标识符构成,通过参数列表可以将参数值传给被调方法。

方法体部分定义了该方法是如何实现的,它由大括号“{}”内的语句序列组成。方法如果有返回值,可通过 return 语句传给调用者,且必须和定义的返回值类型一致。

【例 4.3】 定义二维空间中的点 Point 和基本方法。

```
package code0403;
class Point {
    int x = 0;
    int y = 0;
```

```
    public void move(int dx, int dy)                //移动点坐标的方法
    {
        x = x + dx;
        y = y + dy;
    }
    public void alert() {                           //无返回值的输出方法
        System.out.println("x=" + x + " y=" + y);
    }
    public static void main(String args[]) {
        Point p = new Point();
        p.move(1, 2);
        p.alert();
    }
}
```

程序运行结果：

```
x=1 y=2
```

提示：方法体中的变量均为局部变量，参数列表中的变量也是局部变量，因此只能在本方法的{…}内使用，如方法 move 中的局部变量 dx 和 dy 不能在方法 alert 中使用。但类中成员变量的作用域是整个类，因此变量 x 和 y 可以在方法 alert 中使用。

Java 中有两种特殊的成员方法：构造方法和 main()方法，其定义形式与传统方法稍有不同：

(1) 构造方法要求方法名必须与类名相同，且不能有返回值(void 也不行)。如类 Student 的构造方法可定义为

```
Student (int i_no, String i_name)
```

或

```
Student(String s)
```

(2) main()方法通常是一个 Java 应用程序的执行起点，其定义格式为

```
public static void main(String args[])
```

其中 main 为方法名，args 为形参，类型为字符串数组，void 表示 main 方法无须返回值。

4.3.3 方法重载

Java 允许使用同一个名字去定义多个方法，只要方法的参数列表不同，即参数的数量、类型不完全相同。方法调用时，编译器根据实参列表的个数和类型自动调用匹配的方法。

方法重载使得类中两个相似的方法可以拥有完全相同的名字，也可以更灵活地基于所接收的参数进行不同方法的调用。

【例 4.4】 编程实现对双操作数和三操作数加法运算的重载。

```
package code0403;
public class OverLoad {
    int sum(int a, int b) {                         //定义双操作数的 sum 方法
        return a + b;
    }
```

```
    int sum(int a, int b, int c) {                //重载的三操作数 sum 方法
        return a + b + c;
    }

    public static void main(String[] args) {
        OverLoad o = new OverLoad();
        System.out.println(o.sum(1, 5));          //调用双操作数 sum 方法
        System.out.println(o.sum(3, 5, 8));       //调用三操作数 sum 方法
    }
}
```

程序运行结果:

```
6
16
```

分析:类 OverLoad 对成员方法 sum()进行了重载,两个方法的参数个数不同,调用时 o.sum(1,5)调用第一个 sum()方法,o.sum(3,5,8)调用第二个 sum()方法。

提示:根据方法的返回值类型是无法区分重载方法的,因为在调用方法时,返回值是不参与调用的。

4.4 对　　象

4.4.1 实例化对象

类作为一种抽象的复合数据类型,必须先要实例化(即生成对象),然后才能使用。基本语句格式如下:

```
类名 对象名= new 类名([参数列表])
```

如下实例化类 Point 的两个对象 p1,p2:

```
Point  p1=new Point ();
Point  p2=new Point ();
```

关键字 new 为每个生成的对象分配一片内存区域,并返回该对象的一个引用(可以理解为该对象的内存首地址)。

可以为一个类创建多个不同的对象,每个对象占用不同的内存空间,存储各自的状态信息,改变其中一个对象的状态不会影响其他对象,如图 4-3 所示。

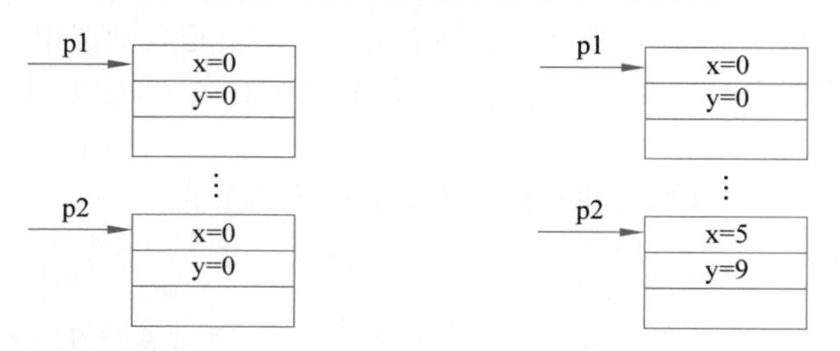

图 4-3　对象 p2 的属性变化与 p1 无关

提示：当运算符==和!=用于两个对象的比较时，它不是比较两个对象的值是否相等，而是判断运算符两边是否是同一个对象（地址空间相同），如上述对象p1、p2显然是两个不同的对象，因为它们的地址空间完全不同，因此表达式p1==p2的值为false。如果要比较两个对象的值是否相等，可以使用equals()方法，如p1. equals(p2)。

如果只想声明某类的实例，而不想为其创建任何对象，则可以使用关键字null（含义为空）。例如：

```
Point p3=null;
```

声明了类Point的对象p3，但并没有为其分配内存空间，在需要时再通过语句p3=new Point()对其实例化。

4.4.2 初始化对象

初始化对象就是为所创建对象的成员变量赋初值，其中最直接的方式是在定义成员变量的同时对其赋值。例如：

```
class Student{
    String name= "zhangshan";
    …
}
```

这样，每个Student对象中变量name的值均为zhangshan。

一种更好的方式是使用4.3节中提到的构造方法，在为对象分配内存空间的同时，实现对象的初始化。具体这项工作包括两个子任务：

(1) 定义一个或多个构造方法，并在方法体中对成员变量赋初值。

(2) 调用或执行相应的构造方法。

【例4.5】 通过构造方法对Student对象进行初始化。

```
package code0404;
class Student {
    int no;
    String name;

    Student(int lno, String lname) {                  //定义构造方法
        this.no = lno;
        this.name = lname;                            //对成员变量 no 和 name 初始化
    }

    public static void main(String args[]) {
        Student s1 = new Student(1, "zhangShan");   //传递参数 lno=1;lname= zhangShan
        Student s2 = new Student(2, "xiaoMing");    //传递参数 lno=2;lname= xiaoMing
        System.out.println("name=" + s1.name + " no=" + s1.no);
        System.out.println("name=" + s2.name + " no=" + s2.no);
    }
}
```

程序运行结果：

```
name=zhangShan  no=1
name=xiaoMing  no=2
```

需要说明的是:构造方法的调用方式和过程非常特别,它不是通过显示指定方法名直接调用的,而是在实例化对象时由系统自动调用的。实际上,当新建一个对象时,系统将自动完成以下 3 项工作:①为每个对象分配不同的内存空间;②如果类定义时有初值则使用该值对成员变量进行初始化,如果没有初值则可以使用默认值;③自动调用构造方法,如果构造方法有多个,则根据参数的类型、个数等决定调用哪个构造方法。

由此可知,当语句 Student s1=new Student(1,"zhangShan")被执行以创建对象 s1 时,构造方法同时被调用,并通过参数传递使得变量 no=1; name= zhangShan。

如果用户在类中没有定义任何构造方法,系统会自动生成一个缺省的构造方法。该方法没有参数,且方法体为空。例如,类 Student 的缺省构造方法为

```
public Student() { }
```

当对该类进行实例化且不指定任何参数时,如执行语句 Student s1=new Student(),系统将自动调用该缺省的构造方法。

与常规方法相同,构造方法也可以被重载,以提供对象不同的初始化过程和属性值。

【例 4.6】 构造方法重载示例。

```
package code0404;
public class StudentOverload {
    int no;
    String name;
    StudentOverload(String l_name) {
        no = 0;
        name = l_name;
    }
    StudentOverload(int l_no, String l_name) {
        no = l_no;
        name = l_name;
    }
    public static void main(String args[]) {
        StudentOverload s1 = new StudentOverload("zhangShan");
        StudentOverload s2 = new StudentOverload(2, "xiaoMing");
        System.out.println("name=" + s1.name + " no=" + s1.no);
        System.out.println("name=" + s2.name + " no=" + s2.no);
    }
}
```

程序运行结果:

```
name=zhangShan no=0
name=xiaoMing no=2
```

分析:当程序中有多个重载的构造方法时,Java 根据 new 语句中的实参个数、类型自动调用与之匹配的构造方法。例如,语句 new Student("zhangShan")中只有一个 String 类型的实参,因此应该匹配第一个构造方法;而语句 new Student(2,"xiaoMing")中第一个实参为整数,第二个实参为 String,与之匹配的是第二个构造方法。

4.4.3 使用对象

Java 中对象的使用包括对成员变量的引用和成员方法的调用,它们都是通过点运算符(.)来实现的,其格式分别如下:

```
对象名.成员变量名
对象名.成员方法名(实参列表)
```

方法调用时,要指定被调用的方法名称、实际参数,并且要求实参的类型、个数和顺序与定义中的形参列表一致。下面通过例子加以说明。

【例 4.7】 对象的基本使用方法示例。

```
package code0404;
public class PassTest {
    float ptValue;
    public void changeInt(int value) {
        System.out.println("In changeInt:" + value);
        value = 55;
    }
    public void changeArrayValue(int xc[]) {
        System.out.println("In changeArrayValue:" + xc[1]);
        xc[1] = 6;
    }
    public static void main(String args[]) {
        int val = 11;
        int sc[] = { 1, 9 };
        PassTest pt = new PassTest();
        pt.ptValue = 20;                          //引用成员变量 ptValue
        pt.changeInt(val);                        //方法调用(传值形式)
        System.out.println("Current Int value is: " + val);
        pt.changeArrayValue(sc);                  //方法调用(引用形式)
        System.out.println(" Current Value in array is: " + sc[1]);
    }
}
```

程序运行结果:

```
In changeInt:11
Current Int value is: 11
In changeArrayValue:9
Current Value in array is: 6
```

分析:本例使用了两种不同的方法调用形式:传值调用和引用。传值调用的特征是形参一般为基本类型的变量(如数值型、字符型、布尔型等)。调用时,实参的值对应传递给形参,这种传递是单向的,即形参的值不会改变实参的值。如上例中,尽管方法 changeInt ()对形参 value 的值做了改变,但这种改变并不影响实参 val。

引用则类似于 C 语言中的传地址调用,直接将实参的地址传递给形参,这样两者使用完全相同的地址空间,因此实参可以改变形参的值,形参也可以改变实参的值。当实参为复合数据类型时(如数组、对象),Java 自动采用引用方式进行参数传递。如例 4.7 中,方法 changeArrayValue()中传入的参数为数组,这时实际传递的是实参数组的引用,如图 4-4 所示,对形参的改变同时也意味着对实参的改变。

4.4.4 使用静态变量和方法

对不同的对象而言,其在内存中的存储空间通常是独立的,对某个对象成员变量值的修改不会影响到其他对象,这种只属于单个对象的成员变量称为实例变量。

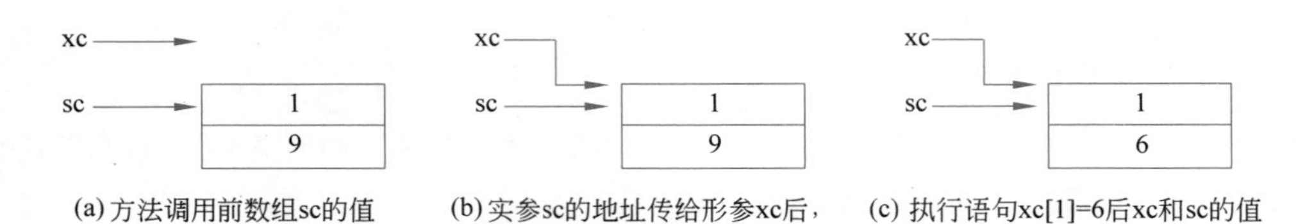

(a) 方法调用前数组sc的值　(b) 实参sc的地址传给形参xc后，它们的地址空间完全相同　(c) 执行语句xc[1]=6后xc和sc的值

图 4-4　基于引用的方法调用过程示例

如果所有对象希望共享同一变量，则可以在变量定义前加上关键字 static，这样即定义了一个类变量，也称静态变量。静态变量不再属于某个实例或对象，而是属于整个类。

类似地，可以通过在方法定义前加上关键字 static，将该方法定义为类方法或静态方法。例如：

```
static int no;                                        //定义静态变量 no
static void alert(){ …}                               //定义静态方法 alert()
```

由于静态变量和静态方法都属于整个类，因此可以直接通过类名访问，而无须创建该类的实例，基本格式如下：

```
类名.静态变量名;
类名.静态方法名([实参列表])
```

重新定义类 StudentStatic 如下：

```
class StudentStatic{
    static int no;
    static String name="test";
    static void alert(){ System.out.println(name); }

}
public class SimpleTest{
    public static void main(String args[]){
        System.out.println(StudentStatic.no); //通过类名引用静态变量 no
        StudentStatic.alert() ;               //通过类名引用静态方法 alert()
    }
}
```

注意：对实例成员变量，每个实例都有一个；而对于静态成员变量，每个类只有一个，该类对应的所有实例均共享该变量，它们的区别可通过下面的示例说明。

【例 4.8】　实例变量和静态变量的区别示例。

```
package code0404;
class StaticTest {
    static int statInt = 4;
    static double statDouble = 16.0;
    int instInt;
    double instDouble;
    public static void staticMethod() {          //输出静态变量的值
        System.out.println("statInt=" + statInt + "; statdouble=" + statDouble);
    }
    public void instMethod() {                   //输出实例变量的值
        System.out.println("instInt=" + instInt + "; instdouble=" + instDouble);
    }
```

```
    public StaticTest(int intArg, double doubleArg) {
        instInt = intArg;
        instDouble = doubleArg;
    }
    public static void changestatic(int newInt, double newDouble) {
                                                    //改变静态变量的值
        statInt = newInt;
        statDouble = newDouble;
    }
    public static void main(String args[]) {
        StaticTest instance1 = new StaticTest(1, 2.0);
        StaticTest instance2 = new StaticTest(3, 4.0);
        instance1.instMethod();                  //调用实例方法
        instance2.instMethod();
        StaticTest.staticMethod();
        StaticTest.staticMethod();               //调用静态方法
        instance2.staticMethod();
        instance1.changestatic(8, 8.0);          //改变静态变量的值
        instance2.staticMethod();                //再次输出静态变量的值
        StaticTest.staticMethod();
    }
}
```

程序运行结果：

```
instInt=1; instdouble=2.0
instInt=3; instdouble=4.0
statInt=4; statdouble=16.0
statInt=4; statdouble=16.0
statInt=4; statdouble=16.0
statInt=8; statdouble=8.0
statInt=8; statdouble=8.0
```

程序分析：

(1) 对静态变量和静态方法，可以通过“类名.*”和“对象名.*”两种方式引用，结果完全相同。

(2) 对实例变量，对象 instance1 和 instance2 提供了不同的内存空间，因此即便同名也是两个不同的变量，其值各不相同。

(3) 静态变量 statInt 和 statDouble 对类 StaticTest 的所有对象都是公共的，因此通过 instance1 对其改变后，通过 instance2 再去访问时其值已经变化了，这进一步证明静态变量不是属于某个对象，而是属于整个类。

4.4.5 清除对象

对象使用完后，如果老占着内存不放，会很快耗尽内存资源，因此必须及时清理。C 和 C++ 中的垃圾清理是由程序员完成的，有时这是一件很困难的事情。因为人们并不总是事先知道垃圾应在何时被释放。Java 语言解除了程序员进行垃圾清理的责任，它提供一种系统级的线程跟踪每一块内存的分配情况。JMV 检测何时一个对象不再有用，然后收回该对象的空间为新对象所用。但如果对象用到任何其他系统资源，如在某个对象的生存期内打开了一些文件，而当这个对象被破坏时，若想确认这些文件是否已经被正确地关闭，那么可以重写

finalize()方法，其格式如下：

```
protected void finalize() throws Throwable{
    …     //撤销对象
}
```

4.4.6 应用程序与命令行参数

一个 Java 应用程序(application)通常由一个或多个类构成，但其中只能有一个类作为整个应用执行的起点，我们称之为主类。

主类的特征之一是类名与 Java 文件名相同，且必须含有 main()方法，此外方法必须被声明为 public、static、void，具体格式如下：

```
public static void main(String args[]){
    …     //方法体
}
```

public 指明该方法可以被所有类使用，static 则表明该方法是一个静态方法。

由于 main()方法是应用程序执行过程中第一个被调用的内容，因此可以在该方法体中启动应用程序的任何其他代码，包括生成其他类的实例等。

与其他方法类似，main()方法也可以接收一个或多个(命令行)参数，这些参数作为字符串依次被传递并存入数组 args[]中，即 args[0]存放第一个参数，args[1]存放第二个参数，以此类推，但要注意 Java 不把文件名作为参数之一。

【例 4.9】 命令行参数使用示例。

```
package code0404;
public class MainTest {
    public static void main(String args[]) {
        int n = args.length;                         //取得参数的个数
        if (n == 0)
            System.out.println(" no parament ! ");
        else {
            System.out.println(" number of paraments : " + n);
            for (int i = 0; i < n; i++)
                System.out.println(" args[ " + i + " ] = " + args[i]);
        }
    }
}
```

编译通过后，执行该程序：

```
java  MainTest  I  Love  China
```

或通过可视化开发工具配置，在 Eclipse 的 Package Explorer 中先找到对应类文件，右击，在弹出菜单上选择 Run as-Run Configurations，在弹出窗口中选择 arguments 标签页，program arguments 栏即为输入命令行参数的位置，如输入：I Love China，然后单击 Run 按钮，如图 4-5 所示。

程序运行结果：

```
number of paraments : 3
args[ 0 ] = I
args[ 1 ] = Love
args[ 2 ] = China
```

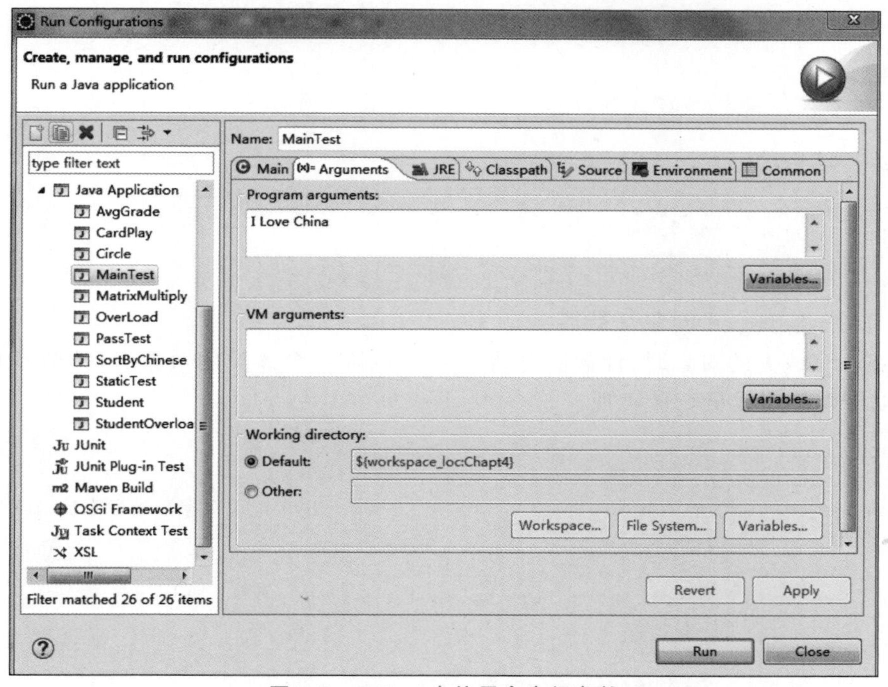

图 4-5　Eclipse 中使用命令行参数

4.4.7　实用案例

【例 4.10】　实现一个冒泡排序算法。

假定用户可以任意指定待排序的数字序列,可以利用 Scanner 类获取用户从控制台的输入,调用 Scanner 的 nextLine()方法即可,详见第 8 章。

```
package code0404;
import java.util.Scanner;
public class BubbleSort {
    public static void main(String args[]) {
        Scanner scanner = new Scanner(System.in);
                                //Scanner 用于获取用户从控制台的输入
        int[] numbers = new int[8];
        System.out.print("Please enter eight number:");
        for (int i = 0; i < numbers.length; i++)
            numbers[i] = scanner.nextInt();
                                //通过 nextInt()方法依次读取用户输入数字并存入数组中
        for (int i = numbers.length - 1; i > 0; i--) {
            for (int j = 0; j < i; j++) {
                if (numbers[j] > numbers[j + 1]) {
                    int temp = numbers[j];
                    numbers[j] = numbers[j + 1];
                    numbers[j + 1] = temp;
                }
            }
        }
        System.out.println("冒泡排序的结果是:");
        for (int i = 0; i < numbers.length; i++) {
```

```
            System.out.print(numbers[i] + " ");
        }
    }
}
```

程序运行结果：

```
Please enter eight number:  24  18  36  30  7  15  9  1
冒泡排序的结果是:
1 7 9 15 18 24 30 36
```

【例 4.11】 实现一个对象计数器。

新创建 1 个该类的对象时，计数器自动加 1，当删除一个该类的对象时，计数器自动减 1。提示：可以借助静态成员变量实现。

```
package code0404;
public class Counter {
    public static int num = 0;                    //用于记录对象数的静态成员变量
    public Counter() {
        num ++;
        System.out.println("after new NO=" + num);
    }

    public static void delete(Counter o) {
        num --;
        System.out.println("after delete NO=" + num);
        System.gc();                              //手工方式回收无用的对象
    }
    public static void main(String [] args) {
        Counter [] objs = new Counter[5];
        for(Counter obj: objs) {
            obj = new Counter();
        }
        Counter.delete(objs[1]);
        Counter.delete(objs[0]);
    }
}
```

程序运行结果：

```
after new NO=1
after new NO=2
after new NO=3
after new NO=4
after new NO=5
after delete NO=4
after delete NO=3
```

4.5 包

包的出现，主要是因为类的种类非常多，类的命名有可能重复，否则只能让类的命名越来越长、越来越复杂，Java 使用包的机制很好地解决了这个问题。在一个包中，类是不可以重名的，但是不同的包中允许相同的类名出现，通过使用“包名＋类名”的方式可以达到有效管理类的目的。

4.5.1 包的定义

创建一个包非常简单，只需要在源文件的起始处添加一条包含关键字 package 的语句，则任何在该文件中定义的类都属于指定的包。package 语句指定了一个类存放的命名空间，如果没有 package 关键字，则类会被放入一个缺省的包中，该包没有名字。

包的定义格式如下：

```
package  [包名 1.[包名 2.[包名 3]]]
```

包定义语句中，用“.”指明包的层次，Java 编译器根据上述定义将包一一映射为文件系统的目录结构。例如，声明 package first.second.third 的包，则该类文件将存储在文件系统的当前目录的 first\second\third 子目录中。

假定当前 project 源码的根目录为 E:\CouresTest\src，在该 project 中生成包 package first.second.third 和文件 packageTest.java 后，其对应的目录结构如图 4-6 所示。

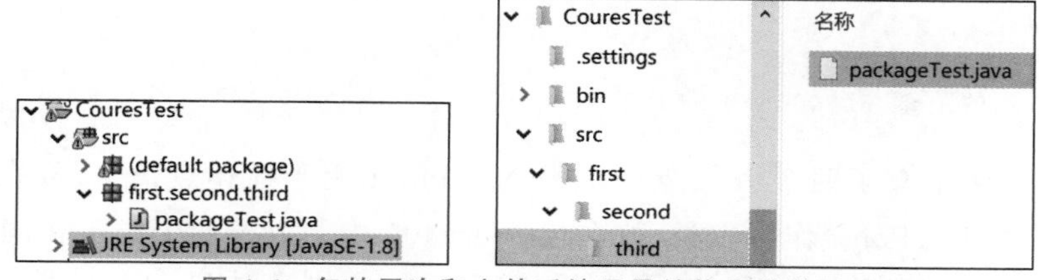

图 4-6　包的层次和文件系统目录结构间的关系

4.5.2 包的引入

如果包已经存在，则可以通过关键字 import 引入包中的任何类，并使得这些类在整个程序中可见。import 语句的基本格式如下：

```
import [包名 1.[包名 2.[包名 3.]]] (类名|*)
```

与 package 相似，[包名 1.[包名 2.[包名 3.]]]为包的层次，实质为文件系统的目录结构，(类名|*)表示只引入指定类名的类，或者引入该包中所有的类(*)。如果要使用 Java 类库中的已有类或其他包中的类，必须先通过 import 语句将其引入后才能使用。例如：

```
import java.util.*;                         //引入 java.util 包中的所有类
import java.applet.Applet;                  //引入 java.applet 包中的类 Applet
Date day1=new Date();
```

【例 4.12】 计算任意两个点之间的距离。

```
package code0405.mypack;
//定义包 code0405.mypack 中的类 NewPoint
public class NewPoint {
    public double x, y;
    public NewPoint(double a, double b) {
        x = a;
        y = b;
    }
```

```
    public double distanceTo(NewPoint p) {     //该点到另一点 p 的距离
        return Math.sqrt((x - p.x) * (x - p.x) + (y - p.y) * (y - p.y));
    }
}
package code0405;
import code0405.mypack.NewPoint;              //引入包 code0405.mypack 中的类 NewPoint
public class UsePoint {
    public static void main(String[] args) {
        NewPoint p1 = new NewPoint(1.0, 2.0); //创建两个 NewPoint 对象
        NewPoint p2 = new NewPoint(2.0, 5.5);
        System.out.println("点 p1 坐标: " + p1.x + ", " + p1.y);
        System.out.println("点 p2 坐标: " + p2.x + ", " + p2.y);
        System.out.println("点 p1 到点 p2 的距离:" + p1.distanceTo(p2));
    }
}
```

程序运行结果：

```
点 p1 坐标: 1.0, 2.0
点 p2 坐标: 2.0, 5.5
点 p1 到点 p2 的距离:3.640054944640259
```

分析：本例定义了两个 Java 类：①类 NewPoint 用于描述点信息，放在包 code0405.mypack 中；②类 UsePoint 作为主类，放在包 code0405 中。由于两个类不在同一包中，因此必须先通过 import 语句将 NewPoint 引入 UsePoint 中，否则它对 UsePoint 是不可见的。

如果使用的两个类都在同一个包中(如将 UsePoint 中起始语句改为 package code0405.mypack)，则无须进行包的引入即可在 UsePoint 中访问 NewPoint。

提示：如果不使用 import 语句，则必须在每个被引用类前加上包名作为前缀，例如 code0405.mypack.NewPoint p1 = new code0405.mypack.NewPoint (1.0, 2.0)；虽然这与使用 import 的方法是等价的，但相比而言，import 语句使程序更简洁。

4.5.3 模块

在 Java 9 中引入了一个新的概念：模块(module)。每个 module 均可以包含若干个子 package(包)。模块、包与类之间的关系如图 4-7 所示。

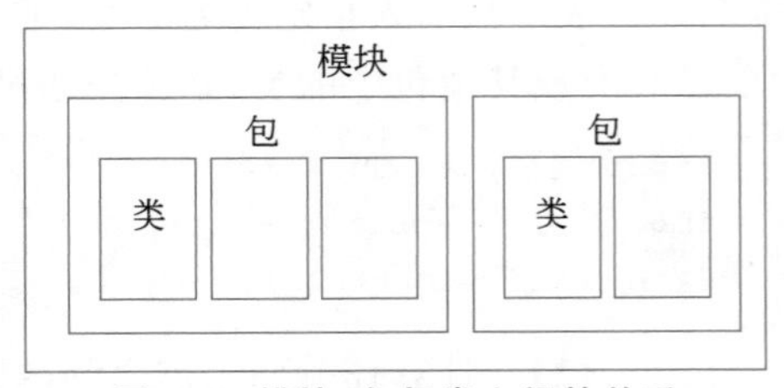

图 4-7 模块、包与类之间的关系

模块由一组密切相关的包和资源以及一个模块描述符文件 module-info 构成。如图 4-8 中的模块 modtest 即由子包 pk1、类 hello 以及 module-info.java 构成。

```
package pk1;
public class hello {
    public void sayHello() {
        System.out.println("hello!!!");
    }
}
```

如果想让模块中的包 pk1 被其他模块访问，则需要在文件 module-info 中用关键字 exports 申明导出包 pk1。

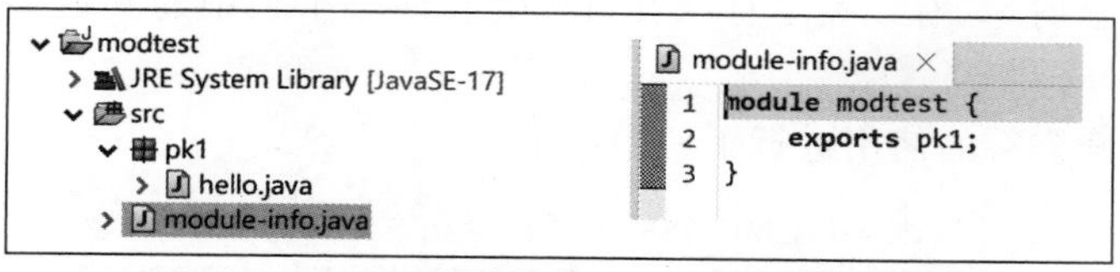

图 4-8　modtest 模块结构以及 module-info 文件

如果另一模块 hellotest(见图 4-9)，想要引用上述模块中的类 hello，则需要在该模块的 module-info 文件中使用 requires 申明依赖的其他模块。

```
module hellotest {
    requires modtest;                                    //依赖模块 modtest
}
```

然后，就可以在模块 hellotest 中引用其他模块 modtest 中的类 hello 了。

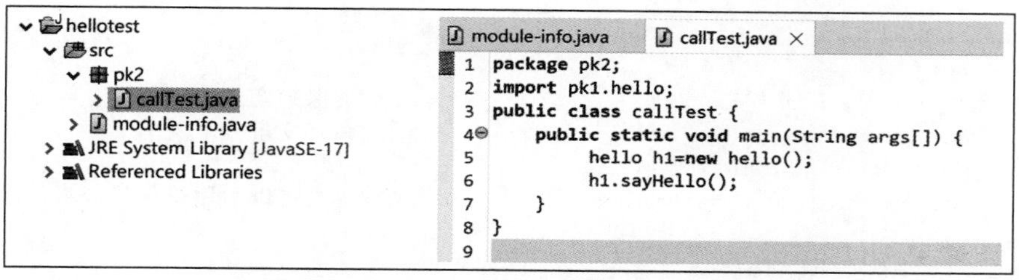

图 4-9　hellotest 模块结构以及 callTest 类

4.6　类及成员修饰符

如果对类及其成员访问有特殊限制，可以引入修饰符对类及其成员进行限定，以说明它们的性质、相互关系和适用范围。常见的修饰符包括 public、protected、private、final、abstract、static 等，本节将重点介绍前 4 种修饰符，abstract 将在第 5 章进行介绍。

1. public

public 可以同时修饰类和成员变量、成员方法。如果 public 修饰一个类名，如 public class A{…}，则表示该类可以被所有的其他类访问或引用，即其他类中可以创建该类的实例，访问该类的所有可见成员变量，调用所有可见方法。

如果类成员(变量和方法)用 public 修饰，表示该类的成员不仅可以被其内部成员访问，而且可以被其他类直接访问，即外界可直接存取公有数据和公有方法。

2. protected

关键字 protected 主要用于修饰类成员，说明该成员是被保护成员，除了可以被类自身访问外，还可以被该类的子类以及与该类在同一个包中的其他类访问。

3. private

关键字 private 主要用于修饰类成员，表示该类成员只能被类自身访问，任何其他类(包括该类的子类)都无权修改或引用。应该将不希望他人随意引用或修改的数据和方法设置为 private，这将使得私有成员对其他类不可见，只有通过声明为 public 的方法才可以对这些数据进行访问,从而达到信息隐藏和封装的目的。

如果类成员前没有 public、protected、private 中的任何一个修饰符,则称它使用了缺省(default)修饰符。这时,只有该类本身以及与该类在同一个包中的其他类才可以直接访问这些缺省成员。

表 4-1 对上述修饰符的使用进行了一个小结。

表 4-1 Java 类的成员变量和方法访问权限

	同一个类	同一个包	不同包中的子类	不同包中的非子类
private	★			
default	★	★		
protected	★	★	★	
public	★	★	★	★

【例 4.13】 修饰符使用示例。

```
package code0406;
class Parent{
    protected int num = 3;                          //受保护变量
    private int day;                                //私有变量
    protected void showNum() {
        System.out.println("number=" + num);        //本类可以引用受保护变量
    }
    private void setDay(int d) {
        day=d;                                      //本类可以引用私有变量
    }
}
class ModifierTest extends Parent {
    public static void main(String args[]) {
        ModifierTest mod= new ModifierTest();
        mod. showNum ();                            //访问父类的被保护成员
        mod.num = 10;                               //访问父类的被保护成员
        mod.setDay(5);                              //错误,引用了父类中的私有方法
    }
}
```

分析:该例中 ModifierTest 是 Parent 的子类,关键字 extends 描述了这种继承关系(继承参见第 5 章)。可以看到,父类中定义的受保护变量(num)、受保护方法(showNum)是可以被该类的子类访问的,但 private 修饰的私有成员(day、setDay 方法)只能父类自己使用,即便对自己的子类也是使用受限的,因此语句 mod.setDay(5)是错误的。

4. final

关键字 final 可用于修饰类、成员变量、成员方法。final 的基本意思是“最终的”,即它所修饰的元素不允许再修改。例如,用 final 修饰的类(叫最终类)不能再有子类;用 final 声明的方法(最终方法)不能再被重写;用 final 声明的成员变量(常量)初始化后,不能再被重新赋值或修改。

【例 4.14】 final 使用示例。

```
package code0406;
final class Circle {
```

```
        final double PI = 3.1416;
        final double area(double r) {
            return (PI * r * r);
        }
    }
    package code0406;
    class FinalTest {
        static public void main(String arg[]) {
            Circle c = new Circle();
            c.PI=2.0;                                   //错误
            System.out.println("area= " + c.area(5.0));
        }
    }
```

分析：类 Circle 中所有成员变量和方法均被 final 修饰，并且类 A 本身也是 final 的。因此，如果生成类 A 的子类或在子类中重写方法 area()是不允许的。与之类似，语句 a.PI＝2.0 对常量进行修改也是错误的。

4.7 类和对象实训任务

【任务描述】

模拟银行 ATM 机，编写一个具有简单操作界面的 Java 应用程序，实现包括存款、取款、查询等功能在内的简单应用。

【任务分析】

(1) 现实世界中的相关操作涉及哪些实体呢？一个是储户，一个是 ATM 机。因此根据面向对象中的抽象原则，可将其抽象为两个类：代表储户的账户信息类，代表银行 ATM 机的 ATM 类，另外加一个主类(负责实例化其他类的对象)。

(2) Account 类记录储户的卡号、姓名、密码和账户余额等信息，并提供 get()方法获取每个属性的值，对账户余额属性提供 subBalance()方法和 addBalance()方法以模拟余额增加、减少的功能。注意，为加强对属性数据的控制访问，考虑将其定义为 private。

(3) ATM 类模拟 ATM 机的主要功能，根据对银行 ATM 机的了解，考虑设计以下主要方法：①welcome()方法，欢迎显示功能；②loadSys()方法，登录功能；③sysOpter()方法，根据用户输入进行任务调度；④inquireInfo()方法，查询用户账户；⑤betBalance()方法，取款；⑥addBalance()，存款；⑦isBalance()方法，判断余额是否足够；⑧isRight()方法，判断卡号、密码是否正确。

(4) ATMTest 为主类，是整个应用执行的入口，主要完成创建 ATM 类实例的功能，并通过对 welcome()和 loadSys()方法的调用显示欢迎界面，提示用户完成系统登录操作。

【任务解决】

【例 4.15】 模拟银行 ATM 机功能。

```
/* Account 类封装储户信息及部分功能 */
package code0407;
class Account {
    private String number = null;                 //卡号
    private String name = null;                   //客户姓名
```

```
    private String password = null;                 //客户密码
    private double money = 0.0;                     //余额
    /***** 构造方法,以生成多个储户信息 ***********/
    public Account(String number, String name, String password, double money) {
        this.number = number;
        this.name = name;
        this.password = password;
        this.money = money;
    }
    protected String getNumber() {
        return number;
    }
    protected String getName() {
        return name;
    }
    protected String getPassword() {
        return password;
    }
    public double getMoney() {
        return money;
    }
    protected void subBalance(double mon) {         //余额减少
        money -= mon;
    }
    protected void addBalance(double mon) {         //余额增加
        money += mon;
    }
}
/ * ATM 类模拟 ATM 机的主要功能 * /
package code0407;
import java.io. * ;

class ATM {
    Account act;
    public ATM() {
        act = new Account("000", "test", "111", 5000);      //生成 Account 实例
    }
    /*********** 欢迎界面 ***********/
    protected void welcome() {
        String str = "-----------------------------------";
        System.out.print(str + "\n");
        System.out.print(" 1.取款." + "\n" + " 2.查询." + "\n" + " 3.存款." + "\n" +
" 4.退出系统." + "\n");
        System.out.print(str + "\n");
    }
    /********** 登录系统 **********/
    protected void loadSys() throws Exception {
        String card, pwd;
        int counter = 0;
        BufferedReader br = new BufferedReader(new InputStreamReader(System.in));
                                                    //创建标准输入输出流,详见第 8 章
        do {
            System.out.println("请输入您的卡号:");
            card = br.readLine();                   //读取键盘输入信息
```

```
            System.out.println("请输入您的密码:");
            pwd = br.readLine();
            if (!isRight(card, pwd)) {
                System.out.println("您的卡号或密码输入有误。");
                counter++;
            } else
                sysOpter();

        } while (counter < 3);
        System.exit(1);                                //应用退出
    }
    /********** 系统操作提示 **********/
    protected void sysOpter() throws Exception {
        int num;
        BufferedReader br = new BufferedReader(new InputStreamReader(System.in));
        System.out.println("请选择您要操作的项目(1-4):");
        num = br.read();                               //num 为 ASCII 码转换的整数
        switch (num) {
        case 49:
            betBalance();
            break;
        case 50:
            inquireInfo();
            break;
        case 51:
            AddBalance();
            break;
        case 52:
            Exit_Sys();
            break;
        }
        System.exit(1);
    }
    /********** 信息查询 **********/
    protected void inquireInfo() throws Exception {
        System.out.print("---------------------\n" + "账号:" + act.getNumber
() + "\n" + "姓名:" + act.getName() + "\n"+ "余额:" + act.getMoney() + "\n" + "----
--------------------\n");
        sysOpter();
    }
    /********** 取款 **********/
    public void betBalance() throws Exception {
        String str = null;
        BufferedReader br = new BufferedReader(new InputStreamReader(System.in));
        do {
            System.out.println("请输入取款数目:");
            str = br.readLine();
            double qu = Double.valueOf(str).doubleValue();
                                                       //将字符串转换为 double 类型
            if (qu > act.getMoney()) {
                System.out.println("余额不足,请重新输入您要取的数目:");
            } else {
                act.subBalance(qu);
```

```
                System.out.println("取款成功,您的账户余额为:" + act.getMoney());
                welcome();
                sysOpter();
            }
        } while (true);

    }
    /********** 存款 **********/
    public void AddBalance() throws Exception {
        String str = null;
        BufferedReader br = new BufferedReader(new InputStreamReader(System.in));
        do {
            System.out.println("请输入存款数目:");
            str = br.readLine();
            double qu = Double.valueOf(str).doubleValue();
            act.addBalance(qu);
            System.out.println("存款成功,您的账户余额为:" + act.getMoney());
            welcome();
            sysOpter();
        } while (true);

    }
    /********** 判断卡内是否有钱 **********/
    protected boolean isBalance() {
        if (act.getMoney() < 0) {
            return false;
        }
        return true;
    }
    /******** 卡号密码是否正确 ********/
    protected boolean isRight(String card, String pwd) {
        if (act.getNumber().equals(card) && act.getPassword().equals(pwd))
            return true;
        else
            return false;
    }
    /********** 结束系统 **********/
    protected void Exit_Sys() {
        System.out.println("感谢您使用本系统,再见!");
        System.exit(1);
    }
}
/* ATMTest 类为主类 */
package code0407;
public class ATMTest {
    public static void main(String[] args) throws Exception {
        ATM atm = new ATM();
        atm.welcome();
        atm.loadSys();
    }
}
```

习题与思考

1. Java类中定义两个或更多方法，它们有相同的方法名而参数不同时称为什么？

A. 继承　　B. 多态性　　C. 构造方法　　D. 方法重载

2. 定义一个时钟类(Clock)，要求如下：

(1) 存储时钟的hour(时,0～23)、minute(分,0～59)、second(秒,0～59)。

(2) 创建新对象时默认为0时0分0秒。

(3) 设置时钟为指定的时间。

(4) 使时钟前进1秒的功能incSecond()。

(5) 以“时:分:秒”的形式输出时钟的时间。

3. 编写一个Java Application程序，使用复数类Complex验证两个复数1+2i和3+4i相加产生一个新的复数4+6i。复数类Complex必须满足如下要求。

(1) 复数类Complex的属性有：①RealPart ,int型，代表复数的实数部分；②ImaginPart,int型，代表复数的虚数部分。

(2) 复数类Complex的方法有：①Complex()，构造函数，将复数的实部和虚部都置0；②Complex(int r , int i)，构造函数，形参r为实部的初值，i为虚部的初值；③Complex complexAdd(Complex a)，将当前复数对象与形参复数对象相加，所得的结果仍是一个复数值，返回给此方法的调用者；④String ToString()，把当前复数对象的实部、虚部组合成a+bi的字符串形式，其中a和b分别为实部和虚部的数据。

4. 给出下列程序运行的结果：

```
public class OverloadDemo {
    void testOverload(int i) {
        System.out.println("int");
    }
    void testOverload(String s) {
        System.out.println("String");
    }
    public static void main(String args[]) {
        OverloadDemo a = new OverloadDemo();
        char ch = 'x';
        a.testOverload(ch);
    }
}
```

5. Java中类成员的访问修饰符有哪些？它们各有什么作用？

第 5 章 Java 继承与高级特性

继承是一种基于已有类创建新类的机制，利用继承可以先创建一个具有广泛意义的类，然后通过派生创建新类，并添加一些特殊的属性和行为。类的继承是实现代码复用最有效的方法。本章将以类的继承为基础并逐渐展开，介绍若干类的高级特征：

(1) 继承的实现。

(2) 方法的重写。

(3) 抽象类与抽象方法。

(4) 接口。

(5) 内部类。

(6) Lambda 表达式。

(7) 泛型。

(8) Java 反射机制。

(9) 注解。

5.1 继承使用初探

当新建一个类时，也许会发现该类与之前的某个类非常相似，如绝大多数的属性和行为都相同。这时，可以选择复制原类中的语句，对其部分修改后加入新类中，但这意味着必须同时维护两个相似的 Java 程序。

另外一种方法是通过继承，让新类自动获得被继承类中已有的属性和方法，同时添加原类中没有的属性和方法即可。例如，根据人的特征定义类 Person 如下：

```
public class Person {
    private String name;
    private int age;
    public void say(){
        System.out.println(name+"can say");
    }
    public void setName(String name) {
        this.name = name;
    }
```

```
    public String getName(){
        return name;
    }
}
```

该类对所有人均适用,但如果根据学生的特点需要定义一学生类,则可以肯定的是学生类中,除了姓名、年龄属性外,还可能有所在学校名称(这是一般人没有的),此外学生的行为中还包括在校学习(study)这一方法(这也是一般人没有的)。因此,可以通过继承的方式建立类 Student 如下:

```
public class Student extends Person{
    String schoolname;                              //增加新的属性 schoolname
    public void study(){                            //增加新的方法 study
        System.out.println("I am studying in school");
    }
    public static void main(String[] args) {
        Student student1 = new Student();
        student1.name="MingM";student1.age=10;//引用继承自父类的变量
        student1.schoolname="CQ";
        student1.say();                             //调用继承自父类的方法
        student1.study();                           //调用子类新增加的方法
        System.out.println("My name is "+student1.name);
        System.out.println("My schoolname is "+student1.schoolname);
    }
}
```

程序运行结果:

```
I can say
I am studying in school
My name is MingM
My schoolname is CQ
```

分析:通过关键字 extends 定义了类 Person 的子类 Student,然后添加了只有学生才有的属性 schoolname 和方法 study()。

在 main()方法中,可以看到尽管 Student 中没有定义变量 name、age 以及方法 say(),但是子类却可以通过继承的方式自动取得,并像访问自己的成员变量和方法一样引用即可。

5.2 类的继承

5.2.1 继承的实现

继承的实现其实非常简单,其格式如下:

```
class 子类名 extends 父类名{
    类体
}
```

extends 是关键字,后跟父类的类名,如果没有父类,则缺省父类是 java.lang.Object。Java 只支持单继承,即只能有一个父类,但类之间的继承可以具有传递性。

子类可以通过继承自动获得父类中访问权限为 public、protected、default 的成员变量和方法，但不能继承权限为 private 的成员变量和方法。

【例 5.1】 类的继承示例。

```
package code0502;
class A {
    int i;
    void showi() {
        System.out.println("i: " + i);
    }
}
class B extends A {
    int k;

    void show() {
        System.out.println("k: " + k);
        showi();
    }
    void sum() {
        System.out.println("i+k: " + (i + k));
    }
}
public class Simple {
    public static void main(String args[]) {
        A superOb = new A();
        B subOb = new B();
        superOb.i = 10;                          //对父类对象的成员赋值.
        System.out.println("Contents in 父类: ");
        superOb.showi();
        subOb.i = 7;                             //对子类中继承得到的变量 i 赋值
        subOb.k = 9;
        System.out.println("Contents in 子类: ");
        subOb.show();
        System.out.println("Sum of i and k in 子类:");
        subOb.sum();
    }
}
```

程序运行结果：

```
Contents in 父类:
i: 10
Contents in 子类:
k: 9
i: 7
Sum of i and k in 子类:
i+k: 16
```

分析：类 B 中虽然没有定义变量 i 和方法 showi()，但却可以通过继承关系获得，因此在类 B 中直接引用 i 和方法 showi()是正确的。但如果在类 A 的语句 int i 前加上修饰符 private，则上述程序会出现编译错误，其原因在于类 B 无法继承到 private 类型的变量 i，因此 i 对类 B 是不可见的。

提示：①尽管一个子类可以从父类继承所有允许的方法和变量，但它不能继承构造函数，掌握这点很重要。一个类要得到构造函数，只有两个办法：重写构造函数；根本不写构造函数，这时系统为每个类生成一个缺省构造函数。②为了防止继承被滥用，Java15 引入了一种特殊的密封类(使用 sealed 修饰 class)，并通过 permits 明确写出能够从该 class 继承的子类名称，如 public sealed class Shape permits Circle{}，这表明只允许类 Circle 继承 Shape。

5.2.2 继承与重写

在类的继承过程中，如果子类中新增的变量和方法与父类中原有的数据和方法同名，则会重写(也称覆盖)从父类继承来的同名变量和方法。重写又分变量重写和方法重写，变量重写是指父类和子类中的变量名相同，数据类型也相同。方法重写与之前介绍的方法重载相似，但更严格，不仅要求父类与子类中的方法名称相同，而且参数列表也要相同，只是实现的功能不同。

【例 5.2】 重写父类中的同名方法和变量。

```
package code0502;
class SuperCla {
    int a = 3, b = 4;
    void show() {
        System.out.println("super result=" + (a + b));
    }
}
class SubCla extends SuperCla {
    int a = 10;                           //重写父类中同名的变量 a
    void show() {                         //重写父类中同名的方法 show
        int c = a * b;
        System.out.println("sub result=" + c);
    }
}
public class OverrideTest {
    public static void main(String args[]) {
        SuperCla sp = new SuperCla();
        SubCla sb = new SubCla();
        sp.show();                        //此处调用的是父类中的方法 show
        System.out.println("In super Class:a=" + sp.a);  //此处引用的是父类中的变量 a
        sb.show();                        //此时子类对象的 show 方法覆盖了父类的同名方法
        System.out.println("In sub Class:a=" + sb.a);
    }
}
```

程序运行结果：

```
super result=7
In super Class:a=3
sub result=40
In sub Class:a=10
```

分析：子类 SubCla 中定义有与父类 SuperCla 同名的变量 a 和方法 show()，因此使用子类对象 sb 时访问变量 a 和方法 show()时，引用的是子类中的成员，父类的同名变量被覆盖，同名的方法被重写。

如果想在子类中访问父类中被覆盖的成员怎么办呢？这时可以使用关键字 super 来解决这一问题，基本格式如下。

访问父类成员：

```
super.成员变量
```

或


```
super.成员方法([参数列表])
```

访问父类构造方法:

```
super([参数列表])
```

【例 5.3】 super 关键字的使用。

```
package code0502;
//定义员工类
class Employee {
    private String name;
    private int salary;
    public String getDetails() {
        return "Name:" + name + "\nSalary:" + salary;
    }
    Employee() {
        name = "Tom";
        salary = 1234;
    }
}
//定义经理类
class Manager extends Employee {
    public String department;
    /*重写 getDetails 方法*/
    public String getDetails() {
        System.out.println("I am in Manager");
        return super.getDetails();                //调用父类的 getDetails 方法
    }
    Manager() {
        super();                                  //访问父类的无参构造方法,即 Employee()
        department = "sale";
    }
}
public class Inheritance {
    public static void main(String arg[]) {
        Manager m = new Manager();
        System.out.println(m.getDetails());
        System.out.println("department:" + m.department);
    }
}
```

程序运行结果:

```
I am in Manager
Name:Tom
Salary:1234
department:sale
```

分析:程序首先对 Manager 实例化,并自动调用无参构造方法 Manager(),主要完成两项主要任务:一是通过 super()调用父类的无参构造函数,即 Employee();二是对子类中新增变量 department 初始化。接着,语句 m.getDetails()调用子类的同名方法 getDetails(),并在方法体中通过 super.getDetails()实现对父类中 getDetails()方法的调用。

注意：生成一个子类的实例时，首先要执行子类的构造方法，但如果该子类继承自某个父类，则执行子类的构造方法前，系统自动调用父类的无参构造函数。因此，例 5.3 的方法 Manager()中去掉语句 super()，效果是完全相同的。

与 super 不同，关键字 this 的主要作用是表示当前对象的引用，当局部变量和类的成员变量同名时，该局部变量作用区域内成员变量就被隐藏了，必须使用 this 来指明。

【例 5.4】 this 关键字的使用。

```
package code0502;
public class ThisTest {
    public static void main(String[] args) {
        Local aa = new Local();
    }
}
class Local {
    public int i = 1;                              //这个 i 是成员变量
    Local(int i) {                                 //这个 i 是局部变量
        System.out.println("this.i ="+ this.i);   //this.i 指的是对象本身的成员变量 i
        System.out.println("i = "+i);             //变量 i 前没有 this,因此是局部变量
    }
    Local(){
        this(6);
    }
}
```

程序运行结果：

```
this.i =1
i = 6
```

分析：关键字 this 的主要作用有：①通过它引用成员变量，如上例中 this.i 即指的是当前对象中的成员变量 i；②通过 this 调用类的构造方法，如上例中的 this(6)将调用对应的构造方法 Local(int i)。

5.2.3 继承与类型转换

和标准类型数据的转换一样，不同类的对象之间也可以相互转换，但前提是源和目标类之间必须通过继承相联系。转换可分为显式和隐式两种，显式转换格式如下：

```
(类名)对象名
```

它将对象转换成类名所表示的其他对象。Java 支持父类和子类对象之间的类型转换，如果是子类对象转换为父类，可进行显式转换或隐式转换；如果是父类对象转换成子类，编译器首先要检查这种转换的可行性，如果可行，则必须进行显式转换。

【例 5.5】 对象类型的转换示例。

```
package code0502;
class CA{
    String s="class CA";
}
class CB extends CA{
    String s="class CB";
```

```
    }
    class Convert{
        public static void main(String args[]){
            CB bb,b=new CB();
            CA a,aa;
            a=(CA)b;                                    //显式转换
            aa=b;                                       //隐式转换
            System.out.println(a.s);
            System.out.println(aa.s);
            bb=(CB)a;                                   //显式转换
            System.out.println(bb.s);
        }
    }
```

程序运行结果：

```
class CA
class CA
class CB
```

分析：b 是子类 CB 的实例，将其转换为父类 CA 的实例时可以进行显式或隐式转换。而父类对象 a 转换为子类 CB 的对象时，必须进行显式转换。

5.2.4 实用案例

【例 5.6】 通过继承定义员工和经理类。

普通员工和经理作为职员有很多共同之处，如都可以取得工资报酬，但经理可能还会获得额外的奖金，因此在成员属性和方法上可能有特殊之处。可以将 NewEmployee 类定义为父类，NewManager 类作为子类，子类继承了父类，并添加了一个新的 setBonus()方法，用于增加奖金，最后打印出结果。参考实现代码如下：

```
package code0502;
import java.util.*;
class NewEmployee {
    private String name;
    private double salary;
    private Date hireDay;

    public NewEmployee(String n, double s, int year, int month, int day) {
        name = n;
        salary = s;
        GregorianCalendar calendar = new GregorianCalendar(year, month - 1, day);
        hireDay = calendar.getTime();
    }
    public String getName() {
        return name;
    }
    public double getSalary() {
        return salary;
    }
    public Date getHireDay() {
        return hireDay;
    }
```

```
    public void raiseSalary(double byPercent) {
        double raise = salary * byPercent / 100;
        salary += raise;
    }
}

package code0502;
class NewManager extends NewEmployee {
    private double bonus;
    public NewManager(String n, double s, int year, int month, int day) {
        super(n, s, year, month, day);
        bonus = 0;
    }
    public double getSalary() {
        double baseSalary = super.getSalary();
        return baseSalary + bonus;
    }
    public void setBonus(double b) {
        bonus = b;
    }
}

package code0502;
public class ManagerTest {
    public static void main(String[] args) {
        NewEmployee e= new NewEmployee("Harry Hacker", 50000, 1989, 10, 1);
        e.getName();
        System.out.println(e.getName()+":"+e.getSalary());
        NewManager boss = new NewManager("Carl Cracker", 80000, 1987, 12, 15);
        boss.setBonus(5000);
        System.out.println(boss.getName()+":"+boss.getSalary());

    }
}
```

程序运行结果：

```
Harry Hacker:50000.0
Carl Cracker:85000.0
```

5.3 多　　态

5.3.1 多态性的概念

简单地说，多态性就是一个名称可以对应多种不同的实现方法。Java 语言的多态性体现在两个方面：编译多态和运行多态。

编译多态是指在程序编译过程中体现出的多态性，如方法重载。尽管方法名相同，但由于参数不同，在调用时系统根据传递参数的不同确定被调用的方法，这个过程是在编译时完成的。例如，定义一个作图的类，它有一个 draw()方法用于绘图，根据不用的使用情况，可以接收字符串、矩形、圆形等参数。对于每一种实现，方法名都为 draw()，只不过具体实现方式不同，不用另外重新起名，这样大大简化了方法的实现和调用，程序员无须记住很多的方法名，只需传递相应的参数即可。

运行多态则是由类的继承和方法重写引起的,由于子类继承了父类的属性和方法,因此,凡是父类对象可以使用的地方,子类对象也可以使用。如例 5.5 中的类 CA 和 CB,可以直接生成子类 CB 的对象,并将该引用赋给父类 CA 的对象,即

```
CA a=new CB();
```

它等价于下面两个子句:

```
CB b=new CB(); CA a=b;          //隐式类型转换
```

现在有一个问题:如果子类重写了父类的成员方法,调用该方法时,到底应该调用父类中的方法还是子类中的方法?这无法在编译时确定,需要系统在运行时根据实际情况来决定,所以这种由方法重写引起的多态叫运行多态。

Java 规定:对重写的方法,Java 根据调用该方法的实例的类型来决定选择哪种方法。对子类的实例,如果子类重写了父类的方法,则调用子类的方法;如果子类没有重写父类的方法,则调用父类的方法。

【例 5.7】 类的多态性示例。

```
package code0503;
class A {
    void callme() {
        System.out.println("inside A");
    }
}
class B extends A {
    void callme() {
        System.out.println("inside B");
    }
}
class Poly {
    public static void main(String args[]) {
        A a = new A();
        B b = new B();
        A c = new B();
        a.callme();
        b.callme();
        c.callme();
    }
}
```

程序运行结果:

```
inside A
inside B
inside B
```

分析:对实例 a 和 b,它们的类型为类 A 和 B,所以分别调用父类和子类中的方法 callme()即可。对实例 c,它被实例化为子类 B 的一个对象,且子类重写了方法 callme(),因此根据转型规则,(由于调用该方法的实例类型为 B)这时应该调用子类 B 的方法,所以输出为 inside B。

多态性在实际应用中有很多好处。

(1) 可替换性:多态对已存在代码具有可替换性。例如,多态对圆 Circle 类有效,对其他

任何圆形几何体(如圆环)也同样有效。

(2) 可扩充性：多态对代码具有可扩充性。增加新的子类不影响已有类的多态性、继承性。实际上，新加子类更容易获得多态功能。例如，在实现了圆锥、半圆锥以及半球体的多态基础上，很容易增添球体类的多态性。

5.3.2 实用案例

【例 5.8】 将员工和经理存入同一数组。

数组一般要求存入的数据必须是同一类型，如整形数组要求每个数组成员均为整数。但之前定义的类 NewEmployee 和 NewManager 显然不是同一类型，那么如何将其存入到同一个数组中呢？实现代码如下：

```
package code0502;
public class EmployeeArray {
    public static void main(String[] args) {
        NewEmployee[] staff = new NewEmployee[3];
        NewManager boss = new NewManager("Carl", 80000, 1987, 12, 15);
        boss.setBonus(5000);
        staff[0] = boss;
        staff[1] = new NewEmployee("Harry", 50000, 1989, 10, 1);
        staff[2] = new NewEmployee("Tommy", 40000, 1990, 3, 15);
        for (NewEmployee e : staff)
            System.out.println("name=" + e.getName() + ",salary=" + e.getSalary());
    }
}
```

程序运行结果：

```
name=Carl,salary=85000.0
name=Harry,salary=50000.0
name=Tommy,salary=40000.0
```

分析：满足上述需求的关键在于数组类型的定义，如本例中将数组申明为父类 NewEmployee 类型，这样将子类对象存入数组时，系统将利用隐式类型转换规则，将其统一到父类类型中，即可实现将不同类对象存入同一数组的目标。

如何测试一个对象是否为某种类型的实例呢？如本例中，我们想知道 staff[0]是否为 NewManager 的实例，可以通过 instanceof 运算符来实现，格式为 variable instanceof TypeName，如果 variable 为 TypeName 或 TypeName 父类型的实例，运算结果为 true，否则返回 false。因此，表达式 staff[0] instanceof NewManager 的结果为 true，staff[0] instanceof NewEmployee 的结果也为 true，staff[1] instanceof NewManager 的结果则为 false。

5.4 抽象类与抽象方法

5.4.1 定义抽象类及实现抽象方法

关键字 abstract 修饰的类称为抽象类，抽象类是一种没有完全实现的类。不能用它实例化任何对象，它的主要用途是用来描述一些概念性的内容，然后在子类中具体去实现这些概念，这样可以提高开发效率，统一用户接口，所以抽象类更多是作为其他类的父类。

抽象类中可以含有抽象方法,抽象方法是用 abstract 修饰的方法,抽象方法只有方法的返回值、名称和参数列表,没有方法体,它必须在子类中具体实现该方法(即给出方法体)。

【例 5.9】 定义抽象类和它的具体实现。

```
package code0504;
abstract class Abs {
    abstract void show();
    abstract void show(int i);
}
/* 定义抽象类 Abs 的子类 Real */
public class Real extends Abs {
    int x;
    void show() {                              //实现抽象方法 abstract void show()
        System.out.println("x=" + x);
    }
    void show(int i) {                         //实现抽象方法 abstract void show(int i)
        x = i;
        System.out.println("x=" + x);
    }
}
```

分析:Abs 是一个抽象类,其中两个方法也是抽象的,没有任何代码。子类 Real 中重写了这两个方法,并给出具体的实现。

注意:如果一个类包含抽象方法,则必须被定义为抽象类,但抽象类不一定要包含抽象方法。

5.4.2 实用案例

【例 5.10】 通过抽象类实现自定义堆栈。

```
package code0504;
abstract class Access{                 //定义一个抽象的类 Access,提供基本的存入、取出操作
    abstract void put(char c);
    abstract char get();
}
package code0504;
class LinkedNode {                                  //定义双向链表节点
    char data;
    LinkedNode back;
    LinkedNode forward;
}
package code0504;
class MyStack extends Access {                      //定义栈类
    private LinkedNode bottom=new LinkedNode();
    private LinkedNode top=bottom;                  //初始化栈顶与栈底

    /* 实现接口 Access 的 put()方法,该方法向栈存一个字符 */
    public void put(char c){
        top.forward=new LinkedNode();
        top.forward.data=c;
        top.forward.back=top;
        top=top.forward;
```

```
    }
    /*实现接口 Access 的 get()方法,该方法从栈中取一个字符*/
    public char get(){
        if(top!=bottom){                                //如果栈不为空,则取数
          char ch=top.data;
          top.back.forward=null;
          top=top.back;
          return ch;
        }
        else{
          System.out.println("The stack is empty!");
          return '\0';
        }
    }
}
package code0504;
public class DataTest {
    public static void main(String args[]){
        MyStack s = new MyStack();
        s.put('x');
        s.put('y');
        s.put('z');                                     //向栈 s 中存入 3 个字符
        System.out.println("In Stack:");
        System.out.println(s.get());
        System.out.println(s.get());
        System.out.println(s.get());                    //从栈 s 中取数并显示
    }
}
```

程序运行结果：

```
In Stack:
z
y
x
```

分析：基于抽象类 Access 的定义，可以通过继承产生更多相似的子类，如队列、树结构等。每个子类虽然在完成数据存入、取出时的具体逻辑不同，但均保持相同的访问接口形式，如方法 get()、put()。

5.5 接　口

5.5.1 接口定义

接口是抽象类的一种变体，与抽象类的定义不同，接口是用关键字 interface 定义的，其格式如下：

```
[修饰符] interface 接口名
{
    //变量和方法声明
}
```

大括号为接口体，其中可以定义常量、抽象方法、默认方法和静态方法等。例如：

```
interface Myinter
{
    double g=9.8;  //等价于 public static final double g=9.8
    void show();   //编译器自动添加 public abstract,等价于 public abstract void show()
}
```

如果接口中定义的是抽象方法（没有 default 或 static 修饰符），则只有方法声明，没有方法体。与抽象类使用相似，不能通过关键字 new 创建接口的实例。

5.5.2 接口实现

由接口生成子类不是通过 extends 实现的，而是用关键字 implements 来实现一个接口的，其格式如下：

```
class 类名 implements 接口名 [extends 类名]
{
    //类体
}
```

被实现的接口可以有多个，之间用逗号分隔。这点是类在继承时是无法做到的，Java 只支持单继承，如果想达到多继承的效果，则可以通过多个接口的方式来实现。例如，根据 5.5.1 节定义的接口 Myinter，可以定义类 MyinterImp 如下：

```
class MyinterImp implements Myinter
{
    public void show()                                    //实现抽象方法 show()
    {
        System.out.println("g="+g);
    }
}
```

与类之间的继承关系相似，接口之间也可以通过关键字 extends 来继承，但其中抽象的方法不能被具体实现。

【例 5.11】 接口之间的继承示例。

```
package code0505;
interface IA {
    int a = 1;
    void showa();
}
interface IB extends IA {                                 //接口 IB 继承自 IA
    int b = 2;
    void showb();
}
interface IC extends IA, IB {                             //接口 C 继承自 IA,IB
    int c = 3;
    void showc();
}
public class InterfaceTest implements IC {
    public void showa() {
        System.out.println("aaaa");
    }
```

```
    public void showb() {
        System.out.println("bbbb");
    }
    public void showc() {
        System.out.println("cccc");
    }
}
```

分析：由于 InterfaceTest 实现了接口 IC，而 IC 又通过继承关系获取了抽象方法 showa()、showb()和 showc()，因此类 InterfaceTest 必须在实现代码中重写这 3 个抽象方法，缺一不可。

5.5.3 接口中的默认方法和静态方法

JDK8 推出以后，允许接口中包含具体实现的两种方法：一是默认方法，用 default 修饰；二是静态方法，使用 static 修饰。

1. default 方法

default 方法可以有方法体，且只能通过接口实现类的对象来调用。实现类可以重写 default 方法。其价值在于：可在不破坏 Java 现有实现架构的情况下，向接口里增加新的方法，子类可以直接调用。

例如，对前述接口 Myinter，如果已经建立了实现类 MyinterImp，再临时向 Myinter 新增抽象方法 methodTest()，系统就会报错。因为类 MyinterImp 中没有对该方法的实现。但可以通过向接口 Myinter 加入默认方法 public default void methodTest ()避免该问题。如下述代码中调用 Myinter 中的默认方法是允许的。

```
class MyinterImp implements Myinter{
    public void show()                              //实现抽象方法 show()
    {
        System.out.println("g="+g);
    }
    public static void main(String[] args) {
        MyinterImp a = new MyinterImp ();
        a.show();
        a. methodTest();                            //调用 Myinter 中默认方法 methodTest
    }
}
```

2. static 方法

static 方法也可以有方法体，且只能通过接口名直接调用，不可以通过实现类的类名或者实现类的对象调用。例如，改写接口 Myinter，添加一 static 方法如下：

```
interface Myinter{
    double g=9.8;
    void show();
    public static void methodStatic(){
        System.out.println("这是接口的静态方法！");
    }
}
```

在其他类中，可以通过 Myinter.methodStatic()直接调用该静态方法。

5.5.4 Comparable 与 Comparator 接口

除可以使用自定义接口外,Java 也提供了一系列公共接口以实现一些常用的功能,如排序。前面可以很方便地比较不同 String 对象大小并排序,其原因就在于 String 类实现了一个重要接口 Comparable。

如果现在比较的不是 String,而是两个自定义类(如 Rectangle)对象的大小,应该怎么做呢?答案是像 String、包装类一样,让 Rectangle 实现 Comparable 接口,并重写 comparaTo()方法;或者实现 Comparator 接口,并重写 compare()方法,通过观察方法的返回值(1、−1 或 0)可以判断对象间的大小。

1. Comparable 接口

Comparable 接口位于包 java.lang 中。一个类如果实现了 Comparable 接口,则意味着该类支持两个对象间比较大小。如果对象数组实现了该接口,则可通过 Arrays.sort 自动排序。

Comparable 接口源代码如下:

```
public interface Comparable<T> {
    public int compareTo(T o);
}
```

该方法的返回值有 3 种情况:①e1.compareTo(e2) > 0 即 e1 > e2;②e1.compareTo(e2) = 0 即 e1 = e2;③e1.compareTo(e2) < 0 即 e1 < e2。

对待比较大小的类,只须实现 Comparable<T>接口中的 compareTo()方法即可。

【例 5.12】 Comparable 接口的使用。

```
package code0505;
public class ComparableRectangle implements Comparable<ComparableRectangle> {
    double width;
    double height;
    public ComparableRectangle(double width, double height) {
        this.width = width;
        this.height = height;
    }
    public double getArea() {
        return width * height;
    }
    @Override
    public int compareTo(ComparableRectangle o) {
        return (int) (getArea() - o.getArea());
    }
    public String toString() {
        return "width" + " = " + width + "  " + "height" + " = " + height + "  " + "
Area: "  + getArea();
    }
    public static void main(String[] args) {
        ComparableRectangle[] rectangles = { new ComparableRectangle(3.0, 5.0),
new ComparableRectangle (12. 0, 6. 0), new ComparableRectangle (7. 0, 15. 0), new
ComparableRectangle(2.0, 25.0) };
        java.util.Arrays.sort(rectangles);
        for (ComparableRectangle rectangle : rectangles) {
            System.out.println(rectangle.toString() + " ");
```

```
        }
    }
}
```

程序运行结果：

```
width = 3.0  height = 5.0   Area: 15.0
width = 2.0  height = 25.0   Area: 50.0
width = 12.0  height = 6.0   Area: 72.0
width = 7.0  height = 15.0   Area: 105.0
```

分析：这里在矩形类 ComparableRectangle 的 compareTo()方法中定义了比较代码，再将 4 个矩形对象放入数组 rectangles 中，并传入 Arrays 类的 sort(List<T> list)方法中，按默认排序规则进行排序。

2. Comparator 接口

Comparator 位于包 java.util 中，接口定义如下：

```
public interface Comparator<T> {
    * @return     返回 = 0  表示 o1 == o2
        返回 > 0  表示 o1 > 02
        返回 < 0  表示 o1 < 02
    int compare(T o1, T o2);
}
```

与 Comparable 接口不同，Comparable 接口通常将比较代码嵌入到需要比较的类体中，而 Comparator 接口则在一个独立的类中实现比较。

因此，如果前期类的设计没有考虑类的 Compare 问题，因而没有实现 Comparable 接口，那么后期可以通过 Comparator 接口来实现比较算法进行排序。

【例 5.13】 Comparator 接口的使用。

```
package code0505;
class User {
    String name;
    int age;
    public User(String name,int age) {
        this.name=name;
        this.age=age;
    }
    public String getName() {
        return name;
    }
    public void setName(String name) {
        this.name = name;
    }
    public int getAge() {
        return age;
    }
    public void setAge(int age) {
        this.age = age;
    }
}
public class ComparatorTest {
```

```
    static class MyComparatorName implements Comparator<User> {
        @Override
        public int compare(User u1, User u2) {
            return u1.getName().compareTo(u2.getName()) * -1;
        }
    }
    static class MyComparatorAge implements Comparator<User> {
        @Override
        public int compare(User u1, User u2) {
            return (u1.getAge() - u2.getAge());
        }
    }
    public static void main(String[] args) {
        User[] auser = new User[3];
        auser[0] = new User("apple", 20);
        auser[1] = new User("orange", 10);
        auser[2] = new User("banana", 30);
        Arrays.sort(auser, new MyComparatorName());
        for (User u : auser) {
            System.out.println(u.getName());
        }
        Arrays.sort(auser, new MyComparatorAge());
        for (User u : auser) {
            System.out.println(u.getAge());
        }
    }
}
```

程序运行结果：

```
orange
banana
apple
10
20
30
```

分析：这里实现了两个比较器：①MyComparatorName 以 name 为依据，通过重写 compare()方法实现对象比较；②MyComparatorAge 则以 age 为依据实现对象比较。与 Comparable 不同，比较器与比较对象(User)之间是独立的。Arrays 类的 sort(List<T> list，Comparator<? super T> c)方法利用接收到的数据集 list 和比较器 c 完成排序。

5.5.5 实用案例

【例 5.14】 人驾驶不同汽车的多态实现。

假定一个司机拥有驾驶小汽车的执照，则通常情况下，他若能驾驶宝马车就应该能驾驶奔驰车，甚至驾驶更多种类的车辆。因此，如何在车型可能不断增加、扩展的情况下，保持代码结构的稳定性是本例中除功能实现外更值得关心的问题。

实现的程序代码如下：

```
package code0505.CarDrive;
public interface IDriver {
```

```
        public void drive(ICar car);            //这里接口 Idriver 构成了对接口 Icar 的依赖
    }
    package code0505.CarDrive;
    public interface ICar {
        public void run();
    }
    package code0505.CarDrive;
    public class BMW implements ICar {
        public void run() {
            System.out.println("宝马车正在运行...");
        }
    }
    package code0505.CarDrive;
    public class Benz implements ICar {
        public void run() {
            System.out.println("奔驰车正在运行...");
        }
    }
    package code0505.CarDrive;
    public class Driver implements IDriver {
        public void drive(ICar car) {           //参数为所有车型的父类 ICar
            car.run();
        }
    }
    package code0505.CarDrive;
    public class Client {
        public static void main(String[] args) {
            IDriver zhangSan = new Driver();
            ICar benz = new Benz();
            zhangSan.drive(benz);               //开奔驰车,调用的是类 Benz 中的 run()方法
            ICar bmw=new BMW();
            zhangSan.drive(bmw);                //开宝马车,调用的是类 BMW 中的 run()方法
            }
    }
```

程序运行结果:

```
奔驰车正在运行...
宝马车正在运行...
```

分析:上述类和接口之间的关系如图 5-1 所示。接口 Idriver 构成了对接口 Icar 的依赖,Driver 是 Idriver 的类实现,BMW 和 Benz 是对 Icar 的类实现。

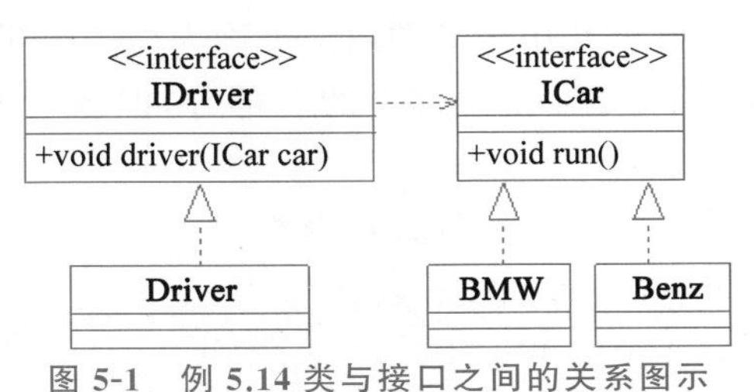

图 5-1 例 5.14 类与接口之间的关系图示

由于 Driver 类的 drive(ICar car)方法中使用了抽象的接口 Icar 作为形式参数,根据 5.3 节的运行多态规则,可以在向其传递实参时直接使用 Icar 的子类对象,如 benz、bmw。当然,如果 Icar 产生有新的车型(即新的子类)了,也可以直接通过 drive(ICar car)方法调用,这样调用的接口对用户是统一的,但实际调用的 run()方法根据传入参数是不同的。

作为对比,如果将上例中的 Driver 类改为

```
public class Driver {
    public void drive(Benz benz) {          //参数类型为 Benz,只能向其传递 Benz 的对象
        benz.run();                         //调用 Benz 的 run()方法
    }
}
```

这时,Driver 类不再依赖抽象的接口 Icar,而是直接依赖具体的实现子类 **Benz**,这会给系统的稳定性造成影响,因为 public void drive(**Benz** benz)是专门针对 Benz 的,如果一个司机现在要驾驶 BMW,就需要修改 Driver,增加一个新的类似方法,如 public void drive(**BMW** bmw),如此反复,Driver 类的稳定性就会很差。

5.6 内 部 类

内部类是指在一个外部类的内部再定义一个类。内部类作为外部类的一个成员,并且依附于外部类而存在的。内部类主要有:成员内部类、局部内部类、静态内部类、匿名内部类。

5.6.1 成员内部类

成员内部类作为外部类的一个成员存在,与外部类的属性、方法并列。

【例 5.15】 成员内部类。

```
package code0506;
public class Outer1 {
    private int k = 20;
    public static void f1() {
    }

    public class Inner1 {
        public void f() {
            System.out.println("In inner Class");
        }
    }
}
```

编译上述代码会产生两个文件:Outer1.class 和 Outer $ Inner1.class。

5.6.2 局部内部类

在方法中定义的内部类称为局部内部类。与局部变量类似,局部内部类不能有访问控制符,因为它不是外部类的一部分,但是它可以访问当前代码块内的常量,和此外部类所有的成员。

【例 5.16】 局部内部类。

```
package code0506;
class Outer2 {
    public void doSomething() {
        class Inner2 {
            public void seeOuter() {
            }
```

```
        }
    }
}
```

5.6.3 静态内部类(嵌套类)

静态的含义是该内部类可以像其他静态成员一样,没有外部类对象时也能够访问它。静态嵌套类不能访问外部类的成员和方法。

【例 5.17】 静态内部类。

```
package code0506;
class Outer3{
    static class Inner3{}
}
public class TestOuter3 {
    public static void main(String[] args){
        Outer3.Inner3 n = new Outer3.Inner3();
    }
}
```

5.6.4 匿名(内部)类

匿名内部类就是没有名字的内部类。什么情况下需要使用匿名内部类?如果满足下面的一些条件,使用匿名内部类是比较合适的:

(1) 只用到类的一个实例。

(2) 类在定义后马上就用到。

(3) 类非常小。

由于匿名类没有名称,所以没办法引用它们,必须在创建时作为 new 语句的一部分来声明它们:

```
new <类或接口>  <类的主体>
```

这种形式的 new 语句声明一个新的匿名类,它继承某个给定的类,或者实现一个给定的接口,并创建该类的一个实例,并把它作为语句的结果返回。要继承的类或要实现的接口一般放在关键字 new 后边,然后是匿名类的主体定义。

【例 5.18】 匿名类。

```
package code0506;
interface Exam {
    void print1();
}
public class Outer4 {
    public Exam test() {
        return new Exam() {
            public void print1() {
                System.out.println("Hello world!!");
            }
        };
    }
    public static void main(String args[]) {
```

```
        Outer4 c = new Outer4();
        Exam e = c.test();
        e.print1();
    }
}
```

5.6.5 实用案例

【例 5.19】 匿名类在可视化界面设计中的应用。

如图 5-2 所示，创建一个可视化界面，界面中含有一个新建(New)按钮(见图 5-2(a))。当用户单击 New 按钮时，将弹出消息对话框，显示“单击了新建按钮”(见图 5-2(b))。

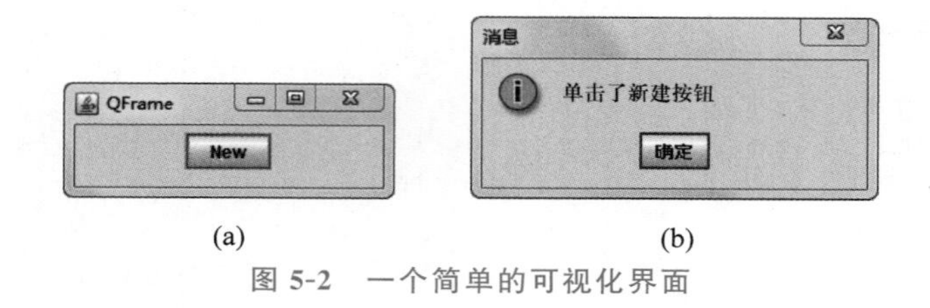

(a)　　(b)

图 5-2　一个简单的可视化界面

Java 中匿名类用的最多的地方就是在可视化界面设计中，特别是将事件监听器注册到某个组件上时，具体代码如下：

```
package code0506;
import javax.swing.*;
import java.awt.event.*;
public class QFrame extends JFrame {
    public QFrame() {
        JButton jbtNew = new JButton("New");
        JPanel panel = new JPanel();
        panel.add(jbtNew);
        add(panel);
        jbtNew.addActionListener(new ActionListener() {
        //新建一匿名类,并将该类对应的事件监听器注册到新建按钮 jbtNew 上
            public void actionPerformed(ActionEvent e) {
                JOptionPane.showMessageDialog(null, "单击了新建按钮");
                System.out.println("Process new");
            }
        });
    }
    public static void main(String[] args) {
        JFrame frame = new QFrame();
        frame.setTitle("QFrame");
        frame.setDefaultCloseOperation(JFrame.EXIT_ON_CLOSE);
        frame.setLocationRelativeTo(null);
        frame.pack();
        frame.setVisible(true);
    }
}
```

关于可视化界面设计与匿名类的更多内容，详见第 10 章。

5.7 Lambda 表达式

5.7.1 初识作用

Lambda 表达式的主要作用：替换之前代码中实现一个匿名内部类的过程，使得代码更加简洁。例如，没有使用 Lambda 表达式之前，Java 对字符串排序的方法如下：

```
String[] names = { "Apple", "Orange", "Banana" };
Arrays.sort(names, new Comparator<String>(){
    @Override
    public int compare(String a,String b){
        return b.compareTo(a);
    }
});
```

这里，sort()方法中需要生成并传入一个匿名的比较器对象。上述代码看似很长，其实最关键的代码就一行：return b.compareTo(a);，其他代码都很冗余，因此 Java8 提出用 lambda 表达式以及函数式接口简化上述过程，可修改代码如下：

```
Arrays.sort(names, (a,b)-> return b.compareTo(a));
```

这里的(a,b)->**return b.compareTo**(a)就是 Lambda 表达式，可以看到相比之前的实现方式，其代码更简洁优雅。

5.7.2 Lambda 表达式定义

Lambda 表达式的主要构成如下：

```
(形式参数) -> {代码块}
```

(1) 形式参数：参数部分可以不用指定参数类型，编译器会根据上下文推导出参数的类型，在仅有一个参数时圆括号可以省略。

(2) ->：由英文中短横线和大于号组成。

(3) 代码块：是我们具体要做的事情，也就是前面常写的匿名函数的方法体。若方法体中只有一行语句，则可以不用大括号包裹，编译器会自动 return 返回值；有大括号和返回值时，需要显式指定 return 表达式。

例如：

```
() -> System.out.print("Java8 lambda.");//无参数,无返回值
x ->  5 * x                           //1 个输入参数,类型为数字,返回值为其值的 5 倍
(x, y) -> x - y                       //2 个输入参数,类型为数字,返回值为其差值
(String str) -> System.out.print(str) //1 个输入参数,类型为 String ,无返回值
```

5.7.3 函数式接口

Lambda 表达式无法单独使用，需要配合相应的函数式接口才能发挥其价值。下面先介绍函数式接口，再结合实例分析 Lambda 表达式如何使用。

函数式接口指的是有且只有一个抽象方法的接口，但是接口中也可以有其他方法(默认、静态、私有)。之前介绍过的 Comparable 接口、Comparator 接口就是函数式接口，它们都只含

有一个抽象方法。

定义函数式接口时可以给接口添加@FunctionalInterface 注解,如果写的接口不符合函数式接口规范,则编译器会报错。例如:

```
@FunctionalInterface
interface MyFunctionalInterface{
    public void method();
}
```

定义好函数式接口后,就可以用作方法的参数类型或返回值类型,配合 Lambda 表达式使用了。

【例 5.20】 函数式接口用作方法的参数类型。

```
public class TestLambda1 {
    public static void show(MyFunctionalInterface myInter){
        myInter.method();
    }
    public static void main(String[] args) {
        show(()->System.out.println("方法实现 one"));
        show(()->{
            System.out.println("方法实现 two");
            Random ran = new Random();
            System.out.println(ran.nextInt(100));
        });
    }
}
```

程序运行结果:

```
方法实现 one
方法实现 two
51
```

分析:Lambda 表达式()->System.out.println("方法实现 one")创建了函数式接口 MyFunctionalInterface 的对象,并向该接口提供了具体实现。可以理解为:在方法 show 被调用时,通过传入参数向函数接口的唯一抽象方法 method()中传递了一段代码,使得原本为空的方法,有了具体的实现(System.out.println("方法实现 one")),而且该实现可以随着不同 Lambda 表达式的传入而灵活改变,如产生新的输出"方法实现 two"和随机数等。

【例 5.21】 函数式接口用作方法的返回值类型。

```
import java.util.Arrays;
import java.util.Comparator;
public class TestLambda2 {
    public static Comparator<String> getComparator(){
        return ((o1, o2) ->o1.length()-o2.length() );    //根据长度排序
    }
    public static void main(String[] args) {
        String[] strs={"aaa","bbbbbb","cc","dddd"};
        System.out.println(Arrays.toString(strs));
        Arrays.sort(strs, getComparator());
        System.out.println(Arrays.toString(strs));
    }
}
```

程序运行结果：

```
[aaa, bbbbbb, cc, dddd]
[cc, aaa, dddd, bbbbbb]
```

分析：本例中函数式接口 Comparator 用作方法 getComparator()的返回值类型，因此需要返回一个 Lambda 表达式。逻辑上，上述 getComparator()方法与下述匿名类实现是等价的。

```
public static Comparator<String> getComparator(){
    return new Comparator<String>() {
        public int compare(String o1, String o2) {
            return o1.length()-o2.length();
        }
    };
}
```

5.7.4 预定义函数式接口

为方便 Lambda 表达式的使用，Java8 提供了多个预定义函数式接口，其位于 java.util.function 中。典型的函数式接口包括 Supplier 接口、Consumer 接口、Predicate 接口、Function 接口。

1. Supplier 接口

Supplier< T >接口仅包含一个无参的方法：T get()，用来获取一个泛型参数指定类型的对象数据。该接口被称为生产型接口，即指定接口的泛型是什么类型，那么接口中的 get()方法就会产生什么类型的数据。

【例 5.22】 使用 Supplier 接口获取数组中的最大值。

```
package code0507;
import java.util.function.Supplier;
public class InterfaceSupplier {
    public static int getMax(Supplier<Integer> sup) {
        return sup.get();
    }
    public static void main(String[] args) {
        int[] arr = { 80, 10, -20, 101, 75, 17, -30};
        int maxValue = getMax(() -> {
            int max = arr[0];
            for (int i : arr) {
                if (i > max)
                    max = i;
            }
            return max;
        });
        System.out.println(maxValue);
    }
}
```

程序运行结果：

```
101
```

分析：Supplier 接口只定义了要通过 get()方法返回一个数据，但“通过什么样的方式，返回何种满足用户要求的数据”，可以通过 Lambda 表达式具体指定。由于方法 get()不接受任何参数，所以例 5.22 中 lambda 表达式的传入参数部分为空。

2. Consumer 接口

consumer< T >接口正好与 Supplier 接口相反，它消费一个数据，其数据类型由泛型决定。它包含抽象方法 void accept(T t)，意为消费一个指定泛型的数据。

【例 5.23】 使用 Consumer 接口。

```
package code0507;
import java.util.function.Consumer;
public class InterfaceConsumer {
    public static void cs(String name, Consumer<String> consumer) {
        consumer.accept(name);     }
    public static void main(String[] args) {
        cs("重庆大学", (name) -> {
            String reName = new StringBuilder(name).reverse().toString();
            System.out.println(reName);
        });
    }
}
```

程序运行结果：

```
学大庆重
```

Consumer 接口中还有一个默认方法 andThen，可以实现连接两个 Consumer 接口，再进行消费的效果。

3. Predicate 接口

Predicate 接口也称断言式接口。有时需要对某种类型的数据进行判断，从而得到一个 boolean 值结果，这时可以用 Predicate<T>接口，接口中包含有一个抽象方法 boolean test(T t)。下面例子展示了对 Predicate 接口中抽象方法 test()的使用。

【例 5.24】 使用 Predicate 接口。

```
public class InterfacePredicate {
    public static void main(String[] args) {
        //数字类型 判断值是否大于 5
        Predicate<Integer> predicate = x -> x > 5;
        System.out.println(predicate.test(10));                        //true
        //字符串为空判断
        Predicate<String> predicateStr1 = x -> null == x || "".equals(x);
        System.out.println(predicateStr1.test(""));                    //true
        //字符串中是否含 b
        Predicate<String> predicateStr2= s -> s.contains("b");
        System.out.println(predicateStr1.and(predicateStr2).test("book"));   //and
        System.out.println(predicateStr1.or(predicateStr2).test("book"));//or
        System.out.println(predicateStr2.negate().test("hello"));//negate
    }
}
```

程序运行结果：

```
true    true    false    true    true
```

此外，Predicate 接口中还提供 3 个默认方法：and(与)、or(或)、negate(非)，以及 1 个静态方法(isEqual)。

4. Function 接口

Function< T , R >接口用来将 T 的数据类型转换为 R 的数据类型，其中的抽象方法为 R apply(T t)。

【例 5.25】 使用 Function 接口中的方法 apply()，把字符串类型的整数转换为 Integer 类型。

```
package code0507;
import java.util.function.Function;
public class InterfaceFunction {
    public static void change(String s , Function<String,Integer> fun){
        int in = fun.apply(s);
        System.out.println(in);
    }
    public static void main(String[] args) {
        String s = "1234567";
        change(s,str->Integer.parseInt(s));
    }
}
```

程序运行结果：

```
1234567
```

与 Consumer 接口类似，Function 接口中也有一个默认方法 andThen，进行组合操作，执行完 Function1 后执行 Function2。

5.7.5 双冒号运算

还可以进一步简化上述 Lambda 表达式，这就要用到 Java 8 中的另一个新特性：双冒号运算。双冒号运算就是 Java 中的“方法引用”，其格式如下：

```
类名::方法名
```

方法引用大体分为以下几种方式：

(1) 静态方法引用语法：classname::methodname。

例如，Math::pow。

(2) 对象的实例方法引用语法：instancename::methodname。

例如，System.out::println。

(3) 特定类型的实例方法引用语法：特定类型::实例方法名。

例如，String::compareToIgnoreCase。

(4) 类构造方法引用语法：classname::new。

例如，ArrayList::new。

(5) 数组构造方法引用语法：typename[]::new。

例如，String[]::new。

方法引用与 Lambda 表达式完全等价，但更简洁。例如：

```
Math::pow              等价于  (x, y) -> Math.pow(x, y);        //静态方法引用
"love"::toUpperCase;   等价于  "love".toUpperCase();            //对象实例方法引用
```

```
String::toUpperCase;     等价于  (str) -> str.toUpperCase();     //类实例方法引用
String::new;             等价于  (str) -> new String(str);       //类构造方法引用
```

【例 5.26】 双冒号运算。

```
package code0507;
public class methodReference {
    public static void main(String[] args) {
        //使用"对象.方法名"形式调用
        Function<String, Integer> fn1 = (String str) ->{
            return str.length();
        };
        Integer aLength = fn1.apply("hello world");
        System.out.println("字符长度为:"+aLength);
        //使用双冒号形式
        Function<String, Integer> fn2 = String::length;
        Integer bLength = fn2 .apply("hello world");
        System.out.println("字符长度为:"+bLength);
    }
}
```

程序运行结果:

```
字符长度为:11
字符长度为:11
```

5.7.6 实用案例

【例 5.27】 使用 Lambda 表达式与函数接口计算斐波那契数列。

```
interface IntCall {
  int call(int arg);
}
public class RecursiveFibonacci {
  IntCall fib;
  RecursiveFibonacci() {
    fib = n -> n == 0 ? 0 :
              n == 1 ? 1 :
              fib.call(n - 1) + fib.call(n - 2);
  }
  int fibonacci(int n) { return fib.call(n); }
  public static void main(String[] args) {
    RecursiveFibonacci rf = new RecursiveFibonacci();
    for(int i = 0; i <= 10; i++)
      System.out.println(rf.fibonacci(i));
  }
}
```

5.8 Java 类的高级特性

5.8.1 泛型

1. 引入泛型的意义

Java 泛型编程是 JDK 1.5 版本后引入的。泛型让编程人员能够使用类型抽象,通常用于集合中。下面是一个不用泛型的简单例子：

```
List myIntList=new LinkedList();                //创建一个列表集合对象
myIntList.add(newInteger(0));                   //向列表集合中加入一个 Integer 对象
Integer x=(Integer)myIntList.iterator().next();
                                                //通过 next 方法从该列表中取出存入对象
```

第 3 行代码比较令人困惑，因为程序员将 Integer 对象放入到 List 中，但是在返回列表元素时，却要通过强制转换类型才能取得 Integer 对象。这是为什么呢？因为编译器只能保证 next()方法返回的是 Object 类型的对象，为保证 Integer 变量的类型安全，所以必须强制类型转换。

如果能保证列表中的元素为一个特定的数据类型，就可以取消类型转换，减少发生错误的机会，这也正是泛型设计的初衷。下面是使用泛型的例子：

```
List<Integer> myIntList=newLinkedList<Integer>();
myIntList.add(newInteger(0));
Integerx=myIntList.iterator().next();
```

在第 1 行代码中指定 List 中存储的对象类型为 Integer，这样在获取列表中的对象时，不必强制类型转换了。

2. 什么是泛型

泛型，即“参数化类型”。一提到参数，最熟悉的就是定义方法时有形参，调用方法时传递实参。与之类似，也可以将“类型”定义成参数形式(称为类型形参)，然后在调用时传入具体的某种类型(称为类型实参)。

如类的方法 method(String str1，String str2)中，参数 str1、str2 的值通常是可变的，泛型也是一样，如下列泛型类定义 class Point＜T1， T2＞中，T1 和 T2 就像方法中的参数 str1 和 str2，也是可通过传递参数改变的。

3. 定义泛型类

【例 5.28】 泛型类的定义和使用。

```
package code0508;
class Point<T1, T2> {
    T1 x;
    T2 y;
    public T1 getX() {
        return x;
    }
    public void setX(T1 x) {
        this.x = x;
    }
    public T2 getY() {
        return y;
    }
    public void setY(T2 y) {
        this.y = y;
    }
}

public class GeneDemo {
    public static void main(String[] args) {
```

```
        //实例化泛型类
        Point<Integer, Integer> p1 = new Point<Integer, Integer>();
        p1.setX(10);
        p1.setY(20);
        int x = p1.getX();
        int y = p1.getY();
        System.out.println("This point is:" + x + ", " + y);
        Point<Double, String> p2 = new Point<Double, String>();
        p2.setX(25.4);
        p2.setY("东经 180 度");
        double m = p2.getX();
        String n = p2.getY();
        System.out.println("This point is:" + m + ", " + n);
    }
}
```

程序运行结果:

```
This point is:10, 20
This point is:25.4, 东经 180 度
```

分析:与普通类的定义相比,上面的代码在类名后面多出了 <T1, T2>,T1 和 T2 是自定义的标识符,也就是我们提到的类型参数(用来传递数据的类型,而非数据的值)。T1 和 T2 只是数据类型的占位符,运行时会被替换为真正的数据类型。

泛型类在实例化时必须指出具体的类型,也就是向类型参数传值,其格式为:

```
className <dataType1, dataType2> variable = new className<dataType1, dataType2>();
```

4. 定义泛型方法

除了定义泛型类,还可以定义泛型方法。定义泛型方法方式:将泛型参数列表用尖括号括起来,放在返回值之前。

【例 5.29】 泛型方法的定义和使用。

```
package code0508;
class NewPoint<T1, T2> {
    T1 x;
    T2 y;
    public T1 getX() {
        return x;
    }
    public void setX(T1 x) {
        this.x = x;
    }
    public T2 getY() {
        return y;
    }
    public void setY(T2 y) {
        this.y = y;
    }
    //定义泛型方法 printPoint()
    public <S1, S2> void printPoint(S1 x, S2 y) {
        S1 m = x;
        S2 n = y;
```

```
        System.out.println("This point is:" + m + ", " + n);
    }
}
public class GeneMethodDemo {
    public static void main(String[] args) {
        NewPoint<Integer, Integer> p1 = new NewPoint<Integer, Integer>();
        p1.setX(10);
        p1.setY(20);
        p1.printPoint(p1.getX(), p1.getY());    //调用泛型方法 printPoint()
        NewPoint<Double, String> p2 = new NewPoint<Double, String>();
        p2.setX(25.4);
        p2.setY("东经 180 度");
        p2.printPoint(p2.getX(), p2.getY());    //调用泛型方法 printPoint()
    }
}
```

分析：上面的代码中定义了一个泛型方法：public <S1, S2> void printPoint(S1 x, S2 y)，与传统方法相比，它在修饰符后、返回值类型前增加了类型参数(< >)。一旦定义了类型参数，就可以在参数列表、方法体和返回值类型中使用了。

与使用泛型类不同，使用泛型方法时不必指明参数类型，编译器会根据传入的参数类型将泛型设置为相关类型。

提示：泛型方法的定义与其所在的类是否是泛型类没有任何关系，在普通类中也可以定义泛型方法，泛型方法可以有自己独立的类型参数。

5. 类型通配符

虽然 Integer 是 Object 的子类，但 Point<Object, Object>和 Point<Integer, Integer>之间其实并没有什么关系，Point<Integer, Integer>也不是 Point<Object,Object>的子类。但有时需要一个在逻辑上可以同时表示 Point<Integer, Integer>和 Point<Double, Double>的父类的一个引用类型，由此类型通配符应运而生。

类型通配符一般是使用问号“?”代替具体的类型实参。注意，此处是类型实参，而不是类型形参。

【例 5.30】 类型通配符的使用。

```
package code0508;
public class WildCardTest {
    public static void main(String[] args) {
        Box<String> name = new Box<String>("Hello");
        Box<Integer> age = new Box<Integer>(12);
        Box<Double> number = new Box<Double>(210.50);
        getData(name);
        getData(age);
        getData(number);
    }

    public static void getData(Box<? > data) {
        System.out.println("data :" + data.getData());
    }
}
class Box<T> {
    private T data;
    public Box() {}
```

```
        public Box(T data) {
            setData(data);
        }
        public T getData() {
            return data;
        }
        public void setData(T data) {
            this.data = data;
        }
    }
```

程序运行结果：

```
data :Hello
data :12
data :210.5
```

5.8.2 Java 类加载机制

JVM 提供的运行时环境中有个模块是 ClassLoader(类加载器)，它主要用于将主类(即包含了 main()方法的类)加载到 JVM 的 code segment(代码区)，然后运行环境找到 main()方法(程序入口)开始执行程序。在整个程序运行的过程中，逐步将更多的 class 动态加载到内存中，如图 5-3 所示。

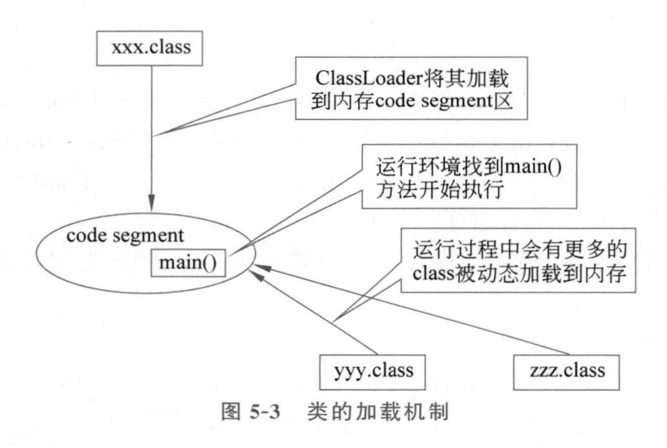

图 5-3　类的加载机制

可以利用 java -verbose：class XXX (XXX 为程序名)可以观察类的具体加载过程。

5.8.3 Java 反射机制

Java 反射机制是在运行状态中，对于任意一个类，都能够知道这个类的所有属性和方法；对于任意一个对象，都能够调用它的任意一个方法和属性。这种动态获取的信息以及动态调用对象的方法的功能称为 Java 的反射机制。正常机制与反射机制对比见图 5-4。

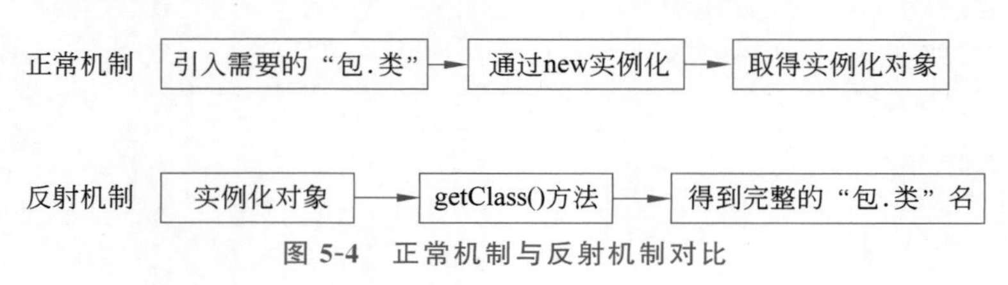

图 5-4　正常机制与反射机制对比

Java 的反射机制可以增加程序的灵活性，避免将程序写死到代码里。

例如，实例化一个 Person()对象，如果不使用反射机制，则可以通过 new Person()实现，但如果想实例化其他类，就必须修改源代码，并重新编译。如果使用反射机制，则代码变为 Class.forName("Person").newInstance()，而且这个类描述还可以写到配置文件中，如 **.xml，这样如果想实例化其他类，只要修改配置文件的类描述即可，不需要重新修改代码并编译。

Java 反射机制提供的主要功能包括：

(1) 加载运行时才能确定的数据类型。

(2) 解析类的结构，分析类的能力，获取其内部信息。

(3) 操作类或其实例(访问属性、调用方法、创建新对象)。

java.lang.Class 类是 Java 反射机制的基础，它是一个比较特殊的类，封装了被装入 JVM 中的类(包括类和接口)的相关信息，可以通过它实现对相关类中详细信息的获取。

1. 获取 Class 类对象

一般可以通过以下 3 种方式获得 Class 类对象：

(1) 使用 Class 类的静态方法 forName，如对类 Person：Class.forName("Person");。

(2) 使用对象的 getClass()方法，如 Person p = new Person();Class c = p.getClass();。

(3) 使用“类名.class”，如 Class c3 = Person.class;。

【例 5.31】 3 种获取 Class 类对象的方法。

```
package code0508;
public class GetClass {
    public static void main(String[] args) {
        checkClass();
    }
    public static void checkClass() {
        try {
            System.out.println("使用对象的 getClass()方法");
            Class cla= new Student().getClass();
            System.out.println(cla.getName());
            System.out.println("使用类名.class 方法");
            Class forClass = Student.class;
            System.out.println(forClass.getName());
            System.out.println("使用 Class 类的静态方法");
            Class forName = Class.forName("code0508.Student");
            System.out.println(forName.getName());
        } catch (Exception e) {
            e.printStackTrace();
        }
    }
}
class Student {
    private int id;
    private String name;
    public Student() {
        id=0;
        name="default";
    }
}
```

程序运行结果：

```
使用对象的 getClass()方法
code0508.Student
使用类名.class 方法
code0508.Student
使用 Class 类的静态方法
code0508.Student
```

2. 获取实例对象

我们不仅可以取得对象所在类的信息，也可以直接通过 Class 类的 newInstance()方法进行对象实例化操作。newInstance 方法原型如下：

```
public T newInstance() throws InstantiationException,IllegalAccessException
```

如对例 5.31 中的类 Student，可以实例化对象如下：

```
Student s=(Student)Class.forName ("code0508.Student").newInstance();
```

newInstance 方法调用默认的构造器初始化新创建的对象，如果这个类没有默认的构造器，就会抛出一个异常。

3. 获取类的构造方法

可以通过 Class 类中的 getConstructors()方法或 getDeclaredConstructors()方法获得本类中的全部构造方法，上述方法的返回类型都是 Constructor(构造器)的数组。

(1) public Constructor＜？＞[] getConstructors()：返回类中所有 public 的构造器集合。

(2) public Constructor＜T＞ getConstructor(Class＜？＞... parameterTypes)：返回指定 public 构造器，参数为构造器参数类型集合。

(3) public Constructor＜？＞[] getDeclaredConstructors()：返回类中所有的构造器，包括私有。

(4) public Constructor＜T＞ getDeclaredConstructor(Class＜？＞... parameterTypes)：返回任意指定的构造器。

【例 5.32】 获取 Person 类中的所有构造方法并实例化对象。

```
package code0508;
import java.lang.reflect.Constructor;
class Person {
    private String name;
    private int age;
    public Person() {
        this.name = "default";
    }
    public Person(String name) {
        this.name = name;
    }
    public Person(int age) {
        this.age = age;
    }
    public Person(String name, int age) {
        this.age = age;
```

```
        this.name = name;
    }
    public String getName() {
        return name;
    }
    public int getAge() {
        return age;
    }
    public String toString() {
        return "[" + this.name + "  " + this.age + "]";
    }
    public int addAge(int a){
        return age+a;
    }
}
public class GetConstructor {
    public static void main(String[] args) {
        Class<?> demo = null;
        try {
            demo = Class.forName("code0508.Person");
        } catch (Exception e) {
            e.printStackTrace();
        }
        //取得全部的构造函数
        Constructor<?> cons[] = demo.getConstructors();
        try {
            Constructor cons0 = demo.getConstructor();
            Constructor cons1 = demo.getConstructor(String.class);
            Constructor cons2 = demo.getConstructor(int.class);
            Constructor cons3 = demo.getConstructor(String.class,int.class);
            Person per1 = (Person) cons0.newInstance();
            Person per2 = (Person) cons1.newInstance("ZhangShan");
            Person per3 = (Person) cons2.newInstance(20);
            Person per4 = (Person) cons3.newInstance("Lisi", 20);
            System.out.println(per1);//[default 0]
            System.out.println(per2);//[ZhangShan 0]
            System.out.println(per3);//[null 20]
            System.out.println(per4);//[Lisi 20]
        } catch (Exception e) {
            e.printStackTrace();
        }
    }
}
```

程序运行结果：

```
[default  0]
[ZhangShan  0]
[null  20]
[Lisi  20]
```

提示：要想调用有参构造方法，必须使用 Constructor 类的 newInstance 方法，因为 Class 类的 newInstance 方法只能调用默认的无参构造方法。

4. 获取类的成员变量

成员变量用 Field 类进行封装,主要的方法包括:

(1) public Field getDeclaredField(String name):获取任意指定名字的成员。

(2) public Field[] getDeclaredFields():获取所有的成员变量。

(3) public Field getField(String name):获取任意 public 成员变量。

(4) public Field[] getFields():获取所有的 public 成员变量。

5. 获取类的成员方法

(1) public Method[] getMethods():获取所有的共有方法的集合。

(2) public Method getMethod(String name,Class<? >... parameterTypes):获取指定公有方法 参数 1:方法名 参数 2:参数类型集合。

(3) public Method[] getDeclaredMethods():获取所有的方法。

(4) public Method getDeclaredMethod(String name,Class<? >... parameterTypes):获取任意指定方法。

通过 Class 类的 getMethod()方法或 getDeclaredMethod()方法取得一个 Method 对象后,即可通过 invoke()方法调用指定的方法。invoke()方法的结构如下:

```
public Object invoke(Object obj, Object … args)
```

其中,obj 从中调用底层方法的对象;args 用于方法调用的参数。

【例 5.33】 通过反射机制获取类的属性和方法。

```
package code0508;
import java.lang.reflect.*;
public class ClassDemo {
    public static void main(String args[]) {
        try {
            Class cls = Class.forName("code0508.Person");
            Constructor ct = cls.getConstructor(int.class);
            Object obj = ct.newInstance(20);
            //获取类中的所有方法及其属性
            Method[] methods = cls.getDeclaredMethods();
            for (Method m : methods) {
                System.out.print(Modifier.toString(m.getModifiers()) + " " +
m.getReturnType().getName() + " "+ m.getName() + "(");
                Class<?>[] paras = m.getParameterTypes();
                for (Class<?> p : paras) {
                    System.out.print(p.getName() + " ");
                }
                System.out.println(")");
            }
            Method meth = cls.getMethod("addAge", int.class);
            //调用指定的方法
            Object retObject = meth.invoke(obj, 5);
            System.out.println("After addAge():" + retObject);
            //获取类中所有的属性
            Field[] field = cls.getDeclaredFields();
            for (Field f : field) {
                //获得属性修饰符
                System.out.print(Modifier.toString(f.getModifiers()) + " ");
```

```
                //获得属性类型
                System.out.print(f.getType().getName() + " ");
                //获得属性名称
                System.out.println(f.getName());
            }
        } catch (Exception e) {
            e.printStackTrace();
        }
    }
}
```

程序运行结果：

```
public java.lang.String toString()
public java.lang.String getName()
public int addAge(int )
public int getAge()
After addAge():25
private java.lang.String name
private int age
```

分析：如果想进一步获取类中方法的具体信息(如返回值类型、参数列表、修饰符)，可以通过类 Method 中的下列方法实现：

① public Class<？ > getReturnType()：取得全部的返回值。

② public Class<？ >[] getParameterTypes()：取得全部的参数。

③ public int getModifiers()：取得修饰符。

同理，如果想进一步了解类中属性的具体信息(如修饰符、类型、属性名)，可以通过类 Field 的下列方法实现：

① public int getModifiers()：取得属性修饰符。

② public Class<？ > getType()：取得属性类型。

③ public String getName()：取得属性名称。

5.8.4 实用案例

【例 5.34】 对 List 对象按不同成员属性排序。

一般对 list 排序可以使用 Collections.sort(list)，但如果 list 中包含的是一个对象，则这种方法是行不通的。如给定对象 UserInfo 如下，现在要求编写一个通用的方法，可以任意指定成员属性及其类型，list 按该成员属性重新排序输出。

```
//类 UserInfo
public class UserInfo implements java.io.Serializable {
    private Integer userId;
    private String username;
    private java.util.Date birthDate;
    private Integer age;
    private SimpleDateFormat formater = new SimpleDateFormat("yyyy-MM-dd");
    public UserInfo() {
    }
    public UserInfo(Integer userId, String username, java.util.Date birthDate,
Integer age) {
```

```
        this.userId = userId;
        this.username = username;
        this.birthDate = birthDate;
        this.age = age;
    }
    public void setUserId(Integer value) {
        this.userId = value;
    }
    public Integer getUserId() {
        return this.userId;
    }
    public void setUsername(String value) {
        this.username = value;
    }
    public String getUsername() {
        return this.username;
    }
    public void setBirthDate(java.util.Date value) {
        this.birthDate = value;
    }
    public java.util.Date getBirthDate() {
        return this.birthDate;
    }
    public void setBirthDatestr(String value) throws Exception {
        setBirthDate(formater.parse(value));
    }
    public String getBirthDatestr() {
        return formater.format(getBirthDate());
    }
    public void setAge(Integer value) {
        this.age = value;
    }
    public Integer getAge() {
        return this.age;
    }
    public String toString() {
        return new StringBuffer().append(getUserId()).append("; " + getUsername()).
append("; " + getBirthDatestr()).append("; " + getAge()).toString();
    }
}
//类 MySort
package code0508.UserInfoSort;
import java.util.Collections;
import java.util.Comparator;
import java.util.List;
import java.util.function.Function;
import java.lang.Comparable;
public class MySort<E,R extends Comparable<R> > {
    public void Sort(List<E> list, Function<E,R> f,  String sort) {
        Collections.sort(list, new Comparator<E>() {
            public int compare(E a, E b) {
                int ret = 0;
                R a1 = f.apply(a);
                R b1 = f.apply(b);
```

```
                if (sort != null && "desc".equals(sort))               //倒序
                {
                    ret = b1.compareTo(a1);
                } else                                                  //正序
                {
                    ret = a1.compareTo(b1);
                }
            return ret;
            }
        });
    }
}
//测试类 Test
package code0508.UserInfoSort;
import java.util.ArrayList;
import java.util.Date;
import java.util.List;
import java.text.SimpleDateFormat;
public class Test{
    public static void main(String[] args) throws Exception {
        List<UserInfo> list = new ArrayList<UserInfo>();
        SimpleDateFormat formater = new SimpleDateFormat("yyyy-MM-dd");
        list.add(new UserInfo(3, "Chang", formater.parse("2017-12-01"), 11));
        list.add(new UserInfo(1, "Li", formater.parse("2016-10-01"), 30));
        list.add(new UserInfo(2, "Zhen", formater.parse("2015-10-01"), 11));
        //按 userId 排序
        MySort<UserInfo,Integer> sortList1 = new MySort<UserInfo,Integer>();
        sortList1.Sort(list, x->x.getUserId(), "desc");
        System.out.println("--------按 userId 倒序-------------------");
        for (UserInfo user : list) {
            System.out.println(user.toString());
        }
        //按 username 排序
        MySort<UserInfo,String> sortList2 = new MySort<UserInfo,String>();
        sortList2.Sort(list,x->x.getUsername(), null);
        System.out.println("---------按 username 排序-----------------");
        for (UserInfo user : list) {
            System.out.println(user.toString());
        }
        //按 birthDate 排序
        MySort<UserInfo,Date> sortList3 = new MySort<UserInfo,Date>();
        sortList3.Sort(list,x->x.getBirthDate(), null);
        System.out.println("---------按 birthDate 排序-----------------");
        for (UserInfo user : list) {
            System.out.println(user.toString());
        }
    }
}
```

分析：排序类 MySort 中没有用到具体的对象和类型，而是使用了两个泛型 E 和 R，这使得该类及方法具有一定的通用性，如果要对 UserInfo 的 userId 排序，则只要将类名(UserInfo)及 userId 的类型名(Integer)作为参数传入即可，以此类推。语句 sortList1.Sort(list，x->x.getUserId()，"desc")实现“按 userId 倒排序”的任务，由于 userId 字段为 Integer

类型,需要先利用 Function 接口将 UserInfo 转为 Integer,并传入 Lambda 表达式 x-> x.getUserId()。

通过 Collections.sort()方法来实现排序,其中 Comparator 是个比较器接口,可以通过重写 compare()或 equals()方法实现比较功能。

5.8.5 枚举类型

枚举是一个被命名的整型常数的集合,用于声明一组带标识符的常数。枚举在日常生活中很常见,例如,一个人的性别只能是"男"或"女",一周的星期数只能是 7 天中的一个。类似这种当一个变量有几种固定可能的取值时,就可以将它定义为枚举类型。

声明枚举时必须使用关键字 enum,其基本语法如下:

```
[修饰符] enum 枚举名{
    //枚举体
}
```

枚举体由一组常量组成,多采用大写形式。例如,定义表示性别的枚举类型 SexEnum 和表示颜色的枚举类型 Color:

```
public enum SexEnum{
    MALE,FEMALE;
}
public enum Color{
    RED,BLUE,GREEN,BLACK;
}
```

Java 中的每一种枚举类型都默认为 java.lang.Enum 的子类。当定义一个枚举类型时,每一个枚举类型常量都代表了该 Enum 类的一个对象,这些枚举常量都默认被 final、public、static 修饰。使用枚举类型常量时,可以通过枚举类型名直接引用,如 SexEnum. male、Color.RED。

下面是 Enum 类中常用的方法。

(1) int compareTo(E e):比较两个枚举常量大小,其实比较的是枚举常量在枚举类中声明的顺序。例如,RED 的下标为 0,BLUE 下标为 1,那么 RED 小于 BLUE。

(2) boolean equals(Object o):比较两个枚举常量是否相等。

(3) int ordinal():返回枚举常量在枚举类中声明的序号,第一个枚举常量序号为 0。

(4) String toString():把枚举常量转换成字符串。

(5) static T valueOf(String name):把字符串转换成枚举常量。

(6) values():以数组形式返回枚举类型的所有成员。

枚举中也可以添加普通方法、静态方法、抽象方法、构造方法等。注意,如果要为 enum 定义方法,那么必须在 enum 的最后一个常量尾部添加一个分号。此外,在 enum 中,必须先定义常量,不能将方法定义在常量前面;否则,编译器会报错。

【例 5.35】 定义并使用枚举类型 Season。

```
package code0508;
enum Season {
    WINTER, SPRING, SUMMER, FALL;
    public void TestMethod(Season s) {
```

```
        System.out.println("索引号是: " + s.ordinal());
    }
}
    public class EnumExample {
    public static void main(String[] args) {
        for (Season s : Season.values()) {
            System.out.print(s);
            s.TestMethod(s);
        }
        System.out.println("WINTER 的 value 值是: " + Season.valueOf("WINTER"));
        System.out.println("比较结果: " + Season.SUMMER.compareTo(Season.WINTER));
    }
}
```

程序运行结果：

```
WINTER 索引号是: 0
SPRING 索引号是: 1
SUMMER 索引号是: 2
FALL 索引号是: 3
WINTER 的 value 值是: WINTER
比较结果: 2
```

5.8.6 Java 注解

Java 从版本 5 开始引入注解，现在很多 Java 框架中也在大量使用注解，如 Hibernate、Jersey、Spring。注解作为程序的元数据嵌入程序中，可以被一些解析工具或编译工具进行解析。

注解是代码的一种附属信息，它遵循一个基本原则：注解不能直接干扰程序代码的运行，无论增加或删除注解，代码都能够正常运行，Java 语言解释器会忽略这些注解，而由第三方工具负责对注解进行处理。

注解的语法比较简单，除了@符号的使用以外，基本与 Java 的固有语法一致，Java 内置了多种注解，定义在 java.lang 包中。

（1）@Override：只能用在方法之上的，用来告诉别人这一个方法是改写父类的。

（2）@Deprecated：建议别人不要使用旧的 API，编译时会产生警告信息，可以设定在程序里的所有的元素上。

（3）@SuppressWarnings：表示关闭一些不当的编译器警告信息。

除使用内置注解外，程序员也可以自定义注解，创建自定义注解与创建接口相似，但是注解的 interface 关键字需要以@符号开头，其基本格式如下：

```
public @interface 注解名 {定义体}
```

使用@interface 自定义注解时，自动继承了 java.lang.annotation.Annotation 接口，由编译程序自动完成其他细节。自定义注解时，不能继承其他注解或接口。下面是定义注解的实例。

```
package code0508;
import java.lang.annotation.ElementType;
import java.lang.annotation.Retention;
```

```
import java.lang.annotation.RetentionPolicy;
import java.lang.annotation.Target;
@Target(ElementType.METHOD)
@Retention(RetentionPolicy.RUNTIME)
public @interface UseCase {
    public int id();
    public String description() default "no description";
}
```

自定义注解时需要注意以下 4 点：

(1) 注解方法不能带有参数，也不能抛出异常。例如，boolean value(String str)、boolean value() throws Exception 等方式是非法的。

(2) 注解方法返回值类型限定为基本类型、String、Enum、Annotation 或者是这些类型的数组。

(3) 注解方法可以有默认值，例如 String level() default "LOW_LEVEL"、int high() default 2 是合法的；也可以不指定默认值。

(4) 注解本身能够包含元注解，元注解被用来注解其他注解。

Java5 中定义了 4 个标准的元注解(meta-annotation)类型：①@Target；②@Retention；③@Documented；④@Inherited。它们可用于提供对其他 annotation 类型的说明。

(1) @Target：表示该注解用于什么地方，可能的 ElemenetType 参数包括：

ElemenetType.CONSTRUCTOR：构造器声明。

ElemenetType.FIELD：域声明(包括 enum 实例)。

ElemenetType.LOCAL_VARIABLE：局部变量声明。

ElemenetType.METHOD：方法声明。

ElemenetType.PACKAGE：包声明。

ElemenetType.PARAMETER：参数声明。

ElemenetType.TYPE：类、接口(包括注解类型)或 enum 声明。

(2) @Retention：表示在什么级别保存该注解信息。可选的 RetentionPolicy 参数包括：

RetentionPolicy.SOURCE：注解将被编译器丢弃。

RetentionPolicy.CLASS：注解在 class 文件中可用，但会被 JVM 丢弃。

RetentionPolicy.RUNTIME：JVM 将在运行时也保留注释，因此可以通过反射机制读取注解的信息。

(3) @Documented：将此注解包含在 javadoc 中。

(4) @Inherited：允许子类继承父类中的注解。

使用注解的基本语法如下：

```
@<注解名>(<成员名 1>=<成员值 1>,<成员名 1>=<成员值 1>,…)
```

下面在类 PasswordUtils 中使用注解 UseCase。

```
package code0508;
public class PasswordUtils {
    @UseCase(id = 47, description = "Passwords must contain at least one numeric")
    public boolean validatePassword(String password) {
        return (password.matches("\\w * \\d\\w * "));
    }
```

```
    @UseCase(id = 48)
    public String encryptPassword(String password) {
        return new StringBuilder(password).reverse().toString();
    }
}
```

当然，使用注解最主要的部分还是在于对注解的处理，这就涉及注解处理器。从原理上讲，注解处理器就是通过反射机制获取被检查方法上的注解信息，然后根据注解元素的值进行特定的处理，从而产生不同的行为。

注解处理器使用 Java 中的反射机制来读取和分析被注解的源代码，使用的主要的包有 java.lang 和 java.lang.reflect。只有当一个 Annotation 类型被定义为运行时的 Annotation 后，该注解才能运行时可见，一旦该 class 文件被装载则保存在 class 文件中的 Annotation 就会被虚拟机读取。

AnnotatedElement 接口是所有程序元素(Class、Method 和 Constructor)的父接口，所以程序通过反射获取了某个类的 AnnotatedElement 对象之后，程序就可以调用该对象的如下 4 个方法来访问 Annotation 信息。

(1) <T extends Annotation> T getAnnotation(Class<T> annotationClass)：返回该程序元素上存在的、指定类型的注解。

(2) Annotation[] getAnnotations()：返回该程序元素上存在的所有注解。

(3) boolean is AnnotationPresent(Class<? extends Annotation> annotationClass)：判断该程序元素上是否包含指定类型的注解，存在则返回 true，否则返回 false。

(4) Annotation[] getDeclaredAnnotations()：返回直接存在于此元素上的所有注释。

下面是一个注解处理器的简单实例。

【例 5.36】 简单注解处理器。

```
package code0508;
import java.util.*;
import java.lang.reflect.Method;
public class UserAnnotation {
    public static void main(String[] args) {
        List<Integer> li = new ArrayList<Integer>();
        Collections.addAll(li, 47, 48, 49, 50);    //将元素 47、48、49、50 插入集合 li 中
        trackUseCases(li, PasswordUtils.class);
    }
    public static void trackUseCases(List<Integer> li, Class<?> cl) {
        for (Method m : cl.getDeclaredMethods()) {
            UseCase uc = m.getAnnotation(UseCase.class);
            if (uc != null) {
                System.out.println("Found Use Case:" + uc.id() + " " + uc.
description());
                li.remove(new Integer(uc.id()));
            }
        }
        for (int i : li) {
            System.out.println("Warning: Missing use case-" + i);
        }
    }
}
```

程序运行结果:

```
Found Use Case:47 Passwords must contain at least one numeric
Found Use Case:48 no description
Warning: Missing use case-49
Warning: Missing use case-50
```

5.9　继承与高级特性实训任务

【任务描述】

假定数据库的每一个表都对应一个 Java 类,表中的每一个字段对应该类中的一个属性,并且类的名字和表的名字相同,属性名和字段名也相同。使用反射类构造一个通用数据库查询应用,实现功能:对用户指定的表名和键值,将其作为参数传入,完成对该表的数据查询功能。如查询 userinfo 表(见图 5-5)中 Id 为 6988 的用户。

Column Name	Datatype	NOT NULL	AUTO INC	Flags
Id	INT(45)	✔		☑ UNSIGNED ☐ ZEROFILL
Name	VARCHAR(45)	✔		☐ BINARY
Pwd	VARCHAR(45)	✔		☐ BINARY
Age	INT(45)	✔		☑ UNSIGNED ☐ ZEROFILL

图 5-5　数据库的 userinfo 表

【任务分析】

针对表 userinfo,首先将其映射为对应的 Java 类,然后建立一个数据库工厂类 DBFactory,实现对数据库的连接功能。最后利用反射类,对作为传入参数的类进行解析,取得类名、属性名、类型等,并将其按照 SQL 语法规范拼接为一个完整的 SQL 语句,提交 DBMS 查询。本例需要用到部分数据库 JDBC 知识,请参见相应章节内容。

【任务解决】

【例 5.37】　将用户指定的表名和键值作为参数传入,完成对该表的查询。

```
//UserInfo类
package code0509.DB;
public class UserInfo {
    private int id;
    private String name;
    private String pwd;
    private int age;
    @Override
    public String toString() {
        return "UserInfo [id=" + id + ", name=" + name + ", pwd=" + pwd + ", age=" +
age + "]";
    }
    public int getId() {
        return id;
    }
    public void setId(int id) {
        this.id = id;
    }
    public String getName() {
        return name;
```

```
    }
    public void setName(String name) {
        this.name = name;
    }
    public String getPwd() {
        return pwd;
    }
    public void setPwd(String pwd) {
        this.pwd = pwd;
    }
    public int getAge() {
        return age;
    }
    public void setAge(int age) {
        this.age = age;
    }
}
//DBFactory 类
package code0509.DB;
public class DBFactory {
    public static Connection getDBConnection() {
        Connection conn = null;
        try {
            Class.forName("com.mysql.jdbc.Driver");
            String url = "jdbc:mysql://localhost:3306/blogsystem";
            String user = "root";
            String password = "yourpass";
            conn = DriverManager.getConnection(url, user, password);
        } catch (Exception e) {
            e.printStackTrace();
        }
        return conn;
    }
}
//DBhandle 类
package code0509.DB;
public class DBhandle {
        public static Object getObject(String className, int Id) {
        //得到表名字
        String tableName = className.substring(className.lastIndexOf(".") + 1,
                className.length());
        //根据类名来创建 Class 对象
        Class<?> c = null;
        try {
            c = Class.forName(className);
        } catch (ClassNotFoundException e1) {
            e1.printStackTrace();
        }
        //拼凑查询 sql 语句
        String sql = "select * from " + tableName + " where Id=" + Id;
        System.out.println("查找 sql 语句:" + sql);
        //获得数据库链接
        Connection con = DBFactory.getDBConnection();
```

```
        //创建类的实例
        Object obj = null;
        try {
            Statement stm = con.createStatement();
            //得到执行查寻语句返回的结果集
            ResultSet rs = stm.executeQuery(sql);
            //得到对象的方法数组
            Method[] methods = c.getMethods();
            //遍历结果集
            while (rs.next()) {
                obj = c.newInstance();
                //遍历对象的方法
                for (Method method : methods) {
                    String methodName = method.getName();
                    //如果对象的方法以 set 开头
                    if (methodName.startsWith("set")) {
                        //根据方法名字得到数据表格中字段的名字
                        String columnName = methodName.substring(3,methodName.
                                length());
                        //得到方法的参数类型
                        Class[] parmts = method.getParameterTypes();
                        if (parmts[0] == String.class) {
                            //如果参数为 String 类型,则从结果集中按照列名取得对应的值,
                            //并且执行该 set 方法
                            method.invoke(obj, rs.getString(columnName));
                        }
                        if (parmts[0] == int.class) {
                            method.invoke(obj, rs.getInt(columnName));
                        }
                    }


                }
            }

        } catch (Exception e) {
            e.printStackTrace();
        }
        return obj;
    }
    public static void main(String args[]) {
        //创建一个 UserInfo 对象
        UserInfo user = new UserInfo();
        //查找对象
        UserInfo userInfo = (UserInfo) getObject(
                "code0509.DB.UserInfo", 6988);
        System.out.println("获取到的信息:" + userInfo);
    }
}
```

注意:本案例中 DBhandle 类中的数据库查询方法是可重用的,只要用户提供对数据库表的任意映射类和 Id,并将其作为参数传入即可。

1. Java 语言中,下面关于类的继承关系描述正确的是哪个?

A. 一个子类可以有多个父类

B. 一个父类可以有多个子类

C. 子类可以使用父类的所有属性和方法

D. 子类一定比父类有更多的成员方法

2. 编写一个类 Student,该类拥有属性:校名、学号、性别、出生日期。方法包含设置姓名和成绩(setName(),setScore()),再编写 Student 类的子类 Undergraduate(大学生)。Undergraduate 类除拥有父类的上述属性和方法外,还拥有附加的属性和方法:属性包括系(department)、专业(major);方法包含设置系别和专业(setDepartment(),setMajor())。

3. 现有以下接口的声明:

```
public interface CalcArea{
  double getArea();
}
```

(1) 定义圆类 Circle 并实现接口 CalcArea,圆类的主要成员变量为半径,构造方法的参数用于初始化半径。

(2) 定义矩形类 Rectangle 并实现接口 CalcArea,矩形类的主要成员变量包括长和宽,构造方法的参数用于初始化长和宽。

(3) 现有如下类 Tester,其功能是求存储在一个数组中的多个图形的面积之和,要求补全其中缺失的代码。

```
public class Tester{
  static ____①____ shapes = { new Circle(1.0), new Rectangle(3.0, 4.0), new Circle
(8.0) };
  public static void main(String[ ] args){
    System.out.println("total area = "+ sumArea(shapes);
  }

  public static double sumArea(____②____ shapes){
    ____③____
  }
}
```

4. 给出下列程序运行的结果:

```
public class Test extends TT{
  public static void main(String args[]){
    Test t = new Test("Tom");
  }
  public Test(String s){
    super(s);
    System.out.println("How do you do? ");
  }
  public Test(){
    this("I am Tom");
```

```
        }
    }
    class TT{
      public TT(){
        System.out.println("What a pleasure!");
      }
      public TT(String s){
        this();
        System.out.println("I am "+s);
      }
    }
```

5. 抽象类可以实例化吗？如何使用抽象类？

6. 定义一个泛型方法，把任意参数类型的集合中的数据安全地复制到相应类型的数组中。

7. 下列关于通过反射机制获取方法并执行的过程说法正确的是哪个？

   A. 通过对象名.方法名(参数列表)的方式调用该方法

   B. 通过 Class.getMethod(方法名，参数类型列表)的方式获取该方法

   C. 通过 Class.getDeclaredMethod(方法名，参数类型列表)获取私有方法

   D. 通过 invoke(对象名，参数列表)方法来执行一个方法

8. 编写类 TestLambda，类中定义一个方法 test()，该方法使用函数接口作为参数，接收一个字符串并将其转换成大写字母，作为方法的返回值。

9. 请使用双冒号运算改写下列 Lambda 表达式：

```
()-> new HashMap<>();
(String string) -> System.out.print(string);
Comparator c = (c1, c2) -> c1.getAge().compareTo (c2.getAge());
```

第 6 章 Java 标准类库

本章学习目标

Java 提供了很多工具类，划分在不同的包中，这些包合称为类库。本章主要介绍常见的工具类和接口的使用，主要内容包括：

(1) 字符串 String、StringBuffer 类的使用，以及简单正则表达式的使用。

(2) 使用数据类型包装器进行数据类型转换。

(3) 数学函数与随机数的使用。

(4) 使用 System 类获得系统时间。

(5) 日期类的使用和格式化。

(6) 常用集合类的使用方法。

(7) 使用 Stream 处理集合元素。

6.1 简　介

Java 提供了许多功能强大的工具类。程序必须首先使用 import 语句引入需要的类或包。本章讲解的类主要来自下面两个包。

(1) java.lang 包：是 Java 语言的核心类库，包含了运行 Java 程序必不可少的系统类，如基本数据类型、基本数学函数、字符串处理、线程、异常处理类等，该包由系统自动引入。

(2) Java.util 包：包括 Java 语言中的一些实用工具，如处理时间的 Date 类，集合类 ArrayList 等，需要手动引入。

6.2 字符串类

字符串类主要包含初始化后不能改变的字符串类 String 和字符串内容可以动态改变的类 StringBuffer。这两个类都是线程安全的。

Java 将字符串作为对象来处理，在对象中封装了一系列方法来进行字符串处理。字符串类还可以对正则表达式进行处理。正则表达式用来描述字符串匹配的模式，可以使用正则表达式来进行字符串匹配、替换和分解。另外，在包 java.util.regex 提供了两个类 Pattern 和 Matcher 对正则表达式进行处理。

6.2.1 String 类

String 类位于 java.lang 包中,用来表示初始化后其内容不能被改变的字符串。

1. 字符串对象的构造

字符串常量是用双引号包含的一系列 Java 合法字符。JVM 自动为每个字符串生成一个 String 类的对象,例如"Java is great"。

字符串变量的声明和其他类一样,格式如下:

```
String s;
```

(1) 调用 String 类的构造方法,使用字符串常量,构造新字符串对象。例如:

```
s=new String("We are students");
```

也可写成:

```
s= "We are students";
```

声明和实例化也可一步完成:

```
String s=new String("We are students");
```

或者

```
String s="We are students";
```

使用没有参数的构造方法可以生成一个空字符串对象,长度为 0。例如:

```
String s=new String();
```

表示 s 为空字符串,与空字符串""等价。

(2) 可以由字符数组构造字符串。例如:

```
char cDem01[]={'1','2','3','4','5','6'};
String strDem01=new String(cDem01);          //使用字符数组构造字符串
String strDem02=new String(cDem01,1,4);      //从字符数组的第 1 个字符开始构造长度
                                             //为 4 的字符串
```

第一个字符串的内容是"123456",第二个字符串的值为"2345"。

(3) 可以由字节数组构造字符串,用于已知字符的编码值构造字符串。例如:

```
byte cDem01[]={65,66,67,68};
String strDem01=new String(cDem01);          //使用字节数组构造字符串
String strDem02=new String(cDem01,1,3);      //从字节数组的第 1 个字节开始,取 3 字节
                                             //构造字符数组
```

第一个字符串的值是"ABCD",第二个字符串的值是"BCD"。

2. String 类的常用方法

String 类提供了很多字符串处理方法,这些方法都不会改变字符串本身的值。按照用途可以分为字符串长度计算、字符串比较、字符串检索、字符串的截取、替换等方法,下面详细介绍这些方法的使用。

1) 字符串长度计算

使用 String 类中的 length()方法可以获取一个字符串的长度,返回值为 int。例如:

```
String s= "we are students",tom= "我们是学生";
int n1,n2;
n1=s.length();                                      //n1 的值是 15
n2=tom.length();                                    //n2 的值 5
```

2) 字符串比较

字符串比较的方法有 equals()、equalsIgnoreCase()、regionMatches()等方法。

equals(String s)方法用来比较当前字符串对象的内容是否与参数指定的字符串 s 的内容相同。例如：

```
String tom=new String("we are students");
String boy=new String("We are students");
String jerry= new String("we are students");
tom.equals(boy)的值是 false;tom.equals(jerry)的值是 true.
```

equalsIgnoreCase(String s)比较当前字符串对象是否与参数指定的字符串 s 相同，比较时忽略大小写。例如：

```
String tom =new String("ABC"),
Jerry=new String("abc");
tom.equalsIgnoreCase(Jerry)的值是 true。
```

3) startsWith()、endsWith()方法

字符串对象调用 **srartsWith(String s)**方法，判断当前字符串对象的前缀是否是参数指定的字符串 s。例如：

```
String tom= "220302620629021",jerry= "21079670924022";
tom.startsWith("220")的值是 true;jerry.startsWith("220")的值是 false。
```

可以使用 **endsWith(String s)** 方法，判断一个字符串的后缀是否是字符串 s。例如：

```
String tom= "220302620629021",jerry= "21079670924022";
tom.endsWith("021")的值是 true;jerry.endsWith("021")的值是 false。
```

4) compareTo，compareToIgnoreCase()方法

compareTo(**String s**)方法，依次按字符的编码值与参数 s 指定的字符串比较大小。如果当前字符串与 s 相同，则该方法返回值 0；如果当前字符串对象大于 s，则该方法返回正值；如果小于 s，则该方法返回负值。例如：

```
String str= "abcde";
str.compareTo("boy");                               //小于 0
str.compareTo("aba");                               //大于 0
str.compareTo("abcde");                             //等于 0
```

按字符的编码值比较两个字符串还可以使用 compareToIgnoreCase(String s)方法，该方法忽略大小写。

【例 6.1】 将字符串数组按字符集顺序重新排列。

```
package code0602;
public class SortStrs{
  public static void main(String args[]){
    String a[]={"Java","Basic","C++","Fortran","SmallTalk"};
```

```java
        for(int i=0;i<a.length-1;i++){                  //使用冒泡排序算法进行数组排序
          for(int j=i+1;j<a.length;j++){
            if(a[j].compareTo(a[i])<0){                 //比较两个字符串的大小
              String temp=a[i];
              a[i]=a[j];
              a[j]=temp;
            }
          }
        }
        for(int i=0;i<a.length;i++) {
          System.out.print(" "+a[i]);
        }
      }
    }
```

程序运行结果：

```
Basic C++ Fortran Java SmallTalk
```

5）字符串检索

搜索指定字符或者字符串在另一个字符串中出现的位置，可以用 indexOf()方法，其定义如下：

```java
public int indexOf(int ch)
public int indexOf(int ch,int fromIndex)
public int indexOf(String str)
public int indexOf(String str,int fromIndex)
```

一些重载方法可以通过 fromIndex 来指定匹配的起始位置。如果没有检索到字符或字符串，则返回值是－1。例如：

```java
String strSource="I love Java";
int nPosition;
nPosition=strSource.indexOf('v');               //nPosition 的值为 4
nPosition=strSource.indexOf('a',9);             //nPosition 的值为 10
nPosition=strSource.indexOf("love");            //nPosition 的值为 2
nPosition=strSource.indexOf("love",0);          //nPosition 的值为 2
```

另外，还可以使用 lastIndexOf()方法搜索一个字符或者字符串在另外一个字符串中的最后位置，其方法如下：

```java
public int lastIndexOf(int ch)
public int lastIndexOf(int ch,int fromlndex)
public int lastIndexOf(String str)
public int lastIndexOf(String str,int fromIndex)
```

定位指定的字符和字符串最后出现的位置。这些方法从后向前搜索，并且可以通过 fromIndex 来指定搜索的起始位置。如果没有检索到字符或字符串，则该方法返回的值是－1。

```java
String strSource="I love Java";
int nPosition;
nPosition=strSource.lastIndexOf('v');           //nPosition 的值为 9,最后一个
nPosition=strSource. lastIndexOf ('a',9);       //nPosition 的值为 8,倒数第二个
```

6）字符串的截取

使用方法 substring(int beginIndex)在字符串中截取子字符串，子串是从 beginIndex 位置开始的所有字符。

方法 substring(int beginIndex,int endIndex)获得的子串是从当前字符串的 beginIndex 位置到 endIndex 位置得到的字符串，但是不包含 endIndex 位置的字符。

例如：

```
String strSource= "Java is interesting";
String strNewl=strSource.substring(5);     //strNew1="is interesting"
String strNew2 =strSource.substring(5,6); //strNew2="i",长度为 1
```

7）字符串的替换

在 String 类中完成字符串替换的方法如下：

（1）public String replace(char oldChar,char newChar)：用新字符 newChar 替换字符串中的所有旧字符 oldChar，得到新字符串，原字符串不改变。

（2）public String replaceAll(String oldStr ,String newStr)：用字符串 newStr 替换所有的旧字符串 oldStr，得到新字符串。oldStr 可以是正则表达式。

（3）public String trim()：去掉字符串的前后空白，得到新字符串。

例如：

```
String s= "I mist theep ";
Strong temp=s.replace('t' , 's' );            //结果是"I miss sheep"
String s="  I am a student  ";                //前后有空格
String temp=s.trim();                         //结果是"I am a student",去掉首尾空格
```

8）大小写转换

使用 toUpperCase()和 toLowerCase()函数可以把字符串中的所有字母转成大写字母或者小写字母。

```
String str1="HelloJava";
String str2=str1.toUpperCase();               //全部大写"HELLOJAVA"
String str3=str1.toLowerCase();               //全部小写"hellojava"
```

【例 6.2】 String 类的使用。

```
package code0602;
public class UseString {
    public static void main(String args[]) {
        int n1, n2, n3;
        String vb = "Visual Baisc", ja = "Java",  s1, s2, s3,s4;
        s1 = vb.toUpperCase();                //转大写
        s2 = s1.substring(4, 10);             //子串长度为 6
        s3 = vb.replace('a', 'x');            //用新字符 x 替换所有旧字符 a
        s4=vb+ja;                             //连接字符串
        n1 = s4.length();                     //长度
        n2 = s4.indexOf(ja);                  //第一个位置
        n3 = s4.lastIndexOf("Visual");        //最后一个位置
        System.out.println("s1:"+s1);
        System.out.println("s2:"+s2);
        System.out.println("s3:"+s3);
        System.out.println("n1:"+n1);
```

```
        System.out.println("n2:"+n2);
        System.out.println("n3:"+n3);
    }
}
```

程序运行结果：

```
s1:VISUAL BAISC
s2:AL BAI
s3:Visuxl Bxisc
n1:16
n2:12
n3:0
```

6.2.2 StringBuffer 类

StringBuffer 类表示的是一个本身内容可变的字符串对象，包含一个缓冲区，主要用于完成字符串的动态添加、插入、替换等操作，这些操作引发字符串本身的变化。另外一个和 StringBuffer 功能相容的类是 StringBuilder，一般在单线程环境替代 StringBuffer 使用。

1. 添加操作 append()方法

将一个字符添加到字符串的后面。在应用中，如果添加字符的长度超过缓冲区的长度，则缓冲区将自动扩充长度。

append()方法有各种重载形式，可以处理不同的数据类型。例如：

```
public StringBuffer append(char c)
public StringBuffer append(char[] str)
```

可用来向字符串缓冲区添加布尔变量、字符、字符数组、双精度数、浮点数、整型数、长整型数、字符串等。

再如：

```
StringBuffer sbf=new StringBuffer("1+2=");
sbf.append(3);                                  //在尾部追加数字 3
System.out.println(sbf);                        //输出结果为 1+2=3
```

2. 插入操作

插入操作主要用于动态地向 StringBuffer 中插入字符。下面为插入数据的方法 insert()的部分重载声明：

```
public StringBuffer insert(intoffset,Boolean b)
public StringBuffer insert(int offset,char[] str)
public StringBuffer insert(int offset,String str)
```

offset 是插入的位置。可以向字符串缓冲区插入布尔变量、字符、字符数组、双精数、浮点数、整型数、长整型数、字符串等。例如：

```
StringBuffer sbf=new StringBuffer("1+=2");
sbf.insert(2,1);                                //在位置 2 插入数字 1
System.out.println(sbf);                        //输出结果为 1+1=2
```

3. 删除字符

删除字符串语句如下：

```
delete(int start,int end)
```

该语句删除字符串缓冲区中起始序号为 start、终止序号为 end－1 的字符，方法的返回类型为 StringBuffer。例如：

```
StringBuffer sbf=new StringBuffer("YouLoveTheJava");
sbf.delete(0,3);                                      //结果是"LoveTheJava"
deleteCharAt(int index)
```

删除字符串缓冲区中指定位置的字符，方法的返回类型为 StringBuffer。

4. 内容替换

内容替换语句如下：

```
public StringBuffer replace(int start,int end,String str)
```

该语句将起始位置为 start、终止位置为 end－1 的字符替换为由字符串 str 指定的内容。例如：

```
StringBuffer sbf = new StringBuffer("YouLoveJava");
String str = "IAlso";
sbf.replace(0, 3, str);                               //结果为 "IAlsoLoveJava"
```

【例 6.3】 使用 StringBuffer 反转字符串。

```
package code0602;
public class ReverseString {
    public static void main(String args[]) {
        StringBuffer sbuf = new StringBuffer("I Love Java");
        System.out.println(sbuf.reverse());          //调用 StringBuffer 的反转方法
        String strDest = reverseIt(sbuf.toString());
        System.out.println(strDest);
    }
    public static String reverseIt(String source) {
        StringBuffer dest = new StringBuffer(source);
        int len=dest.length();
        for (int i=0;i<=len/2;i++)                       //字符串的首尾字符互换
        {
            char c=dest.charAt(i);                      //得到 i 位置的字符
            dest.setCharAt(i, dest.charAt(len-i-1));    //设置 i 位置的字符
            dest.setCharAt(len-1-i, c);                 //设置 len-1-i 位置的字符
        }
        return dest.toString();
    }
}
```

程序运行结果：

```
avaJ evoL I
I Love Java
```

分析：例子中先用 StringBuffer 的 reverse()方法反转了字符串的顺序，然后编写了一个反转方法，又把顺序反转过来。

6.2.3 正则表达式

正则表达式是一个字符串，用来描述字符串匹配模式，可以进行字符串匹配、替换和分解。

一个正则表达式是由普通字符(如字符 a~z)以及特殊字符(元字符)组成的文字模式,用以描述在查找文字主体时待匹配的一个或多个字符串。正则表达式作为一个模板,将某个字符模式与所搜索的字符串进行匹配。例如,下面的正则表达式:

"java":由普通字符组成。

"java*":包含特殊字符 *,表示 java 后面跟有 0 个或者多个字符 a。

"[a-zA-Z]":表示从 a~z 或者 A~Z 当中的任意一个字符。

1. 正则表达式示例

String 类的 matches()、split()、replaceFirst()和 replaceAll()方法都需要传入一个正则表达式作为参数。下面是一个分割字符串为数组的例子。

【例 6.4】 使用多个条件分割字符串。

```
package code0602;
public class SplitString {
    public static void main(String[] args) {
        //构造包含逗号、句号的正则表达式
        String regex="[,。]+";
        //被分割的字符串
        String s="中国,你好。Java。。程序设计。";
        //使用逗号、句号分割字符串
        String[] ss=s.split(regex);              //分隔字符串,返回数组
        for(String a:ss)
        {
            System.out.println(a);
        }
    }
}
```

程序运行结果:

```
中国
你好
Java
程序设计
```

该例子构造了一个正则表达式进行字符串分割。正则表达式"[,。]+"表示任意一个或者多个逗号、句号。然后把正则表达式传递给 split()方法作为分隔符进行字符串分割。从输出结果可以看出,成功把字符串分割成了数组。

2. 正则表达式的特殊字符

所谓特殊字符(元字符),就是一些有特殊含义的字符,要在正则表达式模式中包含元字符以使其不具有特殊含义,必须使用反斜杠 (\) 转义字符。正则表达式的特殊字符如表 6-1 所示。

表 6-1 正则表达式的特殊字符

字符	说明
$	匹配输入字符串的结尾位置。要匹配 $ 字符本身,请使用 \$
()	标记一个子表达式的开始和结束位置。子表达式可以获取,供以后使用
*	匹配前面的子表达式 0 次或多次。要匹配 * 字符,请使用 *
+	匹配前面的子表达式 1 次或多次。要匹配 + 字符,请使用 \+

续表

<table>
<tr><th>字符</th><th>说　　明</th></tr>
<tr><td>.</td><td>匹配除换行符 \n 之外的任何单字符。要匹配.字符，请使用 \.</td></tr>
<tr><td>[</td><td>标记一个中括号表达式的开始。要匹配 [字符，请使用 \[</td></tr>
<tr><td>?</td><td>匹配前面的子表达式 0 次或 1 次，或指明一个非贪婪限定符。要匹配 ? 字符，请使用 \?</td></tr>
<tr><td>\</td><td>将下一个字符标记为或特殊字符或原义字符或向后引用或八进制转义符。例如，'\n' 匹配换行符。序列 '\\' 匹配 "\"，而 '\(' 则匹配 "("</td></tr>
<tr><td>^</td><td>匹配输入字符串的开始位置，除非在方括号表达式中使用，此时它表示不接受该字符集合。要匹配 ^ 字符本身，请使用 \^</td></tr>
<tr><td>{</td><td>标记限定符表达式的开始。要匹配 {字符，请使用 \{</td></tr>
<tr><td>|</td><td>指明两项之间的一个选择。要匹配 |字符，请使用 \|</td></tr>
<tr><td>\s</td><td>空格字符(空格键、Tab、换行、换页、回车)</td></tr>
<tr><td>\S</td><td>非空格字符([^\s])</td></tr>
<tr><td>\d</td><td>一个数字，在 Java 中使用\\d</td></tr>
<tr><td>\D</td><td>一个非数字的字符</td></tr>
<tr><td>\w</td><td>一个单词</td></tr>
<tr><td>\b</td><td>一个单词的边界</td></tr>
<tr><td>\G</td><td>前一个匹配的结束</td></tr>
</table>

除了这些特殊字符，还有限定符。限定符用来指定正则表达式的一个给定组件必须要出现多少次才能满足匹配。有 *、+、?、{n}、{n,}和{n,m}6 种。其中，n 表示出现的最小次数，m 表示出现的最大次数，具体使用时，需要替换成数字。

3. 正则表达式的字符类

使用方括号([ ])定义字符类。可以使用字符类指定字符列表以匹配正则表达式中的一个位置。例如，下面的正则表达式定义了匹配 bag、beg、big、bug 的字符类：

```
"b[aeiu]g"
```

使用连字符指定字符的范围，例如 A～Z、a～z 或 0～9。这些字符必须在字符类中构成有效的范围。例如，下面的字符类匹配 a～z 范围内的任何一个字符或任何数字：

```
"[a-z0-9]"
```

如果在字符类的开头使用尖号 (^) 字符，则将反转该集合的意义，即未列出的任何字符都认为匹配。下面的字符类匹配除小写字母(a～z)或数字以外的任何字符：

```
"[^a-z0-9]"
```

4. 正则表达式的工具类

为了更好地处理正则表达式，包 java.util.regex 里面有两个工具类 Pattern 和 Matcher 可以使用。使用方法分为以下 3 步。

(1) 构造一个模式：

```
Pattern p=Pattern.compile("[a-z]*");
```

(2) 构造一个匹配器:

```
Matcher m = p.matcher(str);
```

str 是需要匹配的字符串。

(3) 进行匹配,得到结果:

```
boolean b = m.matches();
```

【例 6.5】 用正则表达式判断手机号码是否规范。

```
package code0602;
import java.util.regex.Matcher;
import java.util.regex.Pattern;
public class JudgeMobileNumber {
    public static void main(String[] args) {
        Pattern p = null;
        Matcher m = null;
        boolean b = false;
        //正则表达式表示第一位是 1,第二位为 3,4,5,结尾为 9 位数字的 11 位数字
        p = Pattern.compile("^[1][3-5]+\\d{9}");
        String[] numbers = { "13996332243", "1227788", "15676789065",
                            "139abcd1234" };
        for (String s : numbers) {
            m = p.matcher(s);
            b = m.matches();
            System.out.println("手机号码正确:"+s+" " + b);
        }
    }
}
```

程序运行结果:

```
手机号码正确:13996332243 true
手机号码正确:1227788 false
手机号码正确:15676789065 true
手机号码正确:139abcd1234 false
```

6.2.4 实用案例 6.1:使用正则表达式检查 IP 地址

【例 6.6】 使用正则表达式检查 IP 地址。

案例代码如下:

```
package code0602;
import java.util.regex.Matcher;
import java.util.regex.Pattern;
public class CheckIPAddress {
    public static void main(String[] args) {
        Pattern p = null;
        Matcher m = null;
        boolean b = false;
        //构造匹配 IP 地址的模式
        p = Pattern.compile("[1-2]\\d{0,2}+\\.[1-2]\\d{0,2}+\\.
            [1-2]\\d{0,2}+\\.[1-2]\\d{0,2}");
        String[] ips = { "192.168.1.1", "192.168.1.1345",
            "222.168.1.134", "322.168.1.134" };
```

```
        for (String s : ips) {
            m = p.matcher(s);
            b = m.matches();
            System.out.println("IP 正确:"+s+" " + b);
        }
    }
}
```

程序运行结果：

```
IP 正确:192.168.1.1 true
IP 正确:192.168.1.1345 false
IP 正确:222.168.1.134 true
IP 正确:322.168.1.134 false
```

分析：本案例的关键是构造一个检查 IP 地址的正则表达式。每个 IP 地址由 4 字节组成，每个字节用 1～3 位十进制数字表示，并且第一位必须是 1 或 2，后面跟 0～2 位数字。由于字符"."在正则表达式中是特殊字符，必须进行转义。

6.3 数据类型包装器类

java.lang 包中有很多类，其中一些类和前面学过的基本数据类型有关系，这些类型称为包装器类（wrappers）。

Java 使用的简单数据类型，如 int 和 char，不是对象。有时需要用对象的形式表达这些简单类型的值，例如，当把数据放到集合中时需要包装成对象。为了能够用对象的形式表达简单数据类型，Java 提供了与每一个简单类型相对应的类。这些类的对象中包装了（wrap）简单类型的数据。因此，这种数据类型被称作类型包装器或者包装器类。

每种简单类型都有对应的包装器类型。Integer、Long、Short、Byte、Double、Float、Character、Boolean 分别是 int、long、short、byte、double、float、char、boolean 对应的包装器类。注意，包装器类的首字母是大写的。后面将以整型包装器类为例，说明包装器类的使用。

6.3.1 整型包装器类

Byte、Short、Integer 和 Long 类分别是字节型（byte）、短整型（short）、整型（int）和长整型（long）整数类型的包装器。使用数字或者是可以转换为数字的字符串构造其对象，例如：

```
Integer oi=new Integer(20223456);
Long  ol=new Long("123456");
```

另外，Java 还支持这些类型的自动装箱（Autobox）和拆箱（Unbox）。自动装箱就是自动地把简单类型封装成对应的对象类型，拆箱就是自动地把对象里面的值提取出来。例如：

```
Integer oi=123456;                              //自动装箱
int i=oi;                                       //自动拆箱
```

在这些类中定义了两个常量：MAX_VALUE 和 MIN_VALUE，分别表示每种数据类型表示的最大值和最小值。另外还定义了以下一些方法。

(1) valueOf()：把简单类型转换为对象类型，把字符串类型解析为包装器类型。

```
Integer oi=Integer.valueOf(2234);                 //数字转为对象
Integer oi2=Integer.valueOf("2234");              //以十进制把字符串解析为数字
Integer oi3=Integer.valueOf("2234", 16);          //以十六进制把字符串解析为数字
```


(2) 提取包装器里面的值：byteValue()、intValue()、longValue()、shortValue()。例如：

```
int i=oi2.intValue();
byte b=oi2.byteValue();
```

(3) 使用静态方法实现从字符串到数字类型的转换。

public static int parseInt(String s)：将参数 string 解析为有符号的十进制 int。

public static long parseInt(String s, int radix)：按照指定的进制 radix 解析字符串。

还有其他把字符串转换为 long、byte、short 的函数,请自行学习。

(4) 数字类型到字符串的转换。

String toString()：把对象里面的值按字面意思转成字符串。

static String toString(int i, int radix)：按照 radix 进制把整数 i 转换成字符串,符号单独处理,转换结果含符号字符。

static String toBinaryString(int value)：按二进制转换成字符串,转换结果不含符号字符,即最高位为 1 表示负数。一般用这个函数查看一个数的补码二进制表达形式。

static String toOctalString(int value)：按八进制转换成字符串,转换结果不含符号字符。

static String toHexString(int value)：按十六进制转换成字符串,转换结果不含符号字符。

6.3.2 实用案例 6.2：字符串和数字的相互转换

程序设计中一个最常见的问题是实现数字和字符串之间的相互转换。Byte、Short、Integer 和 Long 类分别提供了相互转换的方法。以 Integer 为例,数字和字符串之间的转换,主要使用 parseInt()方法和 toString()方法。下例使用 parseInt()方法将字符串转换成与之相应的整型(int)值。使用 toBinaryString()、toHexString()和 toOctalString()方法,可以分别将一个值以二进制、十六进制和八进制形式转化成字符串。

【例 6.7】 数字和字符串的转换。

```
package code0603;
public class ConvertDigitAndStr {
    public static void main(String[] args) {
        String s1 = "-20227";
        String s2 = "-20228";
        int i1 = 0, i2 = 0, sum = 0;
        i1 = Integer.parseInt(s1);                //把字符串解析成整数
        i2=Integer.valueOf(s2);                   //得到 Integer 对象,并自动拆箱
        sum=i1+i2;
        System.out.println("和:"+sum);
        String s3=Integer.toBinaryString(sum);    //转换成二进制字符串形式
        System.out.println("二进制:"+s3);
        String s4=Integer.toHexString(sum);       //转换成十六进制形式
        System.out.println("十六进制:"+s4);
    }
}
```

程序运行结果：

```
和:-40455
二进制:11111111111111110110000111111001
十六进制:ffff61f9
```

分析：两个负数相加，计算结果为负数，转成二进制形式后，最高位为1，表示负数。

6.4 System 类的使用

System类包含了很多静态方法和变量，包含标准的输入(in)、输出(out)和错误输出(err)，对系统属性和环境变量的访问，加载文件和库的方法，还有快速复制数组的实用方法。

6.4.1 记录程序执行的时间

System类的currentTimeMillis()方法返回自1970年1月1日午夜起到现在的时间，时间单位是毫秒。如果要记录程序中一段程序的运行时间可以在这段程序开始之前存储当前时间，在该段程序结束之际再次调用currentTimeMillis()方法。执行该段程序所花费的时间为其结束时刻的时间值减去其开始时刻的时间值。

【例6.8】 计算程序运行的时间。

```
package code0604;
public class RunElapsed {
    public static void main(String[] args) {
        long start, end, sum = 0, times = 1_000_000_000;      //十亿次
        System.out.println("执行" + times + "次循环需要的时间");
        start = System.currentTimeMillis();
        for (int i = 0; i < times; i++) {
            sum = sum + i * i;
        }
        end = System.currentTimeMillis();
        System.out.println("需要的时间是: " + (end - start) + "毫秒");
    }
}
```

程序运行结果：

```
执行1000000000次循环需要的时间
需要的时间是: 1202毫秒
```

6.4.2 复制数组

使用arraycopy()方法可以将一个任意类型的数组快速地从一个地方复制到另一个地方。下面是一个用arraycopy()方法复制两个数组的例子，将数组a复制给数组b，然后又将b数组的部分元素复制给a数组。arraycopy()函数的定义如下：

```
static void arraycopy(Object src, int srcPos, Object dest, int destPos, int length)
```

【例6.9】 复制数组。

```
package code0604;
public class ArrayCopyDemo {
```

```
    public static void main(String[] args) {
        byte a[] = { 65, 66, 67, 68, 69, 70, 71 };
        byte b[] = { 88, 88, 88, 88, 88, 88, 88, 88, 88, 88 };
        System.out.println("a = " + new String(a));        //使用字节数组构造字符串
        System.out.println("b = " + new String(b));        //转成字符串进行输出
        System.arraycopy(a, 0, b, 0, a.length);            //从 a 全部复制到 b
        System.out.println("b = " + new String(b));        //转成字符串进行输出
        System.arraycopy(b, 5, a, 0, 4);          //从 b 的第 5 个位置开始复制 4 个元素到 a
        System.out.println("a = " + new String(a));        //转成字符串进行输出
    }
}
```

程序运行结果:

```
a = ABCDEFG
b = XXXXXXXXXX
b = ABCDEFGXXX
a = FGXXEFG
```

6.5 Math 和 Random 类的使用

6.5.1 Math 类

Math 类用于数学计算,包含了常用的指数、对数、平方根、三角函数等静态函数。例如,正弦函数 sin()、反正弦函数 asin()等。

一些指数函数,例如,double pow(double y, double x),将返回以 y 为底数、以 x 为指数的幂值。

其他函数,例如,绝对值函数 int abs(int a)、最大值函数 int max(int x, int y)、伪随机函数 double random()等。

Math 定义了两个双精度(double)常数:E(2.718281828459045)和 PI(近似为 3.141592653589793)。

下面通过一个例子来说明 Math 类的使用。

【例 6.10】 Math 类的使用。

```
package code0605;
public class MathDemo
{
    public static void main(String[] args)
    {
        double radians = 2 * Math.PI;
        double d1 = 3.1415655678;
        System.out.println("上取整函数:" + Math.ceil(d1));
        System.out.println("四舍五入函数:" + Math.round(d1));
        System.out.println("下取整函数:" + Math.floor(d1));
        System.out.println("指数函数:" + Math.exp(d1));
        System.out.println("幂函数:" + Math.pow(Math.E, d1));
        System.out.println("正弦函数:" + Math.sin(Math.PI / 6));
        System.out.println("反正弦函数:" + Math.asin(0.5));
        System.out.println("自然对数函数:" + Math.log(Math.E));
        System.out.println("角度转弧度:" + Math.toDegrees(radians));
```

```
        System.out.println("随机函数:" + Math.random());
    }
}
```

程序运行结果：

```
上取整函数:4.0
四舍五入函数:3
下取整函数:3.0
指数函数:23.140065857331336
幂函数:23.140065857331333
正弦函数:0.49999999999999994
反正弦函数:0.5235987755982989
自然对数函数:1.0
角度转弧度:360.0
随机函数:0.024433998339623453
```

6.5.2 Random 类

java.util.Random 类提供了更丰富、更方便获取随机整数和浮点数的方法。使用前，先要创建它的对象：

```
Random random=new Random();
```

以这种形式实例化对象时，Java 编译器以当前系统时间作为随机数生成器的种子。相同的种子产生相同的随机数序列。也可以在实例化 Random 类对象时，设置随机数生成器的种子。

```
Random random = new Random(seedValue);
```

Random 类中的关键方法如表 6-2 所示。

表 6-2 Random 类中的关键方法

方　法	说　明
nextInt()	返回一个随机整数
nextInt(int n)	返回大于或等于 0 且小于 n 的随机整数
nextLong()	返回一个随机长整型值
nextBoolean()	返回一个随机布尔型值
nextFloat()	返回一个随机浮点型值
nextDouble()	返回一个随机双精度型值
nextGaussian()	返回一个概率密度为高斯分布的双精度值

【例 6.11】 Random 类的使用。

```
package code0605;
import java.util.Random;
public class RandomDemo {
    public static void main(String[] args) {
        Random r = new Random(202208);          //实例化一个 Random 类
```

```
        System.out.println("随机产生一个整数:" + r.nextInt());
        System.out.println("随机产生一个大于等于 0 且小于 10 的整数:  " + r.nextInt
(10));
        System.out.println("随机产生一个布尔型的值:" + r.nextBoolean());
        System.out.println("随机产生一个双精度型的值:" + r.nextDouble());
        System.out.println("随机产生一个浮点型的值:" + r.nextFloat());
        System.out.println("随机产生一个概率密度为高斯分布的双精度值:  " + r.
nextGaussian());
    }
}
```

程序运行结果:

```
随机产生一个整数:-1527427849
随机产生一个大于等于 0 且小于 10 的整数:  4
随机产生一个布尔型的值:true
随机产生一个双精度型的值:0.32511450950263365
随机产生一个浮点型的值:0.61597526
随机产生一个概率密度为高斯分布的双精度值:  0.749750051704871
```

6.5.3 实用案例 6.3:随机生成字符数组并排序

【例 6.12】 随机生成字符数组。

```
package code0605;
import java.util.Arrays;
import java.util.Random;
public class SortRandomArray {
    public static void main(String[] args) {
        char[] cc = new char[10];
        Random r = new Random();
        for (int i = 0; i < cc.length; i++) {
            int n=r.nextInt(26);                  //得到 0~25 的随机整数
            cc[i] = (char) ('A' +n );              //构造字符
        }
        System.out.println(new String(cc));       //排序前输出
        Arrays.sort(cc);                           //排序
        System.out.println(new String(cc));
    }
}
```

程序运行结果:

```
QASQHHPCOJ
ACHHJOPQQS
```

6.6 日期时间实用工具类

本节介绍几个和日期时间有关的工具类。

6.6.1 Date 与 LocalDateTime 类

Date 类封装当前的日期和时间,也可以封装一个特定的日期。Date 支持下面的构造函数：

```
Date()
Date(long millisec)
```

其中，第一个构造函数用当前的系统日期和时间初始化对象。第二个构造函数接收一个以毫秒为单位的参数，该参数为从 1970 年 1 月 1 日 0 时起至今的毫秒数。创建 Date 对象后，可以比较两个日期的大小，并获取毫秒数。

Date 类是 Java 最早的日期类，功能比较弱；自 JDK8 起，提供了 java.time 包，重新定义了功能更强大、体系更完整的日期相关类。

其中主要的一个是 LocalDateTime 类，使用此类可以获取本地当前日期时间，精确到纳秒；还可以获取日期时间分量，并进行日期时间计算。该类没有构造函数，使用静态函数 now()获取当前日期和时间，使用静态函数 of()封装一个指定日期时间。

LocalDateTime 表示的是本地日期时间，没有包含时区信息。ZonedDateTime 增加了时区信息表示，其使用方法和 LocalDateTime 类似。

【例 6.13】 LocalDateTime 类的使用。

```
package code0606;
import java.time.LocalDateTime;
import java.time.temporal.ChronoUnit;
public class DateDemo {
    public static void main(String[] args) {
        LocalDateTime one = LocalDateTime.now();            //获得当前日期和时间
        System.out.println("日期时间 1:"+one);
        int year=one.getYear();                             //获取年
        int month=one.getMonthValue();                      //获取月
        int day=one.getDayOfMonth();                        //获取日
        int week=one.getDayOfWeek().getValue();             //获取星期几
        System.out.printf("%d 年%d 月%d 日 周%d\n",year,month,day,week);
        LocalDateTime two=one.plusYears(30);                //30 年后
        two=two.plusMonths(8);                              //8 月后
        two = two.plusDays(100);                            //100 天后
        System.out.println("日期时间 2:"+two);
        long days=one.until(two, ChronoUnit.DAYS);          //计算两个日期之间的天数
        System.out.println("两个日期之间的天数:"+days);
        long years=one.until(two, ChronoUnit.YEARS);        //计算两个日期之间的年数
        System.out.println("两个日期之间的年数:"+years);
        LocalDateTime three=LocalDateTime.of(year+20, 8, 8, 20, 8, 8);
                                                            //封装一个日期时间,20 年后
        System.out.println("日期时间 3:"+three);
    }
}
```

程序运行结果：

```
日期时间 1:2022-08-17T12:19:33.072525500
2022 年 8 月 17 日 周 3
日期时间 2:2053-07-26T12:19:33.072525500
两个日期之间的天数:11301
两个日期之间的年数:30
日期时间 3:2042-08-08T20:08:08
```

分析：本例分别使用 now()和 of()两个静态函数构建了本地日期时间对象。使用了

一些 plus 函数对日期进行了增加计算,还可以使用 minus()函数进行减少计算。另外,使用 util()函数计算了两个日期时间对象之间的距离。LocalDateTime 类的其他功能函数,请自行查看文档。

Date 和 LocalDateTime 对象之间还可以进行相互转换。它们转换的桥梁是 Instant 对象。Instant 表示时间线上的瞬时点。下面通过例子说明它们之间的转换。

【例 6.14】 Date 和 LocalDateTime 的转换。

```
package code0606;
import java.time.Instant;
import java.time.LocalDateTime;
import java.time.ZoneId;
import java.time.ZoneOffset;
import java.util.Date;
public class ConvertDateToLocal {
    public static void main(String[] args) {
        Instant ins=Instant.EPOCH;                              //时间的起始点
        System.out.println("Java 时间起始点:"+ins);
        System.out.println("当前瞬时点:"+Instant.now());     //0 时区的当前瞬时点
        LocalDateTime now = LocalDateTime.now();                //当前日期时间
        Instant ins2=now.toInstant(ZoneOffset.of("+8"));       //转成 Instant
        System.out.println("转换后的瞬时点:"+ins2);
        Date date = Date.from(ins2);                            //从 Instant 得到 Date
        System.out.println("得到的日期对象:"+date);
        Instant ins3=date.toInstant();                          //日期转 Instant
        ZoneId zoneId=ZoneId.of("+10");                         //创建时区信息
        //根据瞬时点和时区信息转成本地日期时间对象
        LocalDateTime ldt=LocalDateTime.ofInstant(ins3, zoneId);
        System.out.println("新时区的日期:"+ldt);
    }
}
```

分析:LocalDateTime 转成 Instant 时需要当前的时区信息,得到 Instant 对象后,可以转成 Date 对象。Date 对象可以直接转成 Instant 对象,然后转 LocalDateTime 对象时,也需要一个时区信息,这里创建了一个新时区。

6.6.2 实用案例 6.4:日期的格式化

当使用 LocalDateTime 替代 Date 类表示和处理日期时,如果需要把日期时间对象转成需要的字符串格式,可以使用工具类 DateTimeFormatter 定义一个转换格式。例如:

```
DateTimeFormatter formatter = DateTimeFormatter.ofPattern("yyyy-MM-dd");
```

函数的参数是一个定义好的字符串格式,其中会用到一些表示特殊含义的字母,例如 y 表示年、M 表示月、d 表示天、H 表示小时、m 表示分钟、s 表示秒。其他更多字母符号可以参考该类的文档。

定义格式化器后,可以使用 format()函数把日期时间转成字符串;可以使用 parse()函数把字符串转成日期时间对象。

【例 6.15】 日期时间的格式化。

```
package code0606;
import java.text.ParseException;
```

```
import java.time.LocalDateTime;
import java.time.ZonedDateTime;
import java.time.format.DateTimeFormatter;
public class FormatDate {
    public static void main(String[] args) throws ParseException {
        LocalDateTime now=LocalDateTime.now();
        System.out.println(now);
        //定义格式化器
        DateTimeFormatter dtf=DateTimeFormatter.ofPattern("yyyy 年 MM 月 dd 日 HH 时
mm 分 ss 秒");
        String ds1=now.format(dtf);                                  //调用格式化函数
        System.out.println("汉字格式:"+ds1);
        ds1=now.format(DateTimeFormatter.BASIC_ISO_DATE);
        System.out.println("基本国标格式:"+ds1);
        ds1=now.format(DateTimeFormatter.ISO_DATE_TIME);
        System.out.println("国标格式:"+ds1);
        ds1=now.format(DateTimeFormatter.ISO_WEEK_DATE);
        System.out.println("国标星期格式:"+ds1);
        ZonedDateTime zdt=ZonedDateTime.now();                       //带时区的时间
        System.out.println("当前时区:"+zdt.getZone());               //获得当前时区
        ds1=zdt.format(DateTimeFormatter.ISO_ZONED_DATE_TIME);
        System.out.println("国标时区格式:"+ds1);
        String ds2="2022 年 11 月 17 日 13 时 51 分 51 秒";
        LocalDateTime ldt2=LocalDateTime.parse(ds2, dtf);            //字符串解析为时间
        System.out.println("字符串转时间:"+ldt2);
    }
}
```

程序运行结果：

```
2022-08-17T13:54:39.964058900
汉字格式:2022 年 08 月 17 日 13 时 54 分 39 秒
基本国标格式:20220817
国标格式:2022-08-17T13:54:39.9640589
国标星期格式:2022-W33-3
当前时区:Asia/Shanghai
国标时区格式:2022-08-17T13:54:39.9920506+08:00[Asia/Shanghai]
字符串转时间:2022-11-17T13:51:51
```

6.7 Java 集合类

与数组类似，Java 提供集合类的主要作用在于"保存多个数据/对象"，提供一种比用数组下标存取方式更灵活的方法。

集合类存在 java.util 包中，主要分为 4 种：Set、List、Queue 和 Map。Set 表示不允许容纳重复元素的集合，List 表示可以容纳重复元素的集合，Queue 表示一个先进先出的队列，Map 表示存储键/值对的集合，每个键/值对称为一项。

6.7.1 集合接口

集合框架定义的接口决定了集合框架各类的基本特性。具体类提供了标准接口的不同实现。主要的集合接口如表 6-3 所示。

表 6-3　主要的集合接口

接　口	描　述
Iterable	实现该接口的对象,可以在 for-each 循环中使用,可以得到迭代器
Collection	集合框架的根接口,继承了 Iterable 接口,定义了操作对象集合的共同方法
List	继承 Collection,表示是有序的,可包括重复元素的列表
Queue	继承 Collection,表示先进先出的队列
Deque	继承 Queue,表示双向队列,可以在两端对等操作队列
Set	继承 Collection,表示是无序的,无重复元素的集合(数学上的含义)
SortedSet	继承 Set,对 Set 中的元素进行排序

1. Collection 接口

Collection 接口是构造集合框架的基础,它声明所有集合类都拥有的核心方法。Collection 接口定义的主要方法总结在表 6-4 中。

表 6-4　Collection 定义的主要方法

方　法	描　述
boolean add(Object obj)	将 obj 加入集合中。如果成功,则返回 true;如果失败,则返回 false
boolean addAll(Collection c)	将 c 中的所有元素都加入类集合中。如果操作成功,则返回 true;否则,返回 false
void clear()	删除所有元素
boolean contains(Object obj)	判断集合是否包含元素 obj
boolean equals(Object obj)	判断 obj 是否与当前集合相同
boolean isEmpty()	判断集合是否为空
Iterator iterator()	返回集合的迭代器对象
boolean remove(Object obj)	删除 obj。如果成功,则返回 true;否则,返回 false
boolean removeAll(Collection c)	删除 c 的所有元素。如果成功,则返回 true;否则,返回 false
boolean retainAll(Collection c)	保留 c 中的元素,删除其他元素
int size()	返回集合中元素的个数
Object[ ] toArray()	把集合转换为数组

2. List 接口

List 接口是有序的 collection,使用此接口能够精确地控制每个元素插入的位置。用户能够使用索引(元素在 list 中的位置,类似于数组下标)来访问 list 中的元素,这类似于 Java 的数组。一个列表可以包含重复元素。

3. Set 接口

Set 接口是一种不包含重复元素的 collection,Set 接口最多有一个 null 元素。

SortedSet 接口继承了 Set,并说明了元素按序排列的集合的特性。

4. Queue 接口

Queue 接口定义了一个先进先出队列,在 Collection 接口基础上,增加了向队列尾部增加元素,从队列首部提取元素等方法。

Deque 接口定义了一个可以双向操作的队列。增加了可以在首部或者尾部对等操作元素的方法。

6.7.2 实现 List 接口的类

实现 List 接口的类如图 6-1 所示,图中斜体表示抽象类,一般不能直接使用,可以直接使用的类是 ArrayList、LinkedList、Vector 和 Stack。这几种类虽然在内部实现上有所不同,但由于实现了相同的接口,其使用方式基本相同。

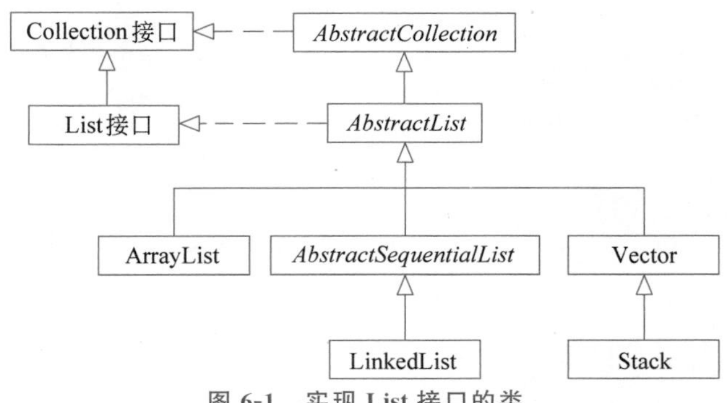

图 6-1 实现 List 接口的类

下面以 ArrayList 为例讲解其使用方法。

1. ArrayList 类

ArrayList 是可随需要而增长的动态数组。在 Java 中,标准数组是定长的,在数组创建之后,长度就不能改变了。ArrayList 能够随着数据元素的增删动态地增加或减小其大小。

ArrayList 有如下的构造函数:

```
ArrayList()
ArrayList(Collection c)
ArrayList(int capacity)
```

其中,第一个构造函数建立一个空的 ArrayList。第二个构造函数由集合 c 中的元素初始化。第三个构造函数指定初始容量(capacity)。容量是用于存储元素的基本数组的大小,需要时容量会自动增加。

ArrayList 在表 6-4 基础上增加了以下主要方法。

void add(int index, Object o):将元素插入列表的指定位置,index 从 0 开始。

Object get(int index):返回指定位置上的元素。

Object remove(int index):移除此列表中指定位置上的元素。

sort(Comparator<? super E> c):按照 c 定义的规则排序。

subList(int fromIndex, int toIndex):获取子列表。

【例 6.16】 ArrayList 类的使用。

```
package code0607;
import java.util.ArrayList;
public class ArrayListDemo {
    public static void main(String[] args) {
        ArrayList<String> al = new ArrayList<String>();   //创建一个 List
```

```
        System.out.println("List 的初始大小: " + al.size());
        for (int i = 0; i < 5; i++) {
            al.add("A"+i);                                    //向 List 中增加元素
        }
        al.add(1, "A2");
        System.out.println("增加元数后的大小: " + al.size());
        //显示其内容
        System.out.println("List 中的内容: " + al);
        al.remove("A2");                                      //删除元素 A2
        al.remove(al.size()-1);                               //删除最后一个位置的元素
        System.out.println("删除元素后的大小:" + al.size());
        System.out.println("删除后的内容: " + al);
        Object[] aa = al.toArray();                           //转化为数组
        for (Object a : aa) {
            System.out.print(a);
        }
    }
}
```

程序运行结果:

```
List 的初始大小: 0
增加元数后的大小: 6
List 中的内容: [A0, A2, A1, A2, A3, A4]
删除元素后的大小:4
删除后的内容: [A0, A1, A2, A3]
A0A1A2A3
```

2. Vector 与 Stack 类

Vector 与 ArrayList 类似,但 Vector 是线程安全的,在多个线程同时访问时安全性更好,但性能稍差。

类 Stack 继承自 Vector,实现一个后进先出的栈。Stack 提供以下额外的方法。

peek():查看栈顶部的元素,但不从堆栈中移除它。

pop():移除栈顶部的元素,返回该元素。

push(E o):把对象压入栈顶部。

int search(Object o):返回元素距离栈顶的距离,以 1 为基数。

6.7.3 实现 Set 接口的类

实现 Set 接口的类如图 6-2 所示,图中斜体表示抽象类,可以直接使用的类是 HashSet、TreeSet、LinkedHashSet。这几个类在内部实现上有所不同。TreeSet 另外实现了 SortedSet 接口,可以对集合中的元素排序。

下面以 HashSet 和 TreeSet 为例进行介绍。

1. HashSet 类

HashSet 使用哈希表存储元素。不保证元素的迭代顺序,可以包含 null 元素;对基本操作提供常量时间性能,是非线程安全的。

HashSet 类的常用构造函数包括:

```
HashSet()
HashSet(Collection c)
```

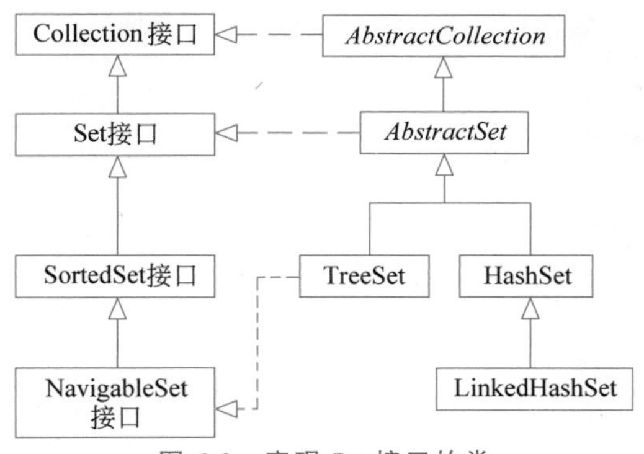

图 6-2 实现 Set 接口的类

其中,第一种形式构造一个默认的哈希集合。第二种形式用集合 c 中的元素初始化哈希集合。

2. TreeSet 类

TreeSet 是使用树结构存储元素的集合,对象默认按升序存储。基本操作提供了 log(n) 时间复杂度。

除包括 Collection 定义的方法,TreeSet 还提供了下列方法。

Object first():返回第一个(最低)元素。

Object last():返回最后一个(最高)元素。

Object floor(Object e):返回小于或等于给定元素的最大元素。如果不存在,则返回 null。

Object higher((Object e):返回大于给定元素的最小元素。如果不存在,则返回 null。

Object lower(Object e):返回小于给定元素的最大元素。如果不存在,则返回 null。

Object pollFirst():获取并移除第一个(最低)元素。如果集合空,返回 null。

Object pollLast ():获取并移除最后一个(最高)元素。如果集合空,返回 null。

【例 6.17】 Set 的使用。

```
package code0607;
import java.util.HashSet;
import java.util.Random;
import java.util.TreeSet;
public class SetDemo {
    public static void main(String[] args) {
        TreeSet<Character> ts = new TreeSet<Character>();
        HashSet<Character> hs = new HashSet<Character>();
        Random r = new Random();
        for (int i = 0; i < 10; i++) {
            char c = (char) ('A' + r.nextInt(25));
            ts.add(c);                              //添加元素
            hs.add(c);
        }
        System.out.println("HashSet:" + hs);
        System.out.println("TreeSet:" + ts);
        System.out.println(ts.lower('T'));      //小于 T 的最大元素
        System.out.println(ts.higher('M'));     //得到大于 M 的元素
        System.out.println(ts.pollFirst());     //删除第一个元素
```

```
        System.out.println(ts.pollLast());       //删除最后一个元素
        System.out.println("TreeSet:" + ts);
        hs.retainAll(ts);                        //操作 HashSet,保留和 ts 相同的元素
        System.out.println("HashSet:" + hs);
    }
}
```

程序运行结果:

```
HashSet:[A, Q, D, V, W, K, L, O]
TreeSet:[A, D, K, L, O, Q, V, W]
Q
O
A
W
TreeSet:[D, K, L, O, Q, V]
HashSet:[Q, D, V, K, L, O]
```

分析:从输出结果可以看出,TreeSet 按自然顺序自动排序,不能有重复元素。HashSet 中的元素无序。

6.7.4 实现 Queue 接口的类

实现 Queue 接口的类如图 6-3 所示。斜体表示抽象类,可以直接使用的类是 PriorityQueue、LinkedList、ArrayDeque。

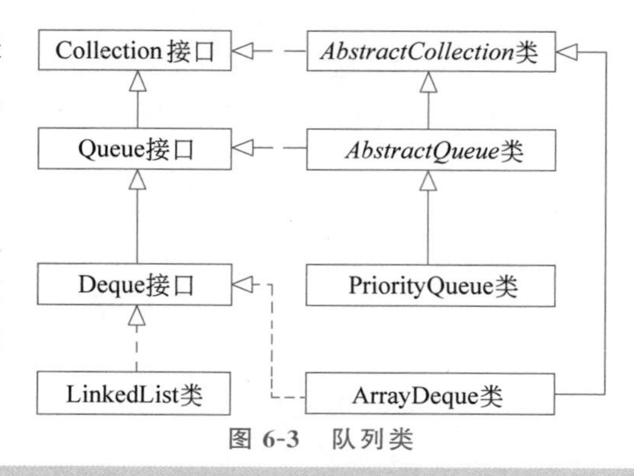

图 6-3 队列类

下面以 ArrayDeque 为例介绍队列类的使用。

ArrayDeque 类在底层用数组实现了双向队列,没有空间限制,可以随着需要动态增加。它是非线程安全的,不允许加入 null 元素,比 LinkedList 的性能好。

ArrayDeque 类的常用构造函数包括:

```
ArrayDeque()
ArrayDeque(Collection c)
```

第一种形式构造一个空队列,第二种形式用集合 c 中的元素初始化队列。

ArrayDeque 主要提供了以下方法。

offerFirst(e):在队列首部加入元素。

offerLast(e):在队列尾部加入元素。

pollFirst():删除队列首部元素并返回该元素。

pollLast():删除队列尾部元素并返回该元素。

peekFirst():获取队列首部元素,不删除该元素,队列空返回 null。

peekLast():获取队列尾部元素,不删除该元素,队列空返回 null。

【例 6.18】 Queue 的使用。

```
package code0607;
import java.util.ArrayDeque;
import java.util.Deque;
```

```
public class QueueDemo {
    public static void main(String[] args) {
        Deque<Character> dl1=new ArrayDeque<Character>();
        for(int i=0;i<10;i++) {
            char c=(char) ('A'+i);
            dl1.offerLast(c);                    //队列尾部加入 10 个数据
        }
        System.out.println("队列尾部加入 10 个数据:"+dl1);
        for(int i=0;i<5;i++) {
            dl1.pollFirst();                     //队列首部取走 5 个数据
        }
        System.out.println("队列首部取走 5 个数据后:"+dl1);
        for(int i=0;i<5;i++) {
            char c=(char) ('A'+i+10);
            dl1.offerFirst(c);                   //队列首部加入 5 个数据
        }
        System.out.println("队列首部加入 5 个数据后:"+dl1);
        for(int i=0;i<5;i++) {
            dl1.pollLast();                      //队列尾部取走 5 个数据
        }
        System.out.println("队列尾部取走 5 个数据后:"+dl1);
    }
}
```

程序运行结果：

```
队列尾部加入 10 个数据:[A, B, C, D, E, F, G, H, I, J]
队列首部取走 5 个数据后:[F, G, H, I, J]
队列首部加入 5 个数据后:[O, N, M, L, K, F, G, H, I, J]
队列尾部取走 5 个数据后:[O, N, M, L, K]
```

分析：使用 ArrayDeque 作为双向队列，分别在首部、尾部进行加入数据和取走数据操作。该类也可以限定为单向队列使用，只需要把变量 dl1 定义 Queue 类型就可以了。

6.7.5 通过迭代接口访问集合类

依次访问集合元素最简单方法是使用迭代器，迭代器是一个实现 Iterator 或者 ListIterator 接口的对象。Iterator 可以遍历集合中的元素，从而获得或删除元素。ListIterator 继承 Iterator，允许双向遍历，并可以修改。Iterator 接口说明的方法总结在表 6-5 中，ListIterator 接口说明的方法总结在表 6-6 中。

表 6-5 Iterator 接口中的主要方法

方　法	描　述
boolean hasNext()	判断是否存在下一个元素
Object next()	返回下一个元素
void remove()	从集合中删除当前元素

表 6-6 ListIterator 接口中的主要方法

方 法	描 述
void add(Object obj)	将 obj 插入下一个用 next()返回的元素之前
boolean hasPrevious()	判断是否存在前一个元素
int nextIndex()	返回下一个元素的下标,如果到达列表尾部,则返回列表的大小
Object previous()	返回前一个元素,如果元素不存在,则引发异常
int previousIndex()	返回前一个元素的下标,如果不存在,则返回 -1
void set(Object obj)	用 obj 替换当前元素。当前元素是上一次调用 next()方法或 previous()方法返回的元素

每一个 Collection 类都提供一个 iterator()函数,该函数返回一个集合的迭代器。使用迭代器循环遍历集合内容的步骤如下:

(1) 通过调用集合的 iterator()方法获得迭代器。

(2) 建立一个含有 hasNext()方法的循环,只要 hasNext()返回 true,就进行循环迭代。

(3) 在循环内部,通过调用 next()方法来得到每一个元素。

对于实现 List 接口的集合,也可以通过调用 ListIterator 来获得迭代器,提供了双向遍历集合的能力,并可修改集合元素。

【例 6.19】 Iterator 接口的使用。

```
package code0607;
import java.util.ArrayList;
import java.util.Iterator;
import java.util.ListIterator;
public class IteratorDemo
{
    public static void main(String[] args)
    {
        ArrayList<Character> al = new ArrayList<Character>();
        for (int i = 0; i < 10; i++) {
            char c = (char) ('A' + i);
            al.add(c);
        }
        System.out.print("原列表内容: ");
        Iterator<Character> itr = al.iterator();
        while (itr.hasNext())                        //用迭代器遍历列表
        {
            Object element = itr.next();
            System.out.print(element + " ");
        }
        System.out.println();
        ListIterator<Character> litr = al.listIterator();
        while (litr.hasNext())                       //使用 List 迭代器
        {
            Character element = litr.next();
            //把当前元素设置为小写
            litr.set(Character.toLowerCase(element));
        }
        System.out.print("修改后后向遍历列表: ");
```

```
        while (litr.hasPrevious())                //后向遍历
        {
            Character element = litr.previous();
            System.out.print(element + " ");
        }
        System.out.println();
    }
}
```

程序运行结果：

```
原列表内容：A B C D E F G H I J
修改后后向遍历列表：j i h g f e d c b a
```

分析：在列表被修改之后，litr 指向列表的末端（当到达列表末端时，litr.hasNext()方法返回 false）。为了以反向遍历列表，程序继续使用 litr，但这一次，程序使用 hasPrevious()检测它是否有前一个元素。只要它有前一个元素，该元素就被获得并被显示出来。

6.7.6 映射接口

映射(map)是一个存储关键字/值对的集合。给定一个关键字（简称键），可以得到它的值。键和值都是对象，每一对键/值作为一项。键必须是唯一的，但值是可以重复的。有些映射可以接收 null 键和 null 值，而有的则不行。映射接口如表 6-7 所示。

表 6-7 映射接口

接口	描述
Map	存储键/值对的集合
Map.Entry	表示键/值对。这是 Map 的一个内部接口
SortedMap	继承 Map，按键的升序存储元素

1. Map 接口

Map 是存储键/值对的集合的接口。给定一个键和值，可以存储到 Map 对象中。然后，就可以使用键获取对应的值。由 Map 定义的方法总结在表 6-8 中，当操作不当时，一些函数会产生异常，详细情况请查看文档。

表 6-8 Map 定义的主要方法

方法	描述
void clear()	删除所有的键/值对
boolean containsKey(Object k)	判断是否包含键 k，若包含，则返回 true；若不包含，则返回 false
boolean containsValue(Object v)	判断是否包含值 v
Set entrySet()	返回键/值对的集合(Set)
Boolean equals(Object obj)	如果 obj 是一个 Map 并包含相同的项，则返回 true；否则，返回 false
Object get(Object k)	返回与键 k 相关联的值
boolean isEmpty()	判断是否为空

续表

方　　法	描　　述
Set keySet()	返回一个键的集合(Set)
Object put(Object k, Object v)	将一个键/值对加入集合
void putAll(Map m)	加入所有来自 m 的项
Object remove(Object k)	删除键等于 k 的项
int size()	返回映射中项的个数
Collection values()	返回包含所有值的集合

映射经常使用两个基本操作方法：get()和 put()。使用 put()方法可以将键/值对加入映射。为了得到值,可以将键传入 get()方法返回该值。

映射可以看作键/值对的集合,可以使用 entrySet()方法得到其内部存储的键/值对的集合。可以使用 keySet()方法得到键的集合;可以使用 values()方法得到值的集合。

2. SortedMap 接口

SortedMap 接口继承了 Map,它确保了项按键升序排列。由 SortedMap 定义的方法总结在表 6-9 中。

表 6-9　SortedMap 定义的方法

方　　法	描　　述
Object firstKey()	返回第一个键
SortedMap headMap(Object end)	返回一个映射,包含了那些键小于 end 的项
SortedMap subMap(Object start, Object end)	返回一个映射,包含了那些键大于或等于 start 同时小于 end 的项
SortedMap tailMap(Object start)	返回一个映射,包含了那些键大于或等于 start 的项

6.7.7　实现 Map 接口的类

有几个类提供了映射接口的实现。可以被用作映射的类如图 6-4 所示,图中斜体表示抽象类,一般不能直接使用,可以直接使用的类是 HashMap 和 TreeMap 等。

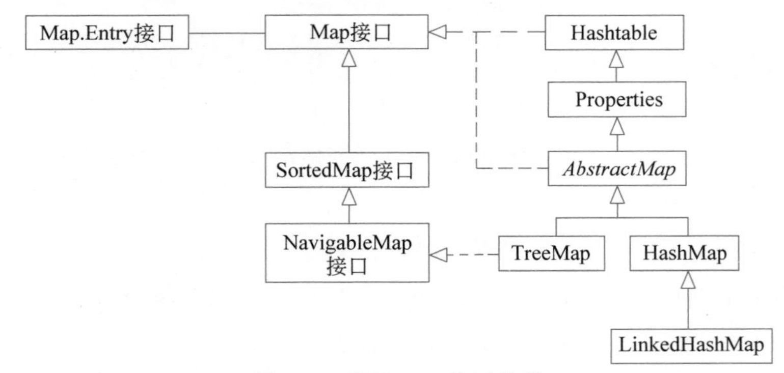

图 6-4　实现 Map 接口的类

1. HashMap 类

HashMap 类使用哈希表实现 Map 接口。一些基本操作方法如 get()和 put()的运行时间保持恒定。哈希表不保证元素的顺序。因此,元素加入哈希表的顺序并不一定是它们被迭代函数读出的顺序。

【例 6.20】 HashMap 的使用。

```
package code0607;
import java.util.HashMap;
import java.util.Iterator;
import java.util.Map;
import java.util.Map.Entry;
import java.util.Random;
import java.util.Set;
public class HashMapDemo
{
    public static void main(String[] args)
    {
        //键的类型为 String,值的类型为 Integer
        HashMap<String, Integer> hm = new HashMap<String, Integer>();
        Random r=new Random();
        hm.put("张三", r.nextInt(10000));                    //把元素加入映射
        hm.put("李四", r.nextInt(10000));
        hm.put("王五", r.nextInt(10000));
        hm.put("刘大",r.nextInt(10000));
        hm.put("钱八", r.nextInt(10000));
        System.out.println("映射的大小:"+hm.size());
        //得到映射项的集合
        Set<Map.Entry<String, Integer>> set = hm.entrySet();
        Iterator<Map.Entry<String, Integer>> i =set.iterator();
        //利用迭代器遍历项
        while (i.hasNext())
        {
            Entry<String, Integer> me =i.next();
            System.out.print(me.getKey() + ": ");
            System.out.println(me.getValue());
        }
        int balance = hm.get("李四");                         //得到键对应的值
        hm.put("李四", balance + 10000);                      //设置键对应的值
        System.out.println("李四的余额: " + hm.get("李四"));
        hm.remove("张三");                                    //删除键和对应的值
        System.out.println("删除后映射的大小:"+hm.size());
    }
}
```

程序运行结果:

```
映射的大小:5
李四: 8860
张三: 8577
刘大: 3990
王五: 6720
钱八: 3188
李四的余额: 18860
删除后映射的大小:4
```

分析:程序开始创建一个 HashMap 对象,然后将名字和账户值构成的项增加到 HashMap 中。接下来,通过使用函数 entrySet()获得项的集合,使用迭代器遍历集合并输出键值对。put()方法自动用新值替换旧值。最后使用 remove()方法删除了一个键值对。

2. TreeMap 类

TreeMap 类使用树实现 SortedMap 接口。在默认情况下,TreeMap 按键升序存储键/值对,同时允许快速检索。

【例 6.21】 TreeMap 的使用。

```
package code0607;
import java.util.Iterator;
import java.util.Random;
import java.util.Set;
import java.util.TreeMap;
public class TreeMapDemo
{
    public static void main(String[] args)
    {
        TreeMap<String, Double>  tm = new TreeMap<String, Double>();
        Random r=new Random();
        int f=10000;
        tm.put("E", f * r.nextDouble());                //放入元素
        tm.put("A", f * r.nextDouble());
        tm.put("B", f * r.nextDouble());
        tm.put("C", f * r.nextDouble());
        tm.put("F", f * r.nextDouble());
        tm.put("D", f * r.nextDouble());
        Set<String> set = tm.keySet();                  //得到关键字集合
        System.out.println("键的集合:"+set.getClass().getName());
        Iterator<String> i = set.iterator();            //得到迭代器
        System.out.println("迭代器:"+i.getClass().getName());
        while (i.hasNext())                             //通过迭代器显示 TreeMap 中的值
        {
            String key = i.next();
            System.out.println(key + ": " + tm.get(key));
        }
        double balance = tm.get("D");                   //存 10000 元到 D 的账户
        tm.put("D", balance + 10000);
        System.out.println("D's new balance: " + tm.get("D"));
    }
}
```

程序运行结果:

```
键的集合:java.util.TreeMap$KeySet
迭代器:java.util.TreeMap$KeyIterator
A: 9658.27150540658
B: 8546.149149735102
C: 5266.864225749369
D: 7300.183608865478
E: 4738.564271102621
F: 2412.9467606789667
D's new balance: 17300.183608865478
```

提示：TreeMap 对关键字进行了排序。键的集合和迭代器使用的是内部类，遍历键的集合时，其顺序就是 TreeMap 中键的顺序。

6.7.8 比较与排序

数据排序是数据处理中的一个重要操作。前面介绍的 List 类型的集合都可以进行数据排序操作。TreeSet 中的元素自动按照元素大小进行升序排列。在 TreeMap 中，默认情况下键按升序排列。

所有的排序功能都依赖于两个元素之间的大小比较。会用到接口 Comparable 或者 Comparator。

对集合排序可以使用工具类 Collections 里面定义的排序函数。还提供了其他集合操作，例如查找、复制、转换、比较、打乱、创建、填充、旋转、同步等操作。详情请查看文档。

下面构建一个学生类，在集合中按照升序排列学生信息。

【例 6.22】 使用 Collections 提供的排序函数。

```
package code0607;
import java.util.ArrayList;
import java.util.Collections;
import java.util.List;
import java.util.TreeSet;
class Student implements Comparable<Student>{
    Integer no;
    String name;
    @Override
    public int compareTo(Student o) {              //定义比较函数
        return this.no.compareTo(o.no);            //利用整数的大小比较
    }
    public Student(Integer no, String name) {
        this.no = no;
        this.name = name;
    }
    @Override
    public String toString() {
        return "(" + no + "," + name + ")";
    }
}
public class ComparableDemo {
    public static <E> void main(String[] args) {
        List<Student> ls=new ArrayList<Student>();
        for(int i=5;i>=1;i--) {
            Student stu=new Student(i,"stu"+i);
            ls.add(stu);
        }
        System.out.println("数组列表:"+ls);
        Collections.sort(ls);                      //排序
        System.out.println("升序排列:"+ls);
        TreeSet<Student> ts=new TreeSet<Student>();
        ts.addAll(ls);                             //所有元素加入树集合
        System.out.println("树集合:"+ts);
    }
}
```

程序运行结果:

```
数组列表:[(5,stu5), (4,stu4), (3,stu3), (2,stu2), (1,stu1)]
升序排列:[(1,stu1), (2,stu2), (3,stu3), (4,stu4), (5,stu5)]
树集合:[(1,stu1), (2,stu2), (3,stu3), (4,stu4), (5,stu5)]
```

分析:初始情况下,ArrayList 集合的元素顺序就是加入元素的顺序,排序后,按照比较函数确定的自然序排列。全部元素加入树集合后,也是按照比较函数确定的自然序排列。

使用 Comparable 接口确定的是一种自然序。一个类实现这个接口后,其对象之间的排序规则就固定下来了。有时需要根据场景灵活的确定两个元素的大小规则。例如,前面的例子中,根据学号确定学生之间的排序规则,不能同时实现根据姓名排序。这时就需要一个第三方的比较器来确定两个对象之间的大小关系。通过实现 Comparator 接口,并实现其中的 int compare(T o1,T o2)函数,构建一个比较器。排序时,使用比较器作为参数,确定对象之间的比较规则。

【例 6.23】 使用 Collections 类和 Comparator 接口进行排序。

```java
package code0607;
import java.util.ArrayList;
import java.util.Collections;
import java.util.Comparator;
import java.util.List;
import java.util.TreeSet;
class ComparatorOfName implements Comparator<Student>{
    //定义比较器,通过学生姓名确定学生之间的大小关系
    public int compare(Student o1, Student o2) {
        return o1.name.compareTo(o2.name);
    }
}
public class ComparatorDemo {
    public static <E> void main(String[] args) {
        List<Student> ls=new ArrayList<Student>();
        for(int i=1;i<=5;i++) {
            Student stu=new Student(i,"stu"+(6-i));
            ls.add(stu);
        }
        System.out.println("姓名降序:"+ls);
        Collections.sort(ls,new ComparatorOfName());    //使用比较器排序
        System.out.println("姓名升序:"+ls);
        TreeSet<Student> ts=new TreeSet<Student>(new ComparatorOfName());
        ts.addAll(ls);                                   //所有元素加入树集合
        System.out.println("树集合:"+ts);
    }
}
```

程序运行结果:

```
姓名降序:[(1,stu5), (2,stu4), (3,stu3), (4,stu2), (5,stu1)]
姓名升序:[(5,stu1), (4,stu2), (3,stu3), (2,stu4), (1,stu5)]
树集合:[(5,stu1), (4,stu2), (3,stu3), (2,stu4), (1,stu5)]
```

分析:程序定义了比较学生姓名的比较器,初次生成列表时,学号升序,姓名降序。使用比较器进行排序后,列表里面按照姓名升序排列。给树集合传入一个比较器后,其中元素也

是按照姓名升序排列。可以发现,通过传入比较器,替代了原来的自然序。

6.7.9 实用案例6.5:学生成绩检索和排序

已知一些学生的姓名、数学和语文成绩,使用类集合存储,并按照总成绩的高低排序。

【例6.24】 学生成绩排序。

```
package code0607;
import java.util.ArrayList;
import java.util.Collection;
import java.util.Collections;
import java.util.Comparator;
import java.util.HashMap;
import java.util.Iterator;
import java.util.Random;
class StuScore {
    String name;
    HashMap<String, Integer> score = new HashMap<String, Integer>();
    int sum = 0;
    public String toString() {
        return name + ":" + sum;
    }
    public void sum()                                          //计算总成绩
    {
        Collection<Integer> c = this.score.values();
        Iterator<Integer> itr = c.iterator();                  //迭代器
        while (itr.hasNext())                                  //遍历
        {
            sum += itr.next();
        }
    }
}
class ScoreDemo{
    public static void main(String[] args) {
        int count = 5;
        ArrayList<StuScore> scores = new ArrayList<StuScore>();
        Random r = new Random();                               //随机初始化学生成绩
        for (int i = 0; i < count; i++) {
            StuScore s = new StuScore();
            s.name = "学生" + i;
            s.score.put("语文", r.nextInt(80, 100));
            s.score.put("数学", r.nextInt(70, 100));
            s.sum();
            scores.add(s);
        }
        System.out.println(scores);                            //打印初始排列
        //使用比较器排序成绩
        Collections.sort(scores, new ComparatorOfSumScore());
        System.out.println(scores);
        //得到一个反向比较器
        Comparator<StuScore> reverseOrder=Collections.reverseOrder(new
ComparatorOfSumScore());
        Collections.sort(scores,reverseOrder);                 //重新排序
        System.out.println(scores);
```

```
    }
}
class ComparatorOfSumScore implements Comparator<StuScore> {
    //定义比较器,确定总成绩升序规则
    public int compare(StuScore o1, StuScore o2) {
        //TODO Auto-generated method stub
        return o1.sum - o2.sum;
    }
}
```

程序运行结果:

```
[学生 0:170, 学生 1:186, 学生 2:162, 学生 3:196, 学生 4:184]
[学生 2:162, 学生 0:170, 学生 4:184, 学生 1:186, 学生 3:196]
[学生 3:196, 学生 1:186, 学生 4:184, 学生 0:170, 学生 2:162]
```

分析:随机产生了学生的总成绩。定义了比较器按照总成绩升序排列,随后又得到了一个反向比较器,按总成绩降序排列。

6.8 Stream 的使用

Stream 将要处理的元素集合看作一种流,流在管道中传输,并且可以在管道的节点上进行各种处理,例如筛选、排序、聚合等。经过一系列的中间处理后,流变为最终需要的结果。

使用一个流的时候,通常包括以下 3 个步骤:

(1) 将集合、数组、文本等转换成管道流。

(2) 对流执行中间操作,如数据过滤/映射等,返回一个新的流,并交给下一个操作。

(3) 执行终结操作,如存储或遍历流元素等。一个流只有一个终结操作,当这个操作执行后,流就被使用"光"了,无法再被使用。

6.8.1 创建 Stream

流分为两类:顺序流和并行流。Stream()函数得到顺序流,由主线程按顺序对流执行操作。parallelStream()函数得到并行流,内部以多线程并行的方式对流进行操作,前提是流中的数据处理没有顺序要求。在多处理器环境下,如果需要提升运算速度,尽量使用并行流。

从数据源中创建顺序流的方式包括 3 种:

(1) 使用 Arrays.stream(T[] array)方法用数组创建流。

```
String[] array = {"h", "e", "l", "l", "o"};
Stream<String> arrayStream = Arrays.stream(array);    //泛型指明元素类型
```

(2) 使用 Collection.stream()方法用集合创建流。

```
List<String> list = Arrays.asList("hello","world","stream");
Stream<String> stream = list.stream();
```

(3) 通过 Stream 的静态方法:of()、iterate()和 generate()。

```
Stream<Integer> stream1 = Stream.of(1, 2, 3, 4, 5, 6);
Stream<Integer> stream2 = Stream.iterate(0, (x) -> x + 2).limit(3);
```

```
stream2.forEach(System.out::println);
Stream<Double> stream3 = Stream.generate(Math::random).limit(3);
stream3.forEach(System.out::println)
```

从数据源中创建并行流的方式有以下两种：

(1) 直接创建并行流。

```
Stream<String> parallelStream = list.parallelStream();
```

(2) 将顺序流转换成并行流。

```
Stream<String> parallelStream=list.stream().parallel();
```

6.8.2 中间操作

中间操作会从流中逐一获取元素并进行处理。所有中间操作都是有惰性的,并不会立即执行,只有等到终结操作被调用时,中间操作才会真正被执行。因此,Stream 具有延迟执行特性。Stream 提供了大量的中间操作和方法,下面给出部分方法。

(1) filter(Predicate predicate)：用于对流中的数据进行过滤。

(2) limit(long maxSize)：截取并返回流中特定数量的元素的数据流。例如：

```
Stream<Integer> stream = Stream.of(1,2,3,4,5,6);
stream.limit(2).forEach(System.out::println);
```

输出：

```
1 2
```

(3) skip(long n)：跳过指定个数的数据,返回后面剩余元素组成的流。例如：

```
Stream<Integer> stream = Stream.of(1,2,3,4,5,6);
stream.skip(3).forEach(System.out::println);
```

输出：

```
4 5 6
```

(4) concat(Stream a, Stream b)：合并 a 和 b 两个流为一个流。

(5) distinct()：返回流中不同元素组成的流(消除重复元素)。例如：

```
Stream<Integer> stream1 = Stream.of(1,2,3,4);
Stream<Integer> stream2 = Stream.of(4,5,6,7);
Stream.concat(stream1, stream2).distinct().forEach(System.out::println);
```

输出：

```
1 2 3 4 5 6 7
```

(6) sorted()：返回由此流的元素组成的流,根据自然顺序排序。例如：

```
List<Integer> list = Arrays.asList(12, 43, 65, 34);
list.stream().sorted().forEach(System.out::println);
```

输出：

```
12 34 43 65
```

(7) sorted(Comparator comparator):返回由该流的元素组成的流,根据提供的 Comparator 进行排序。

(8) map(Function mapper):根据函数 mapper 对流中的值进行某种形式的转换,并将返回值传递到输出流中。例如:

```
String[] strArr = { "abc", "bcd", "def"};
Stream<String>  strList = Arrays.stream(strArr).map(String::toUpperCase);
strList.forEach(System.out::println);
```

输出:

```
ABC   BCD   DEF
```

(9) dropWhile(Predicate predicate):使用一个断言作为参数,舍去断言为真的前序元素,直到断言第一次为 false,返回剩余元素构成的 Stream。例如:

```
long c=Stream.of("a","b","c","","e","f").dropWhile(s->!s.isEmpty()).count();
```

c 的值为 3。

(10) takeWhile(Predicate predicate):使用一个断言作为参数,返回断言为真的前序元素构成的 Stream,直到断言第一次返回 false。例如:

```
Stream.of("a","b","c","","e","f").takeWhile(s->!s.isEmpty()).forEach(System.
out::print);
```

输出:

```
a b c
```

6.8.3 终结操作

终结操作的主要作用是生成一个最终结果,并且启动中间操作的执行。以下是 Stream 中部分终结操作和方法。

(1) forEach(Consumer action):该方法接收一个 Lambda 表达式或者函数,然后在每一个元素上执行该表达式。相当于 for 循环遍历。例如:

```
List<String> strAry = Arrays.asList("Zhang", "Yang", "Liu", "Huang");
strAry.stream().forEach(System.out::println);
```

(2) forEachOrdered(Consumer action):该方法与 forEach 功能相似,主要区别在于遍历元素时按照流定义的元素遭遇顺序(encounter order)进行。

流本身不存储数据,因此流中的数据完成处理后,需要将其重新归集到新的集合里,这时就需要用到 collect 方法。

(3) collect(Collector):聚合操作,封装目标数据,将流转换并收集到结果集合,例如 List、Map、Set 中。根据收集方式的不同,工具类 Collectors 提供了不同的方法以生成不同集合对象。

public static <T> Collector toList():把元素收集到 List 集合中。

public static <T> Collector toSet():把元素收集到 Set 集合中。

public static Collector toMap(Function keyMapper,Function valueMapper):把元素收集到 Map 集合中。

例如：

```
List<Integer> list = Arrays.asList(1, 6, 3, 4, 6, 7, 9, 6, 20);
List<Integer> list2 = list.stream().filter(x -> x % 2 == 0).collect(Collectors.
toList());
```

通过过滤操作把偶数收集到一个新的列表。

(4) allMatch(Predicate)：如果流的每个元素都能匹配断言 Predicate，则返回 true。在第一个 false 时，则停止执行。例如：

```
System.out.println(Stream.of(1, 2, 3, 4, 5).allMatch(n -> n > 2));
```

数组中第一个元素小于 2，则停止匹配返回结果 flase。

(5) anyMatch(Predicate)：如果流中有任意一个元素能匹配到断言 Predicate，则返回 true。在第一个 true 时停止执行。例如：

```
System.out.println(Stream.of(1, 2, 3, 4, 5).anyMatch(n -> n > 2));
```

数组中第三个元素大于 2，则停止匹配返回结果 true。

(6) noneMatch(Predicate)：如果流中所有元素都无法匹配到断言 Predicate，则返回 true。

(7) findFirst()：返回满足条件的第一个元素(该方法的返回类型为 Optional 类)。例如：

```
System.out.println(Stream.of(1, 2, 3, 4, 5).filter(n -> n > 2).findFirst().get());
```

根据条件过滤后取第一个元素，返回结果为 3。

(8) max(Comparator comparator)：根据提供的比较器返回此流的最大元素。例如：

```
System.out.println(Stream.of(1,2,3,4,5).max((x,y)->x-y).get());;
```

输出取流中最大值，结果为 5。

(9) min(Comparator comparator)：根据提供的 Comparator 返回此流的最小元素。

(10) count()：返回此流中的元素个数。例如：

```
Stream<String> stringStream=Stream.of("Zhang","Yang","Liu");
System.out.println(stringStream.count());
```

获取流中元素数量，返回结果为 3。

(11) toArray()：返回包含流元素的数组。

在使用 Stream 过程中需要注意以下几点：

(1) stream 不存储数据。

(2) stream 不会改变数据源，通常情况下会产生一个新的集合。

(3) stream 执行终结操作后被关闭了，不能再执行其他操作。

(4) 为了避免不必要的拆箱和装箱，提供了 IntStream、LongStream 和 DoubleStream 专门处理数字，不需要使用 Stream<Integer>等。

6.8.4 实用案例 6.6：使用 Stream 处理成绩

在本章实用案例 6.5 的基础上，使用 Stream 处理学生的成绩。可以比较两个案例的不同处理方法。

【例 6.25】 学生成绩排序和检索。

```
package code0608;
import java.util.ArrayList;
import java.util.HashMap;
import java.util.List;
import java.util.Random;
import java.util.stream.Collectors;
class StuScore {
    String name;
    int sum = 0;
    HashMap<String, Integer> score = new HashMap<String, Integer>();
    public String toString() {
        return name + ":" + sum;
    }
    public void sum()                                          //使用 Stream 计算总成绩
    {
        this.sum = score.values().stream().mapToInt(x->x.intValue()).sum();
    }
}
public class ScoreStreamDemo{
    public static void main(String[] args) {
        int count = 5;
        ArrayList<StuScore> scores = new ArrayList<StuScore>();
        Random r = new Random();                               //随机初始化学生成绩
        for (int i = 0; i < count; i++) {
            StuScore s = new StuScore();
            s.name = "学生" + i;
            s.score.put("语文", r.nextInt(80, 100));
            s.score.put("数学", r.nextInt(70, 100));
            s.sum();
            scores.add(s);
        }
        System.out.println("初始:"+scores);                   //打印初始排列
        //使用 Stream 升序排序成绩
        List<StuScore> sortedAsc=scores.stream().sorted((x,y)->x.sum-y.sum).
collect(Collectors.toList());
        System.out.println("升序:"+sortedAsc);
        int total=scores.stream().mapToInt(x->x.sum).sum();  //计算总分
        int avg=total/count;
        System.out.printf("总分%d,平均分%d\n",total,avg);
        //过滤大于平均成绩的学生
List<StuScore> grt=scores.stream().filter(x->x.sum>avg).collect(Collectors.
toList());
        System.out.println("大于平均分:"+grt);
    }
}
```

分析：和本章实用案例 6.6 相比，在计算每个学生总分时，构建了 Stream，并转成了 IntStream，直接调用 sum 函数得到总分。在排序时，将一个 lambda 表达式表示的比较器传入 sorted 函数。使用 Stream 的 filter 操作过滤满足条件的学生。

6.9 标准类实训任务

【任务描述】

设一个学生的信息包含学号、姓名、出生日期、性别等。把 n 个学生的信息放入一个集合中，可以根据学号对学生信息进行检索，并且可以根据出生日期对学生进行排序输出。

【任务分析】

对于每个学生的个人信息，可以定义一个类来表示，命名为 Student，其每个对象表示一个学生。题目要求按照学号对学生进行检索，这样可以以学号为键值，把学生的信息保存到一个 Map 中，出生日期可以用 LocalDate 类表示。

为来了完成检索和排序功能，定义一个工具类 StuManager：

(1) 初始化学生信息。

(2) 检索学生信息。

(3) 按照出生日期对学生进行排序。

(4) 统计某一性别人数。

【任务解决】

完整程序代码如下：

```
package code0609;
import java.time.LocalDate;
import java.util.List;
import java.util.Optional;
import java.util.Random;
import java.util.TreeMap;
class Student {                                          //定义学生类
    int id;                                              //学号
    String name;                                         //姓名
    String gender;                                       //性别
    LocalDate birth;                                     //出生日期
    public Student(int id, String name, String gender, LocalDate birth) {
        this.id = id;
        this.name = name;
        this.gender = gender;
        this.birth = birth;
    }
    public String toString() {
        return id + "," + name + "," + gender + "," + birth;
    }
}
//定义学生管理类
public class StuManager {
    static int STU_NUM = 10;
    static TreeMap<Integer, Student> stuMap = new TreeMap<Integer, Student>();
    static void initStudents() {                         //初始化学生信息
        Random r=new Random();
        for (int i = 0; i < STU_NUM; i++) {
            String name = "Name" + i;
            String gender =r.nextInt() % 2 == 0 ? "女":"男";
            //随机生成出生日期
```

```
            LocalDate birth=LocalDate.of(2000+r.nextInt(5), r.nextInt(1,13), r.
nextInt(1,29));
            Student stu = new Student(i, name, gender, birth);
            stuMap.put(i, stu);
        }
    }
    public static void main(String args[]) {
        StuManager.initStudents();
        Random r=new Random();
        int id = r.nextInt(STU_NUM * 2);           //随机生成一个学号进行检索
        System.out.println("检索学号:" + id);
        Student stu=StuManager.stuMap.get(id);
        if (stu == null) {
            System.out.println("未检索到学生信息");
        } else
            System.out.println("检索到的学生信息是:" + stu);
        System.out.println("按照出生日期排序后的结果:");
        //构建 Stream,按照出生日期排序
        List<Student> sortedAsc=StuManager.stuMap.values().stream().sorted((x,
y)->x.birth.compareTo(y.birth)).toList();
        //把学生信息使用 map 操作转成字符串,然后使用 reduce 操作合并成一个字符串
        Optional<String> result=sortedAsc.stream().map(x->x.toString()).reduce
((x,y)->x.toString()+"\n"+y.toString());
        System.out.println(result.get());
        //使用过滤操作得到所有男生
        long count=sortedAsc.stream().filter(x->x.gender=="男").count();
        System.out.println("男生数量:"+count);
    }
}
```

程序可能的运行结果(使用了随机数,每次运行可能不同):

```
检索学号:2
检索到的学生信息是:2,Name2,男,2000-07-14
按照出生日期排序后的结果:
2,Name2,男,2000-07-14
1,Name1,女,2001-01-01
6,Name6,女,2002-02-06
4,Name4,女,2002-03-23
8,Name8,女,2002-08-16
7,Name7,男,2003-03-22
3,Name3,男,2003-09-19
9,Name9,女,2004-02-18
5,Name5,女,2004-04-22
0,Name0,女,2004-09-15
男生数量:3
```

1. 将一个字符串中的小写字母转换成大写字母,并将大写字母转换成小写字母。
2. 找出两个字符串中所有共同的字符。
3. 找出两个字符串的最长子串。

4. 定义一个正则表达式，识别字符串的邮件地址。

5. 编写一个程序，用 Map 实现学生成绩单的存储和查询，并且对成绩进行排序，存储到 TreeSet 中，求出平均成绩、最大值、最小值。

6. 给定一个整数－1234567，输出它的二进制、八进制和十六进制表示形式。

7. 编写一个程序，检查当前系统的 Java 版本和类路径。

8. 编写一个 Java 程序，在其中调用外部程序 cmd，并显示其输出结果。

9. 一个箱子既可以放面包，也可以放苹果，创建一个泛型化类型，表示只放其中一种物品。

10. 计算距离下一届奥运会开幕还有多少天。

11. 编写一个形如图 6-5 的日历程序。

2022 年 8 月

日	一	二	三	四	五	六
**	1	2	3	4	5	6
7	8	9	10	11	12	13
14	15	16	17	18	19	20
21	22	23	24	25	26	27
28	29	30	31			

图 6-5　习题 11 的日历

12. 分析以下 Stream 流代码，最终输出的结果是什么？

```
public class Test {
    public static void main(String[] args) {
        ArrayList<String> list = new ArrayList<>();
        Collections.addAll(list, "解放碑", "沙坪坝", "杨家坪", "江北", "渝北");
        long count = list.stream().filter(s -> s.length() == 3).skip(2).count();
        System.out.println(count);
    }
}
```

第二篇　提 高 篇

第 7 章 异常处理

在程序运行过程中,经常出现由于硬件设备问题、软件设计错误、用户输入错误等原因引起的程序运行错误。为了能处理这些问题,Java 引入了异常处理机制。本章主要讲述如何利用异常处理机制实现对 Java 应用中各种类型异常进行发现、捕获和有效处理。通过本章学习,应该重点掌握以下主要内容:

(1) 异常处理机制。

(2) 异常类及其层次关系。

(3) 异常的捕获和处理。

(4) 自定义异常类。

7.1 为什么需要异常处理

【例 7.1】 假设你正在盘点库存的盒子数量,现在需要将盒子数量输入到程序中以计算盒子的总价值,编写这样一个盘点程序。

```
package code0701;

import java.io.*;
import java.text.NumberFormat;

public class NoException {
    public static void main(String[] args) throws IOException {
        String numBoxesIn;
        int numBoxes;
        double boxPrice = 3.25;
        NumberFormat currency = NumberFormat.getCurrencyInstance();
        System.out.println("有多少个盒子:");
        BufferedReader in = new BufferedReader(new InputStreamReader(System.in));
        //构建输入流以接收用户的输入数据
        numBoxesIn = in.readLine();
        numBoxes = Integer.parseInt(numBoxesIn);
        System.out.println("盒子的总价值是:");
        System.out.println(currency.format(numBoxes * boxPrice));
    }
}
```

程序可能正常运行，也可能发生异常，其可能出现如图 7-1 所示的运行结果。

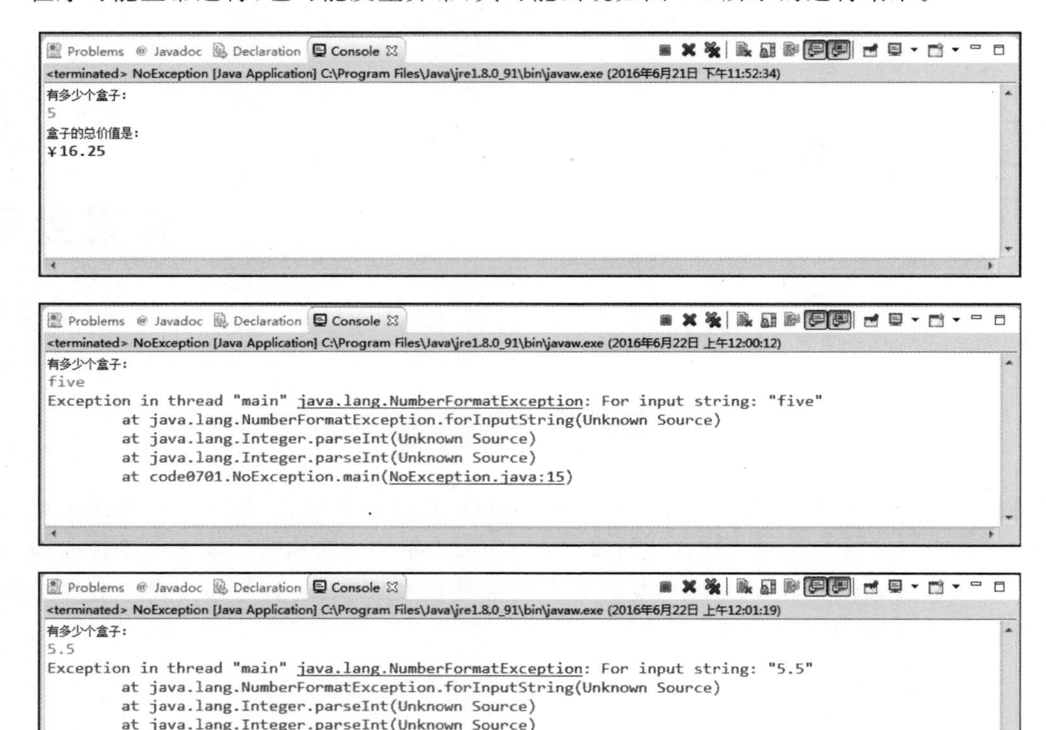

图 7-1　例 7.1 可能的运行结果及异常示例

分析：该软件有一个缺陷，即当用户输入一个整数值时，一切正常；但当用户输入其他值时，程序就会崩溃。为了在出现程序运行错误时，使其他所有代码能够继续完成其正常的工作，或者使程序能够体面地退出，最好能够对出现的错误进行处理，这时就需要引入异常处理。

可以对盘点软件进行如下修改。

【例 7.2】　修改后的盘点程序。

```
package code0701;

import java.io.*;
import java.text.NumberFormat;
import java.util.logging.FileHandler;
import java.util.logging.Logger;
import java.util.logging.SimpleFormatter;

public class NoExceptionModified {
    public static void main(String[] args) throws IOException {
        Logger logger = Logger.getLogger("code0701.NoExceptionModified");
        FileHandler fh = new FileHandler("log.log", 1000, 2, true);
        fh.setFormatter(new SimpleFormatter());
        logger.addHandler(fh);

        String numBoxesIn;
        int numBoxes;
        double boxPrice = 3.25;
```

```
        NumberFormat currency = NumberFormat.getCurrencyInstance();

        System.out.println("有多少个盒子:");
        BufferedReader in = new BufferedReader(new InputStreamReader(System.in));
        numBoxesIn = in.readLine();
        try {
            numBoxes = Integer.parseInt(numBoxesIn);
            System.out.println("盒子的总价值是:");
            System.out.println(currency.format(numBoxes * boxPrice));
        } catch (NumberFormatException e) {
            logger.warning("输入的不是一个整数。");
        }
    }
}
```

程序可能的运行结果以及项目日志文件 log.log.0 中的内容如图 7-2 所示。

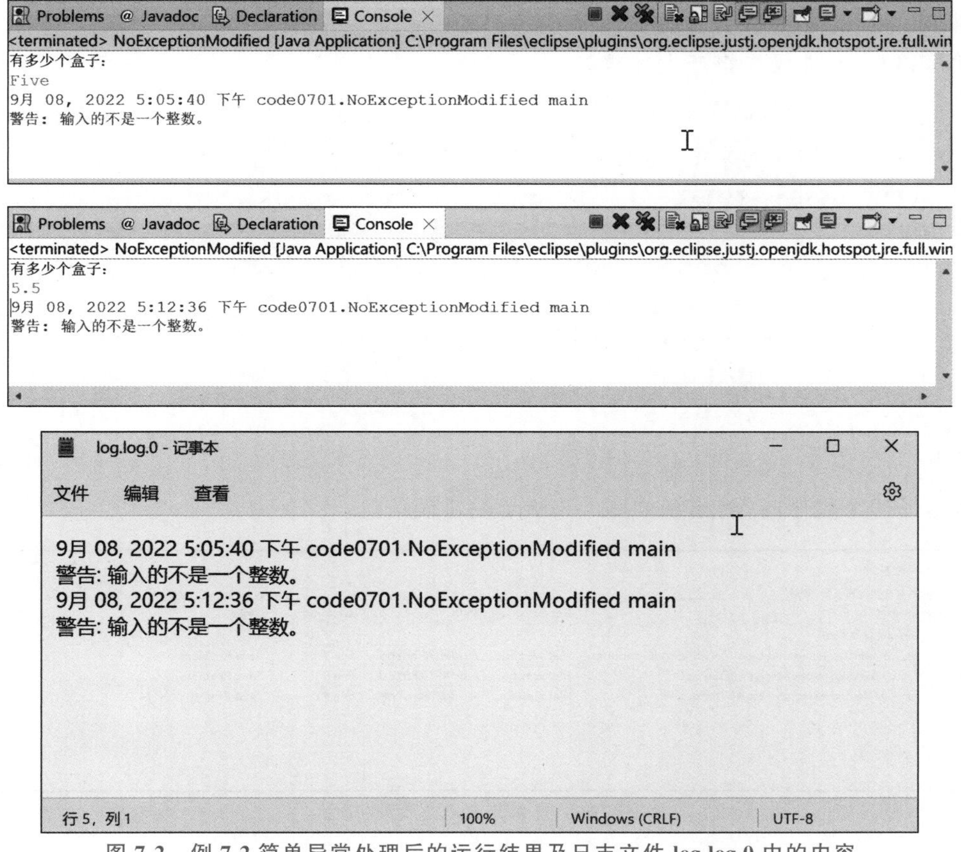

图 7-2　例 7.2 简单异常处理后的运行结果及日志文件 **log.log.0** 中的内容

分析：当用户输入错误时，程序将给出提示“输入的不是一个整数”。上述程序能够处理程序运行错误的关键是把 Integer.parseInt()方法调用封装在 try 语句块中。在执行 Integer.parseInt()方法时，计算机将监视该语句。如果该语句抛出了一个异常，那么计算机就从 try 语句块跳到下面的 catch (NumberFormatException e)语句块，并执行该语句块中的写日志操作，然后继续执行 catch 语句块后面的语句。

修改后的程序涉及 4 个关键问题：①什么情况下会产生异常；②如何使用 try-catch 处理异常；③需要处理的异常类有哪些；④日志功能如何实现。

此外，程序中还可能涉及如何从标准输入流中读取数据(System.in、InputStreamReader、BufferedReader)的问题，相关内容将在第 8 章中介绍。

7.2 异常概述

7.2.1 什么是异常

通常所说的异常(exception)指的是异常事件(exception event)。在软件运行时，很多情况都将导致异常事件的发生。例如：想打开的文件不存在，网络连接中断，操作数超出预定范围，访问的数据库打不开，正在装载的类文件丢失。

可见，在程序中出现异常是普遍存在的。在 Java 编程语言中，对异常的处理有非常完备的机制，如下面的案例所示。

【例 7.3】 文件操作将产生异常。

```
package code0702;

import java.io.FileInputStream;

public class FileException {
    public static void main(String[] args) {
        FileInputStream fis = new FileInputStream("test.txt");
        int b;
        while((b=fis.read())!=-1){
            System.out.print(b);
        }
        fis.close();
    }
}
```

当编译这个程序时，将出现如图 7-3 所示的错误信息。

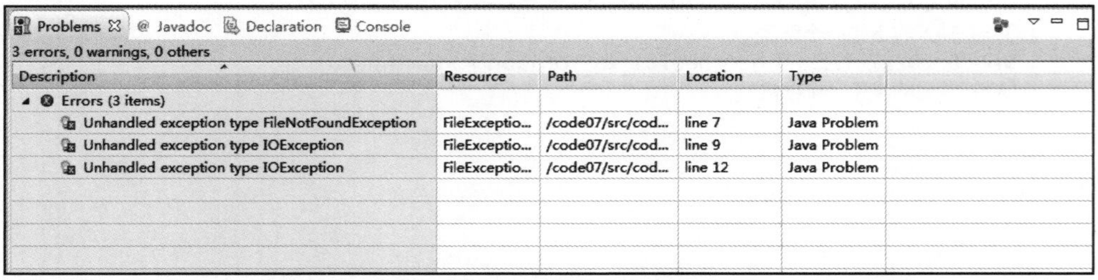

Problems ⊠ @ Javadoc Declaration Console

3 errors, 0 warnings, 0 others

Description	Resource	Path	Location	Type
Errors (3 items)				
Unhandled exception type FileNotFoundException	FileExceptio...	/code07/src/cod...	line 7	Java Problem
Unhandled exception type IOException	FileExceptio...	/code07/src/cod...	line 9	Java Problem
Unhandled exception type IOException	FileExceptio...	/code07/src/cod...	line 12	Java Problem

图 7-3 异常未处理的错误提示

分析：上述程序中出现的 3 个编译错误是由于文件 test.txt 不存在或无法找到引起的，因此导致异常 java.io.FileNotFoundException 和 java.io.IOException 的出现。

【例 7.4】 数组下标越界引发异常。

```
package code0702;

public class ArrayIndexException {
    public static void main(String[] args) {
```

```
        String words[] = {"I","love","Java"};
        for(int i=0; i<4; i++){
            System.out.println(words[i]);
        }
    }
}
```

程序运行结果出现如图 7-4 所示的数组越界异常。

```
Problems @ Javadoc Declaration Console
<terminated> ArrayIndexException [Java Application] C:\Program Files\eclipse\plugins\org.eclipse.justj.openjdk.hotspot.jre.full.win32.x86_64_17.0.3.v202205
I
love
Java
Exception in thread "main" java.lang.ArrayIndexOutOfBoundsException: Index 3 out of bounds for length 3
	at code0702.ArrayIndexException.main(ArrayIndexException.java:7)
```

图 7-4 运行时数组越界异常

分析：上述程序编译可以通过，但在运行时出现了异常。数组 words 的长度为 3，当循环语句执行第 4 次时，导致数组下标越界，引发异常，并使程序终止。

例 7.3 和例 7.4 中都遇到了异常。屏幕上所显示的 java.io.IOException、java.io.FileNotFoundException、java.lang.ArrayIndexOutOfBoundsException 分别指明了异常的类型和异常所在的包。对于某些异常，必须在应用程序中对它进行处理，否则编译程序会指出该错误(如例 7.3)；而对另外一些异常，在应用程序中可以不做处理，而是直接交给运行时系统来处理(如例 7.4)。7.3 节将详细介绍异常的种类及其处理方法。

7.2.2 异常处理带来的好处

异常处理比传统的错误管理技术更具优势。使用异常处理可以将错误处理代码与正常代码分离。在传统的编程技术中，错误的检测、报告和处理代码通常使程序的处理逻辑变得复杂。

【例 7.5】 读取文件的 readFile()方法。

readFile()方法处理逻辑的伪代码如下：

```
readFile {
    打开文件;
    得到文件的长度;
    分配足够大小的内存空间;
    读取文件内容到内存;
    关闭文件;
}
```

上述处理逻辑虽然简单，但却忽略了对可能出现错误的处理。这些可能出现的错误包括不能打开文件、无法得到文件的长度、不能分配足够大小的内存空间、读取文件内容失败、不能关闭文件等。

因此，需要增加错误的检测、报告和处理代码。下面分别基于传统编程技术和异常处理技术，编写读取文件的 readFile()方法的伪代码，并比较两者的不同。

1)基于传统编程技术的伪码形式

```
errorCodeType readFile {
    初始化 errorCode = 0;
    打开文件;
    if (文件正确打开) {
        得到文件的长度;
        if (已得到文件的长度) {
            分配足够大小的内存空间;
            if (内存分配成功) {
                读取文件内容到内存;
                if (读取失败) {
                    errorCode = -1;
                }
            } else {
                errorCode = -2;
            }
        } else {
            errorCode = -3;
        }
        关闭文件;
        if (文件没有正确关闭 && errorCode == 0) {
            errorCode = -4;
        } else {
            errorCode = errorCode and -4;
        }
    } else {
        errorCode = -5;
    }
    return errorCode;
}
```

2)使用异常处理技术之后的伪码形式

```
readFile {
    try {
        打开文件;
        得到文件的长度;
        分配足够大小的内存空间;
        读取文件内容到内存;
        关闭文件;
    } catch (fileOpenFailed) {                          //文件打开失败
      doSomething;
    } catch (sizeDeterminationFailed) {                 //读取文件长度失败
        doSomething;
    } catch (memoryAllocationFailed) {                  //分配内存空间失败
        doSomething;
    } catch (readFailed) {                              //读取文件内容失败
        doSomething;
    } catch (fileCloseFailed) {                         //关闭文件失败
        doSomething;
    }
}
```

分析:通过比较上述两段代码,可以看到,采用传统编程技术编写的代码存在非常多的错误检测、报告和处理代码,使得正常的处理逻辑变得模糊,不便于程序的维护和修改。而采用异常处理技术编写的代码，则在不同的地方编写正常的处理逻辑代码和错误处理代码。虽然异常处理技术并没有减少程序员进行错误处理的工作量，但却可以帮助程序员将错误处理工作组织得更加有效。

7.3 异常处理机制

7.3.1 Java 的异常处理机制

当一个方法在运行时出现异常事件时，该方法就会创建一个对象，并将该对象抛给运行时系统，这个对象称为异常对象(exception object)。异常对象存储了与错误有关的信息，包括错误类型以及异常事件出现时应用程序的状态和调用过程。创建异常对象并抛给运行时系统的过程称为抛出异常(throwing an exception)。能够处理异常的语句块称为异常处理程序(exception handler)。

Java 语言提供了两种处理异常的机制。

1. 捕获异常

当方法抛出异常后，运行时系统将沿着方法的调用栈逐层回溯，去查找能处理该异常的异常处理程序。这一过程称为捕获异常(catching exception)。

在图 7-5 中形成了一个方法调用栈，即 main()方法调用 f1()方法、f1()方法调用 f2()方法、f2()方法调用 f3()方法。如果 f3()方法抛出了异常，则运行时系统首先查看 f2()方法是否能处理；此时 f2()方法不包含异常处理程序，因此运行时系统就继续查看 f1()方法是否能处理；f1()方法包含异常处理程序，因此异常被 f1()方法捕获下来并进行处理，此过程如图 7-6 所示。这是一种积极的异常处理机制。如果运行时系统在遍历了调用栈中所有的方法后，都没有找到合适的异常处理程序，则运行时系统将终止，Java 程序也将相应地退出。

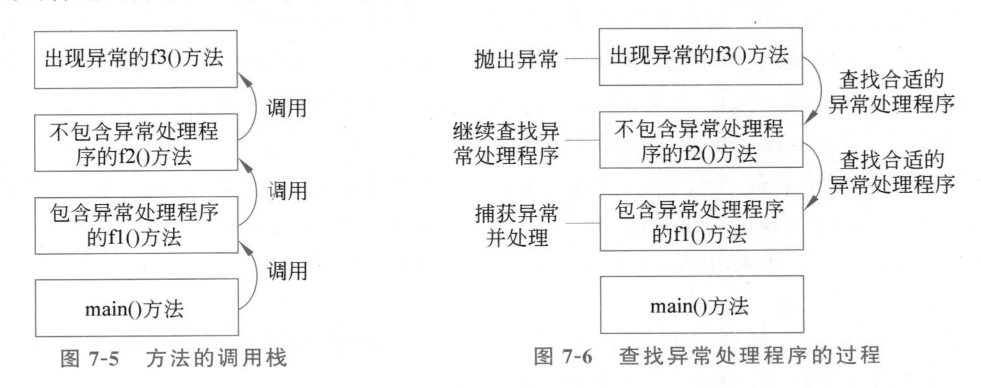

图 7-5 方法的调用栈　　图 7-6 查找异常处理程序的过程

2. 声明抛弃异常

如果一个方法并不知道如何处理所出现的异常，则可以在该方法声明时，声明抛弃异常(Specifying the Exceptions Thrown by a Method)，具体声明方式参见 7.3.3 节。

7.3.2 异常类的类层次

Java 是采用面向对象的方法来处理错误的，一个异常事件是由一个异常对象来表示的。在 Java 类库的每个包中都定义了自己的异常类，所有这些类都直接或间接地继承于 Throwable 类，如图 7-7 所示。Java 中的异常可以分为错误类(Error)、运行时异常类

(RuntimeException)和非运行时异常类(CheckedException)。

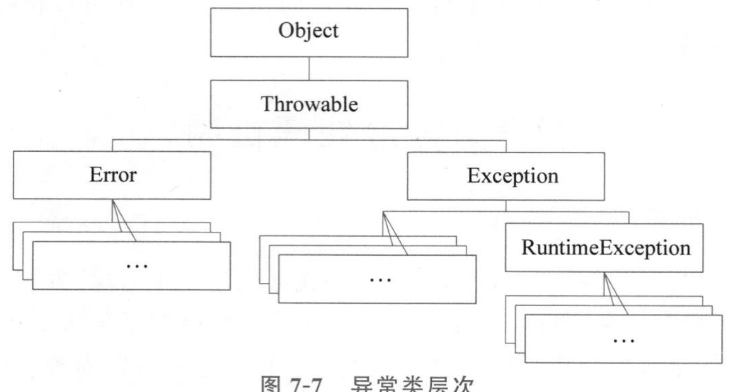

图 7-7　异常类层次

1. 错误类

错误类继承于 Error 类。错误类往往是由程序外部的运行环境出现错误引起的,因此程序通常不必去处理这类异常。例如,一个应用程序成功地打开了一个输入文件,但因为硬件故障或者操作系统故障使得应用程序无法读取文件的内容,此时将产生 java.io.IOError。有时应用程序为了通知用户出现了软硬件故障,也会捕获该类异常。

常见的错误类有 AnnotationFormatError、AssertionError、AWTError、CoderMalfunctionError、FactoryConfigurationError、IOError、LinkageError、ServiceConfigurationError、ThreadDeath、TransformerFactoryConfigurationError、VirtualMachineError 等,用于标识动态链接失败、线程死锁、图形界面错误、虚拟机错误等。

2. 运行时异常类

运行时异常类继承于 RuntimeException 类,它代表了 JVM 在运行时所生成的异常。由于这类异常的出现很普遍,要求程序对这类异常全部都做出处理可能对程序的可读性和高效性带来不好的影响,因此 Java 允许程序可以不对运行时异常做出处理。如例 7.4 中,程序对运行时异常 ArrayIndexOutOfBoundsException 并没有做出任何处理,而是直接交给运行时系统。当然在需要时,程序也可以处理运行时异常。

常见的运行时异常有 ArithmeticException(算术异常)、ClassCastException(类型转换异常)、ArrayIndexOutOfBoundsException(数组越界异常)、NegativeArraySizeException(指定数组维数为负值异常)和 NullPointerException(空指针异常)等。

3. 非运行时异常类

除了运行时异常以外的继承于 Exception 类的子类统称为非运行时异常。对于这类异常,Java 编译器要求程序必须捕获或者声明抛弃异常。

常见的非运行时异常有 ClassNotFoundException(找不到类或接口所产生的异常)、IllegalAccessException(非法访问异常)和 IOException(输入输出异常)等。

4. Throwable 类

Throwable 类是所有异常类的父类,常用的方法有

(1) String getMessage():返回该 Throwable 对象详细信息。

(2) void printStackTrace():把该 Throwable 对象和它的跟踪情况打印到标准错误流。

(3) void printStackTrace(PrintStream s):把该 Throwable 对象和它的跟踪情况打印到指定的打印流。

(4) void printStackTrace(PrintWriter s)：把该 Throwable 对象和它的跟踪情况打印到指定的打印流。

(5) String toString()：返回该 Throwable 对象的简短字符串描述。

7.3.3 异常的处理

对于非运行时异常，程序必须捕获或者声明抛弃异常，否则程序无法通过编译。

1. 捕获异常

一个方法中如果对某种类型的异常对象提供了相应的处理代码，则这个方法可捕获该种异常。捕获异常是通过 try-catch-finally 语句实现的，其语法如下：

```
try{
  …
}catch(exceptionType e){
  …
}catch(exceptionType e){
  …
}
  …
}finally{
  …
}
```

说明：(1) try 语句块：捕获异常的第一步是用 try 语句块选定捕获异常的范围。在 try 语句块中的语句在执行的过程中可能会抛出异常。

(2) catch 语句块：每个 try 语句块可以伴随一个或多个 catch 语句块，用于处理 try 代码块中产生的异常事件。catch 语句只需一个形式参数来指明它所能够捕获的异常类型，这个类必须是 Throwable 的子类。运行时系统通过参数值把抛出的异常对象传递给 catch 语句块。

捕获异常的顺序与 catch 语句的顺序有关。当捕获到一个异常时，剩下的 catch 语句就不再进行匹配。因此，在安排 catch 语句的顺序时，首先应该捕获最特殊的异常，然后再捕获一般化的异常。也就是说，一般先安排子类，再安排父类。例如，在捕获异常时如果写成如下形式：

```
}catch(IOException e){
    e.printStackTrace();
}catch(FileNotFoundException e){
    e.printStackTrace();
}
```

由于第一个 catch 语句首先得到匹配，第二个 catch 语句将不会被执行。编译时将会出现 Unreachable catch block 的错误提示信息。

(3) finally 语句块：finally 语句块在 try 语句块退出后总会被执行，这保证了不论 try 语句块中是否有异常发生，finally 语句块都会被执行。finally 语句块在 try-catch-finally 结构中是可选的。finally 语句块最大的作用是防止资源泄露。在进行文件关闭或者其他资源回收操作时，一般都将相关代码放在 finally 语句块中，这样可以确保资源总是能够被正确回收，防止资源泄露。

此外，try-catch-finally 可以嵌套。

【例 7.6】 捕获异常。

```
package code0703;

import java.io.FileInputStream;
import java.io.FileNotFoundException;
import java.io.IOException;
import java.util.logging.FileHandler;
import java.util.logging.Logger;
import java.util.logging.SimpleFormatter;

public class CatchException {
    public static void main(String[] args) throws SecurityException, IOException {
        Logger logger = Logger.getLogger("code0703.CatchException");
        FileHandler fh = new FileHandler("log.log", 1000, 2, true);
        fh.setFormatter(new SimpleFormatter());
        logger.addHandler(fh);

        FileInputStream fis = null;
        try {
            fis = new FileInputStream("test.txt");
            int b;
            while ((b = fis.read()) != -1) {
                System.out.print(b);
            }
        } catch (FileNotFoundException e) {
            logger.warning("Error: " + e.getMessage());
        } catch (IOException e) {
            logger.warning("Error: " + e.getMessage());
        } finally {
            try {
                if (fis != null) {
                    fis.close();
                }
            } catch (IOException e) {
            }
        }
    }
}
```

程序运行结果如图 7-8 所示。

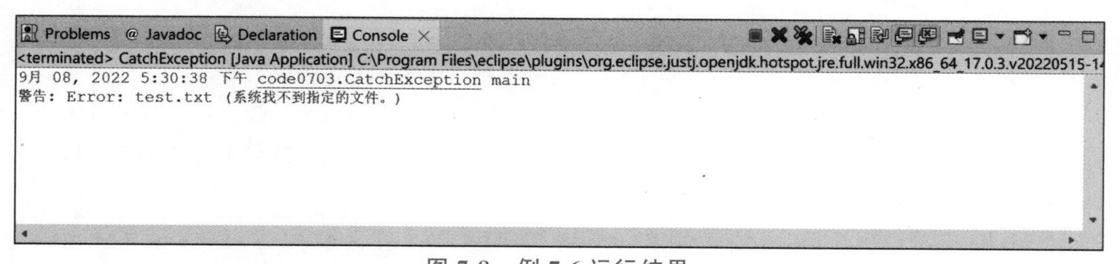

图 7-8 例 7.6 运行结果

分析：由于需要打开的 test.txt 文件不存在，程序在创建文件输入流时会抛出 FileNotFoundException 的异常，该异常被 catch 语句块捕获并处理，将相应的错误描述信息记录到日志文件中。为了保证打开的文件能够被正确地关闭，将 fis.close 语句放在了 finally 语句块中。

程序中的 java.util.logging.Logger 类是 Java 自带的日志系统，用来记录特定系统或应用程序组件的日志消息。一般使用圆点分隔的层次命名空间来命名 Logger。Logger 名称可以是任意的字符串，但是它们一般应该基于被记录组件的包名或类名，如例 7.6 中的程序代码中，使用了 code0703.CatchException 作为 Logger 的名字。通过调用 Logger 类的 getLogger 工厂方法来获得 Logger 对象，该方法要么创建一个新 Logger，要么返回一个合适的现有 Logger。

日志消息被转发到已注册的 Handler 对象，该对象可以将消息转发到各种目的地，包括控制台(ConsoleHandler)、文件(FileHandler)等。在例 7.6 中，通过注册 FileHandler，将日志消息写入指定的 log.log 文件中。java.util.logging.SimpleFormatter 可以为日志信息做标准日志格式化处理。每个 Logger 都有一个与其相关的 Level(日志级别)，这反映了此 logger 所关心的最低 Level。Java 日志级别有 7 级，按照优先级从高到低分别是：SEVERE、WARNING、INFO、CONFIG、FINE、FINER、FINEST，其中 SEVERE 指示严重失败的消息级别，WARNING 指示潜在问题的消息级别，INFO 是报告消息的消息级别，CONFIG 是用于静态配置消息的消息级别，FINE 是提供跟踪信息的消息级别，FINER 指示一条相当详细的跟踪消息，FINEST 指示一条最详细的跟踪消息。另外，还有 ALL 和 OFF 两种日志级别设置，ALL 指示应该记录所有消息，OFF 是一个可用于关闭日志记录的特殊级别。如果将日志级别设置在某个级别上，那么比此级别优先级高的日志都会打印出来。例如，如果设置日志级别为 CONFIG，那么 SEVERE、WARNING、INFO、CONFIG 这 4 个级别的日志能正常输出，而 FINE、FINER、FINEST 这 3 个级别的日志会被忽略。FileHandler 默认的日志级别是 ALL，ConsoleHandler 默认的日志级别是 INFO。通过 Logger 对象的 severe()、warning()、info()、config()、fine()、finer()、finest()方法，可以记录一条相应级别的日志消息。在例 7.6 的 catch 语句块中，就记录了一条 WARNING 级别的错误信息。Logger 上的所有方法都是多线程安全的。

2. 声明抛弃异常

如果在一个方法中发生了异常，但是该方法并不知道该如何处理该异常，此时该方法就应该声明抛弃异常，使得异常对象可以沿着方法的调用栈向后传播，直到有合适的方法捕获它为止。

声明抛弃异常是在一个方法声明的 throws 子句中指明的。例如：

```
public void writeList() throws IOException, ArrayIndexOutOfBoundsException {
    …
}
```

throws 子句中可以同时指明多个异常，之间由逗号隔开。

【例 7.7】 声明抛弃异常。

```
package code0703;

import java.io.FileInputStream;
import java.io.IOException;

public class ThrowException {
    public static void main(String[] args) throws IOException {
        FileInputStream fis = new FileInputStream("test.txt");
        int b;
        while ((b = fis.read()) != -1) {
```

```
            System.out.print(b);
        }
        fis.close();
    }
}
```

程序运行结果如图 7-9 所示。

```
Problems  @ Javadoc  Declaration  Console ×
<terminated> ThrowException [Java Application] C:\Program Files\eclipse\plugins\org.eclipse.justj.openjdk.hotspot.jre.full.win32.x86_64_17.0.3.v20220515-1
Exception in thread "main" java.io.FileNotFoundException: test.txt (系统找不到指定的文件。)
        at java.base/java.io.FileInputStream.open0(Native Method)
        at java.base/java.io.FileInputStream.open(FileInputStream.java:216)
        at java.base/java.io.FileInputStream.<init>(FileInputStream.java:157)
        at java.base/java.io.FileInputStream.<init>(FileInputStream.java:111)
        at code0703.ThrowException.main(ThrowException.java:8)
```

图 7-9　例 7.7 运行结果

分析:由于需要打开的 test.txt 文件不存在,程序在创建文件输入流时会抛出 FileNotFoundException 的异常,由于 main()方法采用了抛弃异常的处理,因此程序出现的异常就被提交给 Java 虚拟机处理。

3. 报告异常情况

程序可以通过抛出异常的方式来报告异常情况。抛出异常就是产生并抛出异常对象的过程。异常对象可以由虚拟机生成并抛出,可以由某些类的实例生成并抛出,也可以由程序员自主创建生成并通过 throw 语句将其抛出。可以抛出的异常对象必须是 Throwable 类或者子类的实例。例如:

```
IOException e = new IOException();
throw e;
```

7.3.4　实用案例:找出数据文件中的最大值

假定程序要从 data.txt 文件中读入一组整数,然后输出这组数据中的最大值。利用异常处理机制,对数据输入可能存在的异常情况进行处理,增强程序的鲁棒性。

```
package code0703;

import java.io.File;
import java.io.FileReader;
import java.io.BufferedReader;
import java.io.FileNotFoundException;
import java.io.IOException;
import java.util.logging.FileHandler;
import java.util.logging.Logger;
import java.util.logging.SimpleFormatter;

public class FindMaxInteger {
    public static void main(String[] args) throws SecurityException, IOException {
        Logger logger = Logger.getLogger("code0703.FindMaxInteger");
        FileHandler fh = new FileHandler("log.log", 1000, 2, true);
        fh.setFormatter(new SimpleFormatter());
        logger.addHandler(fh);
```

```
        String path = FindMaxInteger.class.getClassLoader().getResource("").
getPath();
        File file = new File(path+"data.txt");
        BufferedReader in = null;
        try {
            in = new BufferedReader(new FileReader(file));
            String line;
            int maxValue=0;
            while((line = in.readLine())!=null)
            {
                int inputValue = Integer.parseInt(line);
                if(inputValue>maxValue)
                {
                    maxValue = inputValue;
                }
            }
            System.out.println("Max int is: "+maxValue);
        } catch (FileNotFoundException e) {
            logger.warning("文件不存在。");
        } catch (IOException e) {
            logger.warning("输入输出错误。");
        } catch (NumberFormatException e) {
            logger.warning("数字格式错误。");
        }
    }
}
```

程序可能的运行结果如图 7-10 所示。

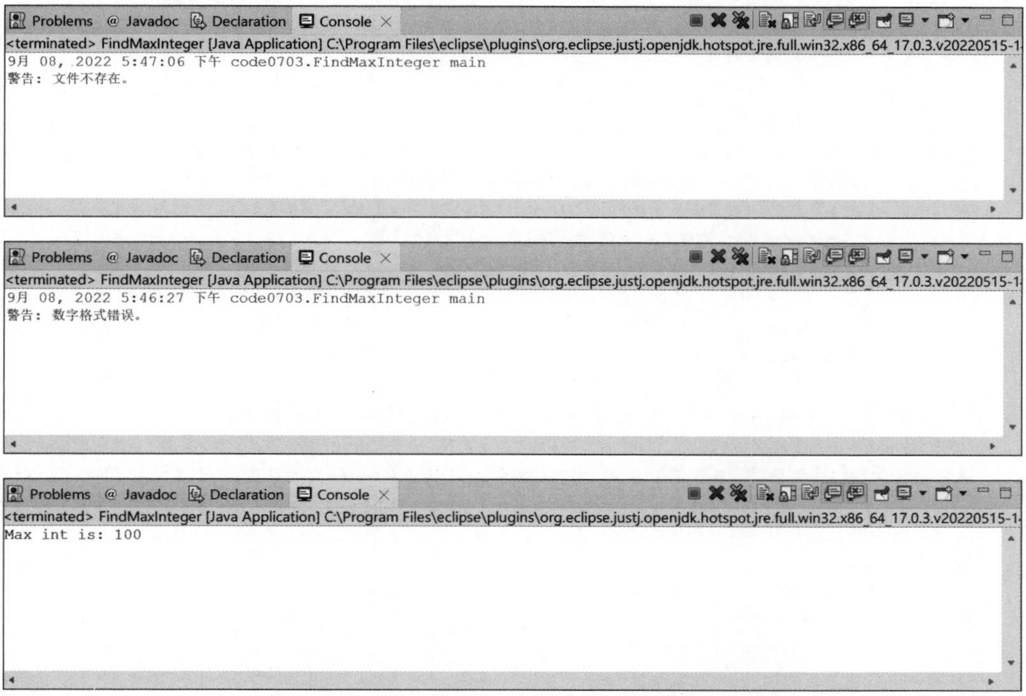

```
<terminated> FindMaxInteger [Java Application] C:\Program Files\eclipse\plugins\org.eclipse.justj.openjdk.hotspot.jre.full.win32.x86_64_17.0.3.v20220515-1
9月 08, 2022 5:47:06 下午 code0703.FindMaxInteger main
警告: 文件不存在。
```

```
<terminated> FindMaxInteger [Java Application] C:\Program Files\eclipse\plugins\org.eclipse.justj.openjdk.hotspot.jre.full.win32.x86_64_17.0.3.v20220515-1
9月 08, 2022 5:46:27 下午 code0703.FindMaxInteger main
警告: 数字格式错误。
```

```
<terminated> FindMaxInteger [Java Application] C:\Program Files\eclipse\plugins\org.eclipse.justj.openjdk.hotspot.jre.full.win32.x86_64_17.0.3.v20220515-1
Max int is: 100
```

图 7-10　实用案例可能的运行结果

分析：由于程序需要读取外部文件中的数据，对于指定的文件是否存在、文件中的数据是否合法等问题，编程时并不能保证这些问题不会出现。因此，通过异常处理机制对这些异常情况加以处理，在遇到异常情况时给用户适当的错误提示，可以提高程序的鲁棒性和人机交互的友好性。

7.4 自定义异常类

如果 Java 提供的系统异常类型不能满足程序设计的需求，那么可以设计自己的异常类型。

从 Java 异常类的结构层次可以看出，Java 异常的公共父类是 Throwable。在程序运行中可能出现两种问题：一种是由硬件系统或者 Java 虚拟机导致的故障，Java 定义该故障为 Error，这类问题是用户程序不能解决的；另外一种问题是程序运行时错误，Java 定义为 Exception，这种情况下，可以在程序中实现异常处理。

因此，用户自定义的异常类型必须是 Throwable 的直接子类或者间接子类。Java 推荐用户的异常类型以 Exception 为直接父类。创建用户异常的方法如下：

```
class UserException extends Exception{
    …     //语句
}
```

【例 7.8】 改写例 7.1 中的盘点程序，使之能够处理用户输入盒子数量为负数的错误。

```
import java.io.*;
import java.text.NumberFormat;
import java.util.logging.FileHandler;
import java.util.logging.Logger;
import java.util.logging.SimpleFormatter;

public class Stocktaking {
    public static void main(String[] args) throws IOException {
            Logger logger = Logger.getLogger("code0701.NoExceptionModified");
            FileHandler fh = new FileHandler("log.log", 1000, 2, true);
            fh.setFormatter(new SimpleFormatter());
            logger.addHandler(fh);

        String numBoxesIn;
        int numBoxes;
        double boxPrice = 3.25;
        NumberFormat currency = NumberFormat.getCurrencyInstance();
        System.out.println("有多少个盒子:");
        BufferedReader in = new BufferedReader(new InputStreamReader(System.in));
                                                    //构建输入流以接收用户的输入数据
        numBoxesIn = in.readLine();
        try {
            numBoxes = Integer.parseInt(numBoxesIn);
            if (numBoxes < 0) {
                throw new NegativeNumberException();
            }
            System.out.println("盒子的总价值是:");
            System.out.println(currency.format(numBoxes * boxPrice));
```

```
        } catch (NumberFormatException e) {
            logger.warning("输入的不是一个整数。");
        } catch (NegativeNumberException e) {
            logger.warning("盒子数量不可能为负数。");
        }
    }
}
class NegativeNumberException extends Exception {
}
```

程序可能的输出结果如图 7-11 所示。

```
Problems @ Javadoc Declaration Console
<terminated> Stocktaking [Java Application] C:\Program Files\eclipse\plugins\org.eclipse.justj.openjdk.hotspot.jre.full.win32.x86_64_17.0.3.v20220515-1416\
有多少个盒子:
-20
9月 08, 2022 5:59:32 下午 code0704.Stocktaking main
警告: 盒子数量不可能为负数。
```

图 7-11　例 7.8 运行结果

分析：通过引入用户自定义异常类 NegativeNumberException，对用户输入盒子数量为负数的错误情况进行了标识，使得程序可以利用异常处理对该错误进行处理。

7.5　异常处理实训任务

【任务描述】

编写一个应用程序，接收用户输入的一个正整数。如果用户输入的不是一个正整数，则提示用户重新输入，直到输入正确为止。

【任务分析】

为了确保用户输入的是一个正整数，需要在程序中判断用户输入的类型，如果不是一个正整数，则抛出异常，由异常处理程序提示用户重新输入。因为需要不断地测试用户输入数据的类型，直到用户输入正整数为止，因此需要在程序中引入循环控制结构。

【任务解决】

完整示例代码如下：

```
package code0705;
import java.io.BufferedReader;
import java.io.IOException;
import java.io.InputStreamReader;
public class GetPositiveInteger {
    public static void main(String[] args) throws IOException {
        String inputStr;
        int inputNum;
        boolean gotGoodInput = false;
        do {
            try {
                System.out.println("请输入一个正整数:");
```

```
                BufferedReader in = new BufferedReader(new InputStreamReader(
                        System.in));
                inputStr = in.readLine();
                inputNum = Integer.parseInt(inputStr);
                if (inputNum <= 0) {
                    throw new NegativeIntegerException();
                }
                gotGoodInput = true;
            } catch (NumberFormatException e) {
                System.out.println("输入的不是一个整数,请重新输入。");
            } catch (NegativeIntegerException e) {
                System.out.println("输入的不是一个正整数,请重新输入。");
            }
        } while (!gotGoodInput);
    }
}
class NegativeIntegerException extends Exception {
}
```

程序可能的运行结果如图 7-12 所示。

```
Problems  @ Javadoc  Declaration  Console
<terminated> GetPositiveInteger [Java Application] C:\Program Files\eclipse\plugins\org.eclipse.justj.openjdk.hotspot.jre.full.win32.x86_64_17.0.3.v20220515
请输入一个正整数:
five
输入的不是一个整数，请重新输入。
请输入一个正整数:
-5
输入的不是一个正整数，请重新输入。
请输入一个正整数:
5
```

图 7-12　实训任务运行结果

1. 什么是异常？简述 Java 的异常处理机制。

2. 下面的代码是否正确？

```
try {

} finally {

}
```

3. 下面的 catch 语句可以捕获哪些类型的异常？如果这样来使用，会有什么不足之处？

```
catch (Exception e) {

}
```

4. 修改下面的代码，使之能够通过编译。

```
public static void cat(File file) {
    RandomAccessFile input = null;
    String line = null;
```

```
    try {
        input = new RandomAccessFile(file, "r");
        while ((line = input.readLine()) != null) {
            System.out.println(line);
        }
        return;
    } finally {
        if (input != null) {
            input.close();
        }
    }
}
```

第 8 章 输入输出处理

本章学习目标

绝大部分程序都需要数据的输入和输出(I/O)。例如,从键盘上读取数据、在网络上交换数据、打印报表、读写文件信息等。在 Java 中,输入输出数据的处理都是通过数据流实现的。本章主要介绍各种输入输出流的使用,应该重点掌握以下主要内容:

(1) 字节流和字符流的不同应用场景。

(2) 文件输入输出字节流的应用。

(3) 标准输入输出流的应用。

(4) 工具类 Scanner 的使用。

(5) 理解部分 NIO 的使用。

8.1 流的作用

在前面的 Java 程序中,程序的运行结果一般输出在控制台,这是一种临时输出措施。有时需要把程序的运算结果长久保存,可以运用 Java 提供的输入输出流把结果保存到文件中。

【例 8.1】 把程序运行结果保存到文件中。

分析:如果把运行结果输出到控制台,可以使用 System.out,如果长久保存,就应输出到文件中,使用文件输出流。问题解决如下:

```
package code0801;
import java.io.FileWriter;
public class SimpleFileStream {
    public static void main(String[] args) throws Exception {
        int sum = 0;
        for (int i = 1; i <= 100; i++) {
            sum += i;
        }
        //输出控制台
        System.out.println("计算结果 1:" + sum);
        //构建文件输出流,输出到文件 f1.txt 中
        FileWriter fout = new FileWriter("f1.txt");
        fout.write("计算结果 2:" + sum);
        fout.close();
    }
}
```

程序的运行结果有两部分,一部分如下输出在控制台:

```
计算结果 1:5050
```

另一部分的输出结果在文件 f1.txt 中,打开文件 f1.txt 可以看到下面的结果:

```
计算结果 2:5050
```

该问题的解决涉及下面 3 个关键点:

(1) 选择并创建合适的文件输出流对象。

(2) 使用输出流的输出方法存储数据,本例使用的是 write()方法。

(3) 关闭输出流,本例使用的是 close()方法。

8.2 流的划分

数据流(Stream)是一组有顺序、有起点和终点的字节集合,是对输入和输出的总称和抽象。

一般地,数据流分为输入流(InputStream)和输出流(OutputStream)。输入流只能读,不能写,而输出流只能写不能读。

Java 程序通过流来连接设备完成输入或输出。所有流具备相同或类似的操作方式。一个输入流能够抽象多种不同类型的输入:从磁盘文件、从键盘或从网络套接字。同样,一个输出流可以输出到控制台、磁盘文件或相连的网络。

Java 主要定义了两种类型的流:字节流和字符流。字节流以字节作为基本处理单位。使用字节流读写二进制数据。字符流以字符作为基本处理单位,一般需要指定字符集编码确定字符和字节的关系。在某些场合,字符流比字节流更方便。字符流也可以由字节流转化而来。

1. 字节流

字节流在顶层有两个抽象类:InputStream 和 OutputStream 定义了所有流的核心读写方法。最重要的两种方法是 read()方法和 write()方法,它们分别用于读字节和写字节。它们被不同的子类重写,可应用于不同的场景。使用完毕后,记得调用 close()方法关闭流。一般名称以 Stream 结尾的都是字节流。

输入字节流的描述如表 8-1 所示。

表 8-1 输入字节流的描述

流	描　述
InputStream	表示输入字节流的抽象类,是其他流的父类
FileInputStream	文件字节输入流
FilterInputStream	过滤字节输入流,使用其他流作为数据来源
BufferedInputStream	缓冲输入流,可以在其他流的基础上构建缓存
DataInputStream	包含读取 Java 标准数据类型方法的输入流
ObjectInputStream	对象输入流,用于对象串行化

输出字节流的描述如表 8-2 所示。

表 8-2　输出字节流的描述

流	描　述
OutputStream	表示输出字节流的抽象类,是其他流的父类
FileOutputStream	文件字节输出流
FilterOutputStream	过滤字节输出流,使用其他流作为数据目的地
BufferedOutputStream	缓冲输出流,在其他流的基础构建缓冲功能
DataOutputStream	包含写 Java 标准数据类型方法的输出流
PrintStream	打印流,包含 print()和 println()的输出流,是标准输出流 System.out 使用的流
ObjectOutputStream	对象输出流,用于对象串行化

2. 字符流

字符流顶层有两个抽象类：Reader 和 Writer。定义了所有字符流的核心读写方法。其中两个最重要的方法是 read()方法和 write()方法,它们分别进行字符数据的读和写。这些方法被不同的子类重写,可应用于不同的场景。

输入字符流的描述如表 8-3 所示。

表 8-3　输入字符流的描述

流	描　述
Reader	表示字符输入流的抽象类,是其他流的父类
BufferedReader	带缓冲功能的字符输入流
InputStreamReader	字节流向字符流转化的流
FileReader	文件字符输入流,是 InputStreamReader 的子类
StringReader	把字符串作为输入流

输出字符流的描述如表 8-4 所示。

表 8-4　输出字符流的描述

流	描　述
Writer	描述字符输出流的抽象类,是其他流的父类
BufferedWriter	带缓冲功能的字符输出流
OutputStreamWriter	字符输出流向字节输出流的转换
PrintWriter	打印字符输出流,属于常用流。可以输出到控制台和文件
FileWriter	输出字符到文件
StringWriter	输出字符到字符串缓冲区

由于涉及的输入输出流比较多,本章主要介绍文件输入输出字节流、标准输入输出流、文件字符流、Scanner 类的使用。其他流的使用请参考 API 文档。

8.3 标准输入输出流

系统类 System 定义了 3 个静态流变量 in、out 和 err，可以直接调用。System.in 表示标准输入，通常情况指键盘数据输入；System.out 表示标准输出，通常指把数据输出到控制台或者屏幕；System.err 表示标准错误输出，通常指把数据输出到控制台或者屏幕。

8.3.1 标准输入

System.in 作为字节输入流实现标准输入，通过 read()方法从键盘接收数据。下述方法遇到流末返回−1。

int read()：读取一个字节，把字节放入整数的最低字节，然后返回整数。

int read(byte b[ ])：读取字节数组，返回实际读取的字节数。

int read(byte b[ ], int offset, int len)：读取 len 个字节放入数组的 offset 位置，返回字节数。

【例 8.2】 从标准输入读取数据。

```
package code0803;
import java.io.IOException;
public class StdInput
{
    public static void main(String[] args) throws IOException
    {
        System.out.println("输入:");
        byte b[] = new byte[512];
        int count = System.in.read(b);          //读取字节数组
        System.out.println("输出:");
        for (int i=0;i<count;i++)
        {
            System.out.print(b[i] + " ");
        }
        System.out.println();
        System.out.println("count=" + count);
    }
}
```

程序运行结果：

```
输入:
abcd 中
输出:
97 98 99 100 -28 -72 -83 13 10
count=9
```

分析：程序运行时，从键盘输入 5 个字符“abcd 中”并按 Enter 键。保存在缓冲区 b 中的元素个数 count 为 9，Enter 占用 2 字节，汉字占用 3 字节，汉字字节的最高位一般为 1，所以是负数。

8.3.2 Scanner 类封装标准输入流

System.in 作为标准输入流，是一个 InputStream 类的对象，其 read()方法的主要功能是读取字节和字节数组，不能直接得到需要的数据(如整型、浮点型、字符串等)。此时，需要另外一个类 java.util.Scanner 的配合。Scanner 类可以对标准输入流 System.in 的数据进行解析，得到需要的数据。

【例 8.3】 从标准输入读取一个整数和浮点数，并计算它们的乘积。

```
package code0803;
import java.util.Scanner;
public class ScannerSystemIn {
    public static void main(String[] args) {
        Scanner sin = new Scanner(System.in);    //封装标准输入流
        int a = sin.nextInt();                   //读取整数
        double b = sin.nextDouble();             //读取浮点数
        double result = a * b;
        System.out.println("a * b=" + result);
    }
}
```

程序运行结果：

```
10
123.456
a * b=1234.56
```

分析：结果中的第一行和第二行是从控制台输入的数据，第三行是程序的输出结果。今后从标准输入读取数据都可以参考这个例子的用法。Scanner 类还提供了很多获得其他类型数据的方法，如 nextBoolean()、nextFloat()、nextByte()等方法，next()方法返回字符串。

8.3.3 标准输出和格式化输出

System.out 作为打印流 PrintStream 的对象实现标准输出。其中定义了 print 和 println 方法，支持 Java 任意基本类型作为参数。例如：

```
public void print(int i);
public void println(int i);
```

两者的区别在于 println 在输出时加一个换行符，前面已经多次使用过。

另外，增加了 printf()方法，支持数据的格式化输出。

```
public PrintStream printf(String format, Object...args)
public PrintStream printf(Locale l, String format, Object...args)
```

该方法支持可变参数，即方法的参数个数是可变的。format 参数定义输出格式，args 是输出参数列表。

format 字符串的格式如下：

```
%[argument_index$][flags][width][.precision]conversion
```

其中，argument_index 用十进制整数表示参数在参数列表中的位置，第一个参数用%1$表示，第二个用%2$表示，以此类推，如果不关心参数的位置，可以只保留%；flags 是调整输出格式的字符集合；width 是一个非负整数，表示输出的最小字符数；precision 是一个非负整数，用于限制字符个数，如果输出浮点数，则是小数点后面的位数。conversion 是一个转换符，表示参数被如何格式化。该方法在格式化输出时，使用了类 java.util.Formatter 的功能。另外,String类的format()函数和这里的printf()使用相同的方法格式化字符串。

下面是常用的转换符。

(1) d:十进制整数。

(2) x:十六进制整数。

(3) f:浮点数。

(4) s:字符串。

(5) c:字符。

(6) t:格式化日期,后面还可以跟其他日期转换符。

详细的转换符参见类java.util.Formatter的文档。

8.3.4 实用案例8.1:数据的格式化输出

【例8.4】 格式化输出。

```
package code0803;
import java.time.LocalDateTime;
import java.util.Date;
public class PrintfDemo {
    public static void main(String[] args) {
        double d1 = 23456789.567;
        int i=65;
        //用逗号作为分隔符,格式化浮点数,字符,使用位置参数
        System.out.printf("%1$,.2f\n%2$d的十六进制=%2$x,对应字符 %2$c\n", d1,i);
        //不使用位置参数
        System.out.printf("pi=%.2f,e=%.5f\n",Math.PI,Math.E);
        Date c =new Date();
        //格式化日期
        System.out.printf("日期:%1$tF 时间:%1$tT %1$tA\n", c);
        //使用字符串的格式化函数
        String s=String.format("日期:%1$tF 时间:%1$tT %1$tA\n", LocalDateTime.now().
plusMinutes(10000));
        System.out.println(s);
    }
}
```

程序运行结果:

```
23,456,789.57
65的十六进制=41,对应字符 A
pi=3.14,e=2.71828
日期:2022-08-20 时间:11:43:18 星期六
日期:2022-08-27 时间:10:23:18 星期六
```

分析:格式化数字时,%表示此处有参数,逗号表示数字的千位分隔符,是一个Flag;“.2”表示数据精度,f表示格式化浮点数,x表示十六进制整数;c表示格式化字符,可以把整数输出为字符;格式化日期时,%1表示第一个参数,t表示格式化日期,F表示输出日期的ISO完整格式(如2015-07-22),T表示输出时间的ISO格式(如16:43:38),A表示提取星期几。其他转换符的使用请参考类java.util.Formatter的文档。

8.4 字节流使用

最典型的字节流是文件字节流。文件字节流包含文件输入流 FileInputStream 和输出流 FileOutputStream。在介绍文件字节流之前,先介绍一个表示文件对象的类 java.io.File,代表文件系统中的一个文件,它不能读取文件的内容,但是可以获取文件信息和对文件进行操作。

8.4.1 File 类

在进行文件操作时,需要知道一些关于文件的信息。File 类提供了一些方法来操作文件和获取文件的信息。对于目录,Java 把它当作一种特殊类型的文件,即文件名列表。

通过 File 类的方法,可以得到文件或者目录的描述信息,包括名称、所在路径、读写性、长度等,还可以创建目录、创建文件,改变文件名、删除文件、列出目录中的文件等。

1. 构造方法

下面的构造函数可以用来生成 File 对象:

```
File(String path)
File(String dir, String filename)
File(File dir, String filename)
File(URI uri)
```

这里,dir 是文件所在的目录,filename 是文件名,path 是文件的路径名。uri 是一个统一资源标识符。

下面的例子创建了 3 个 File 对象:f1、f2 和 f3。第一个 File 对象是由仅有一个路径参数的构造函数生成的。第二个对象有两个参数:目录和文件名。第三个 File 对象的参数包括指向 f1 文件的目录及文件名,f3 和 f2 指向相同的文件。第四个 File 对象用一个 uri 参数构造一个文件。

```
File f1 = new File("D:/Java");                    //计算机中存在此目录,f1 表示目录
File f2 = new File("D:/Java","test.txt");
File f3 = new File(f1,"test.txt");
File f4 = new File("file://D:/Java/test.txt");
```

注意:Java 能正确处理 UNIX 和 Windows/DOS 约定的路径分隔符。如果在 Windows 下的 Java 中用斜线(/),路径处理依然正确。记住,如果在 Windows/DOS 使用反斜线(\),则需要在字符串内使用它的转义序列(\\)。Java 约定用 UNIX 和 URL 风格的斜线作为路径分隔符。

File 定义了很多获取 File 对象标准属性的方法。例如,getName()方法返回文件名,getParent()方法返回父目录名,exists()方法在文件存在的情况下返回 true,反之返回 false。

2. File 类提供的方法

创建一个文件对象后,可以用 File 类提供的方法获得文件相关信息,对文件进行操作。File 类的主要方法如表 8-5 所示。

表 8-5 File 类的主要方法

方　法	描　述
boolean canRead()	测试文件是否可读
boolean canWrite()	测试文件是否可写

续表

方　法	描　述
boolean createNewFile()	创建文件
static File createTempFile(String prefix, String suffix)	创建临时文件
boolean delete()	删除文件
boolean equals(Object obj)	比较两个文件对象是否相等
boolean exists()	测试文件是否存在
File getAbsoluteFile()	返回绝对文件名
String getAbsolutePath()	返回绝对路径
String getName()	返回文件名(不包括路径)
String getParent()	返回父目录
String getPath()	返回路径
boolean isAbsolute()	是否是绝对路径
boolean isDirectory()	是否是目录
boolean isFile()	是否是文件
boolean isHidden()	是否隐藏文件
long lastModified()	上次修改时间,从 1970 年 1 月 1 日开始的标准时间(UTC)的毫秒数
long length()	文件长度
boolean renameTo(File dest)	重命名文件

【例 8.5】 File 类的使用。

```
package code0804;
import java.io.File;
import java.io.IOException;
public class FileDemo {
    public static void main(String[] args) throws IOException {
        File dir = new File("src/code0804");
        File f1 = new File(dir, "FileDemo.java");
        System.out.println(f1);
        System.out.println("exist: " + f1.exists());
        System.out.println("name:  " + f1.getName());
        System.out.println("path:  " + f1.getPath());
        System.out.println("abosolute path:" + f1.getAbsolutePath());
        System.out.println("parent:  " + f1.getParent());
        System.out.println("is a file :  " + f1.isFile());
        System.out.println("is a directory:     " + f1.isDirectory());
        System.out.println("length:   " + f1.length());
        File temp=File.createTempFile("临时文件", ".tmp");
        System.out.println("abosolute path:     " + temp.getAbsolutePath());
        System.out.println("length:  "+temp.length());
    }
}
```

程序运行结果：

```
src\code0804\FileDemo.java
exist: true
name:  FileDemo.java
path:  src\code0804\FileDemo.java
absolute path:C:\Users\yangrl\eclipseworkspace\Ch08\src\code0804\FileDemo.java
parent: src\code0804
is a file : true
is a directory:     false
length:   904
absolute path: C:\Users\yangrl\AppData\Local\Temp\临时文件 3468493380817714287.tmp
length:  0
```

3. 目录

File 对象也可以表示一个目录，目录是一个包含其他文件和路径列表的 File 类。当创建一个 File 对象且它是目录时，isDirectory()方法返回 true。这种情况下，可以调用该对象的 list()方法来提取该目录内部其他文件和目录的列表。该方法有两种形式，第一种形式如下：

```
String[ ] list()
```

文件名在一个 String 对象数组中返回。

第二种形式如下：

```
File[ ] listFiles()
```

文件对象在一个 File 对象的数组中返回。

有时需要列出目录下指定类型的文件，如.java、.class 等扩展名的文件。可以使用 File 类的下述 3 个方法，列出指定类型的文件。

```
String[ ] list(FilenameFilter FFObj)
File[ ] listFiles(FilenameFilter FFObj)
File[ ] listFiles(FileFilter FObj)
```

第一种方法用文件名过滤器返回目录下的文件名数组；第二种方法用文件名过滤器，返回文件对象数组；第三种方法用文件对象过滤器返回符合条件的文件对象数组。

参数 FFObj 是一个实现 FilenameFilter 接口的类的对象。

FilenameFilter 仅定义了一个方法，即 accept()。用 list()方法列出文件时，将调用 accept()方法检查该文件名称是否符合要求。它的通常形式如下：

```
boolean accept(File directory, String filename)
```

如果返回 true，名称为 filename 的文件包含在返回列表中；否则，不包含在返回列表中。

FObj 是一个实现了 FileFilter 接口的类的对象。针对目录中的每个文件调用一次 accept()方法，它不仅可根据文件名过滤文件，也可根据文件的其他信息过滤文件。其通常形式如下：

```
boolean accept(File path)
```

如果返回值是 true，path 代表的文件包含在返回列表中；否则不包含在返回列表中。

【例 8.6】 列出目录中特定类型的文件。

```
package code0804;
import java.io.File;
import java.util.stream.Stream;
public class DirDemo
{
    public static void main(String[] args)
    {
        File dir = new File("bin/code0803");
        System.out.println("列出目录" + dir + "中的 class 文件");
        //使用 lambda 表达式实现文件名过滤器
        File fs1[] = dir.listFiles((path,name)->name.endsWith("class"));
        Stream.of(fs1).forEach(System.out::println);
                                                //使用 Stream 的终结操作 forEach
        dir=new File("./bin");
        System.out.println("列出目录" + dir + "中的子目录");
        //在创建一个子目录
        File f=new File(dir,"_newDir");
        System.out.println(f.mkdir());          //创建目录成功返回 true
        //使用 lambda 表达式构建了文件对象过滤器
        File fs3[] = dir.listFiles(pathname->pathname.isDirectory());
        Stream.of(fs3).forEach(System.out::println);
                                                //使用 Stream 的终结操作 forEach
    }
}
```

程序运行结果：

```
列出目录 bin\code0803 中的 class 文件
bin\code0803\PrintfDemo.class
bin\code0803\ScannerSystemIn.class
bin\code0803\StdInput.class
列出目录.\bin 中的子目录
true
.\bin\code0801
.\bin\code0803
.\bin\code0804
.\bin\_newDir
```

8.4.2 文件字节流

文件字节流 FileInputStream 和 FileOutputStream 可用于对文件的输入输出处理。

1. FileInputStream

FileInputStream 用于顺序读取本地文件。从超类继承了 3 个 read()方法等。它的两个常用的构造函数如下：

```
FileInputStream(String filepath)
FileInputStream(File fileObj)
```

当指定的文件在文件系统中不存在时，引发 FileNotFoundException 异常。这里，filepath 是文件的全称路径，fileObj 是描述该文件的 File 对象。假设有一个文件，名称为 Test.java，可以用下面的代码构造文件输入流。

直接使用文件名构造输入流:

```
FileInputStream f1 = new FileInputStream("Test.java")
```

或者先构造 File 对象,再用 File 对象构造输入流:

```
File f = new File("Test.java");
FileInputStream f2 = new FileInputStream(f);
```

FileInputStream 重写了抽象类 InputStream 的读取数据的方法。使用方式和标准输入流一致。

2. FileOutputStream

FileOutputStream 用于向一个文件写数据。它从父类中继承 3 个 write()方法等。它常用的构造函数如下:

```
FileOutputStream(String filePath)
FileOutputStream(File fileObj)
FileOutputStream(String filePath, boolean append)
FileOutputStream(File fileObj, boolean append)
```

这里 filePath 是文件的全称路径,fileObj 是描述该文件的 File 对象。如果 append 为 true,则文件以追加方式打开,不覆盖已有文件的内容;如果为 false,则覆盖原有的内容。默认是覆盖文件的内容。

如果 filePath 表示的文件不存在,FileOutputStream 在打开之前先创建它;如果文件已经存在,则打开它,准备写。试图打开一个只读文件,会引发一个 IOException 异常。

FileOutputStream 重写了抽象类 OutputStream 的写数据方法:

```
public void write(byte[] b)
```

将字节数组 b 写入文件输出流。

```
public void write(byte[] b, int off, int len)
```

将字节数组 b 中从 off 开始的 len 个字节写入文件输出流。

```
public void write(int b)
```

将整数 b 的最低位字节写入文件输出流。

b 是 int 类型时,占用 4 字节,只有最低的一个字节被写入输出流,忽略其余字节。

下面的文件复制程序使用 FileOutputStream 创建一个输出流,实现源文件到目标文件的内容复制。分别使用了文件流的 3 种重载的 read()和 write()方法,在实际使用过程中,选择一种方法就可以了。

【例 8.7】 文件复制。

```
package code0804;
import java.io.FileInputStream;
import java.io.FileOutputStream;
import java.io.IOException;
public class FileStreamCopy {
    public static void main(String[] args) throws IOException {
        int size;
        //构造输入输出流对象
```

```
        FileInputStream f = new FileInputStream("src/code0804/FileStreamCopy.java");
        FileOutputStream fout = new FileOutputStream("copy-of-file.txt");
        System.out.println("总长度: " + (size = f.available()));
        int n = size / 10;
        System.out.print("使用单字节方法读取后:");
        //使用 read()和 write()方法
        for (int i = 0; i < n; i++) {
            fout.write(f.read());
        }
        System.out.println("剩余长度: " + f.available());
        System.out.print("读取一个字节数组后:");
        //使用 read(byte[]b )和 write(byte[] b);
        byte b[] = new byte[n];
        f.read(b);
        fout.write(b);
        System.out.println("剩余长度:" + f.available());
        //使用 read(b,offset,len)和 write(b,offset,len)
        System.out.print("读取余下数据:");
        int count = 0;
        while ((count = f.read(b, 0, n)) != -1)
            fout.write(b, 0, count);
        System.out.println("剩余长度: " + f.available());
        //最后注意关闭流
        f.close();
        fout.flush();
        fout.close();
    }
}
```

程序运行结果:

```
总长度: 1175
使用单字节方法读取后:剩余长度: 1058
读取一个字节数组后:剩余长度:941
读取余下数据:剩余长度: 0
```

可以用记事本打开 copy-of-file.txt 文件,检查其内容和本程序是相同的。

8.4.3 字节过滤流

过滤流在读写数据的同时可以对数据进行处理,它提供了同步机制,使得某一时刻只有一个线程可以访问一个 I/O 流,以防止多个线程同时对一个 I/O 流进行操作所带来的意想不到的结果。

为了使用一个过滤流,必须首先把它连接到某个输入输出流上。过滤流扩展了输入输出流的功能,典型的扩展是缓冲、字符字节转换和数据转换。

当向缓冲流写入数据时,数据先发送到缓冲区,而不是直接发送到外部设备,缓冲区自动记录数据,当缓冲区满时,系统将数据全部发送到外部设备。

当从一个缓冲流中读取数据时,系统实际是从缓冲区中读取数据。当缓冲区空时,系统会自动从相关设备读取数据。

因为有缓冲区可用,缓冲流支持跳过(skip)、标记(mark)和重新设置流(reset)等方法。

常用的缓冲输入流有 BufferedInputStream、DataInputStream;常用的缓冲输出流有

BufferedOutputStream、DataOutputStream、PrintStream。

缓冲流的操作方法和普通流的操作方法是类似的,主要使用其 read()方法和 write()方法。下面以 BufferedInputStream/BufferedOutputStream 为例进行说明。

1. BufferedInputStream/BufferedOutputStream

BufferedInputStream 类允许把任何字节输入流"包装"成缓冲流并提高它的性能。BufferedInputStream 有两个构造函数:

```
BufferedInputStream(InputStream inputStream)
BufferedInputStream(InputStream inputStream, int bufSize)
```

第一种形式生成了一个包含默认缓冲区的输入流。第二种形式,缓冲区大小是由 bufSize 传入的。使用内存页或磁盘块等的若干倍的缓冲区可以给执行性能带来很大的正面影响。

BufferedOutputStream 用 flush()方法把数据缓冲区写入实际的输出设备。

下面是两个可用的构造函数:

```
BufferedOutputStream(OutputStream outputStream)
BufferedOutputStream(OutputStream outputStream, int bufSize)
```

第一种形式创建了一个包含默认缓冲区输出流。第二种形式,缓冲区的大小由 bufSize 参数传入。

下面再用缓冲流来实现文件的复制。

【例 8.8】 使用缓冲流的文件复制。

```
package code0804;
import java.io.BufferedInputStream;
import java.io.BufferedOutputStream;
import java.io.FileInputStream;
import java.io.FileOutputStream;
import java.io.IOException;
public class BufferedStreamCopy {
    public static void main(String[] args) throws IOException {
        //构造输入输出流对象
        FileInputStream fin = new FileInputStream("src/code0804/
BufferedStreamCopy.java");
        FileOutputStream fout = new FileOutputStream("copy-of-file.txt");
        //使用缓冲流
        BufferedInputStream bis = new BufferedInputStream(fin);
        BufferedOutputStream bos = new BufferedOutputStream(fout);
        System.out.println("开始复制...");
        int size=fin.available();
        System.out.println("文件大小:"+size);
        int n = size / 5;
        byte b[] = new byte[n];
        int count = 0;
        int finished=0;
        while ((count = bis.read(b, 0, n)) != -1)
        {
            bos.write(b, 0, count);
            finished+=count;                    //记录完成数量
            System.out.print("复制"+finished+" ");
        }
```

```
        System.out.println("完成复制");
        bis.close();
        bos.close();
        fin.close();
        fout.close();
    }
}
```

当文件比较大时,使用缓冲流能够提高输入输出的效率。

2. DataInputStream/DataOutputStream

DataInputStream 和 DataOutputStream 它们不仅能使用一般的 read()方法读取数据流,一般的 write()方法写数据流,而且能直接读写各种各样 Java 语言的基本数据类型:如 boolean、int、float、double 等。这些基本数据类型在文件中的表示方式和它们在内存中的一样,无须多余转换。详情参考 API 文档。

8.4.4 实用案例 8.2:文件加密解密

给定一个密钥,读取文件内容,加密后,输出到另外一个文件。

【例 8.9】 文件加密。

```
package code0804;
import java.io.FileInputStream;
import java.io.FileOutputStream;
import java.io.IOException;
public class EncryptFile {
    public static void main(String[] args) throws IOException {
        byte pwd = 123;
        FileInputStream fin = new FileInputStream("src/code0804/
BufferedStreamCopy.java");
        FileOutputStream fout = new FileOutputStream("encrypted.txt");
        System.out.println("开始加密...");
        EncryptAndDecrypt(pwd, fin, fout);
        System.out.println("完成加密");
        fin.close();
        fout.close();
        fin = new FileInputStream("encrypted.txt");
        fout = new FileOutputStream("unencrypted.txt");
        System.out.println("开始解密...");
        EncryptAndDecrypt(pwd, fin, fout);
        System.out.println("完成解密");
        fin.close();
        fout.close();
    }
    //使用异或算法时,加密、解密方法相同
     private  static  void  EncryptAndDecrypt ( byte  pwd,  FileInputStream  fin,
FileOutputStream fout) throws IOException {
        int n;
        byte[] buf;
        int count;
        n = fin.available() / 5;
        buf = new byte[n];
        count = 0;
```

```
            while ((count = fin.read(buf, 0, n)) != -1) {
                for (int i = 0; i < count; i++) {
                    buf[i] = (byte) (buf[i] ^ pwd);  //密码与值进行异或运算
                }
                fout.write(buf, 0, count);
            }
        }
    }
```

分析：使用文件输入字节流读取文件内容，然后对每个字节和密码进行异或加密。加密完成后，写入另外一个文件。解密方法和加密相同。可以用记事本打开加密后的文件进行查看，发现无法阅读。另外，Java在javax.crypto中提供了专用的加密流。

8.5 字符流使用

字节流不能直接操作字符。一般文本文件(可以看作字符串)适合用字符流。

8.5.1 字节流向字符流的转化

字节流以单个字节为读写单位，字符流以字符为读写单位。一般字符由多个字节组成，具体由使用的字符集确定，如当前的汉字编码集是GB18030。

InputStreamReader和OutputStreamWriter用来在字节和字符之间作为中介，可以把以字节形式表示的流转化为特定编码集上的字符表示。为将字节流构造为字符流对象，可以在构造流对象时指定字符编码，也可以用当前平台的缺省编码。

InputStreamReader的构造函数如下：

```
public InputStreamReader(InputStream in)
public InputStreamReader(InputStream in, String charsetName)
```

第一种形式使用当前平台缺省的编码，从字节流in构造一个字符流对象；第二种形式使用特定的字符集编码，从字节流in构造一个字符流对象。

如果使用了不支持的字符集，那么会产生一个UnsupportedEncodingException异常。

OutputStreamWriter的构造函数如下：

```
public OutputStreamWriter(OutputStream out)
public OutputStreamWriter(OutputStream out, String charsetName)
```

第一种形式使用当前平台缺省的编码，从字节流out构造一个字符流对象；第二种形式使用特定的字符集编码，从字节流out构造一个字符流对象。

下面用例子说明InputStreamReader和OutputStreamWriter的使用。

【例8.10】 字节流转换成字符流，实行文件复制功能。

```
package code0805;
import java.io.File;
import java.io.FileInputStream;
import java.io.FileOutputStream;
import java.io.IOException;
import java.io.InputStreamReader;
import java.io.OutputStreamWriter;
```

```java
public class StreamToReaderWriter
{
    public static void main(String[] args) throws IOException
    {
        File file = new File("src/code0805/StreamToReaderWriter.java");
        FileInputStream fin = new FileInputStream(file);
        FileOutputStream fout = new FileOutputStream("copy-of-file.txt");
        //把文件输入字节流向字符流转换
        InputStreamReader isr = new InputStreamReader(fin, "UTF-8");
        //把文件输出字节流向字符流转换
        OutputStreamWriter osw = new OutputStreamWriter(fout, "UTF-8");
        System.out.println("当前输入流编码是 :"+ isr.getEncoding());
        System.out.println("当前输出流编码是:"+ osw.getEncoding());
        int n = (int) (file.length() / 30);
        char b[] = new char[n];
        System.out.println("复制开始...");
        int count = 0;
        while ((count = isr.read(b, 0, n)) != -1)
            osw.write(b, 0, count);
        isr.close();
        fin.close();
        osw.close();
        fout.close();
        System.out.println("复制完成。");
    }
}
```

在从字节流构造字符流时，指定了字符集 UTF-8，因为在操作系统中 Java 程序文件是按照 UTF-8 编码存储的。只有指定了正确的字符集，才能正确地从字节流构造字符流。如果在输出时指定了不同的字符集，则在输出文件中会产生乱码。

Java 还提供了其他字符流，例如 CharArrayReader、CharArrayWriter、StringReader、StringWriter、PrintWriter、FileReader、FileWriter 等。一般地，名称以 Reader 结尾的类都是字符输入流；名称以 Writer 结尾的类都是字符输出流。这些字符流的详细使用说明请参考 Java 文档，在此不再详述。

8.5.2 读写文本文件

对文本文件内容的解析和计算结果的格式化输出是程序设计过程中经常要碰到的问题。例如，一个 source.txt（见图 8-1(a)）文件中存放着两行整数，请分别计算每一行的和，追加到每一行数字的末尾，并输出到文件 dest.txt 中（见图 8-1(b)）。

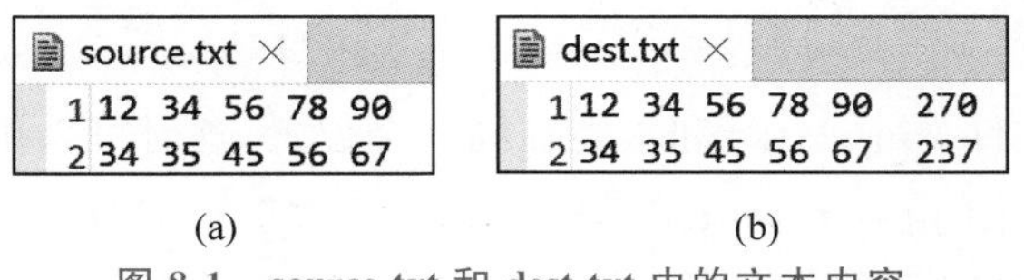

(a) (b)

图 8-1 source.txt 和 dest.txt 中的文本内容

读写文本文件时，一般不使用文件字符流，常用的方法是使用工具类 java.util.Scanner 封装字节流读取数据；使用输出流 java.io.PrintWriter 输出字符或字符串。PrintWriter 向字符（文本）输出流打印格式化表示形式。

Scanner 是一个可以使用正则表达式解析字符串的简单文本扫描器。Scanner 使用分隔符将其输入分解为标记,默认情况下该分隔符与空白匹配。然后,可以使用不同的 next()方法将得到的标记转换为不同类型的值,在读取下一个标记之前可以使用 hasNext()方法检测是否有剩余标记。

【例 8.11】 解析文本文件的数据并格式化输出。

```
package code0805;
import java.io.FileNotFoundException;
import java.io.IOException;
import java.io.InputStream;
import java.io.PrintWriter;
import java.util.Scanner;
public class ScannerPrintWriterDemo {
    public static void main(String[] args) throws IOException {
        Scanner s = null;
        PrintWriter pw = null;
        //装载和本类在同一个目录下的文件
        InputStream in = ScannerPrintWriterDemo.class.getResourceAsStream
("source.txt");
        //使用 Scanner 解析文件字符输入流的内容
        s = new Scanner(in);
        pw = new PrintWriter("dest.txt");        //使用 PrintWriter 进行格式化输出
        while (s.hasNextLine()) {                //判断是否还有未读行
            String strLine = s.nextLine();       //读取一行
            //调用方法求每一行的和
            int sum = getLineSum(strLine);
            //输出每一行
            pw.println(strLine + "\t" + sum);
        }
        s.close();
        in.close();
        pw.close();
    }
    //计算每一行的和的方法
    private static int getLineSum(String strLine) {
        //使用 Scanner 分割每一行为多个整数
        Scanner s = new Scanner(strLine);        //用 Scanner 解析字符串
        int sum = 0;
        while (s.hasNextInt()) {                 //判断是否还有数据
            sum += s.nextInt();                  //获得下一个整数
        }
        return sum;
    }
}
```

分析:数据源文件 source.txt 和本类在同一个位置,这时可以用类路径文件装载方式。使用当前类的 Class 对象的 getResourceAsStream()函数得到字节输入流。用工具类 Scanner 把字节流中的字节转换成字符流,按行读取数据。在解析每一行的数据时,再一次使用 Scanner 工具类,可以分割字符串,并获取数据。输出时,使用打印流 PrintWriter,一次打印一行数据。今后读写文本文件时,参照此方法处理。

8.5.3 实用案例 8.3：文本替换

查找某一个文件中的字符串，全部替换成另外的字符串。

【例 8.12】 替换文件中的字符串。

```
package code0805;
import java.io.File;
import java.io.PrintWriter;
import java.util.Scanner;
public class ReplaceText {
    public static void main(String[] args) throws Exception {
        File f;
        String filename;
        System.out.println("输入文件名:");
        Scanner console=new Scanner(System.in);
        filename=console.nextLine();
        f=new File(new File("src/code0805"),filename);
        System.out.println("输入要替换的字符串:");
        String  pattern=console.nextLine();
        System.out.println("输入要替换的结果:");
        String  result=console.nextLine();
        PrintWriter pw=new PrintWriter("replace.txt");
        Scanner reader=new Scanner(f);
        while(reader.hasNextLine())
        {
            String s1=reader.nextLine();
            String s2=s1.replaceAll(pattern, result);
            pw.println(s2);
        }
        console.close();
        pw.close();
    }
}
```

程序运行结果：

```
输入文件名:
ReplaceText2.java
输入要替换的字符串:
String
输入要替换的结果:
Chongqing
```

分析：该程序使用 Scanner 分行读取文件，然后进行替换，使用 PrintWriter 进行输出。打开输出文件，可以看出所有的字符串“String”被替换成了“Chongqing”。

8.6 高级流处理

Java 在包 java.nio 中提供了更高级的文件和流处理功能，称为 Java NIO。使用 java.nio.file.Path 接口表示文件对象和工具类 java.nio.file.Files 进行文件处理，功能比前面介绍的 File 类更丰富、更方便。

标准 IO 基于流读写数据，NIO 基于通道（Channel）和缓冲区（Buffer）进行操作，数据总是从通道读取到缓冲区中,或者从缓冲区写入通道中。NIO 引入了选择器(Selector)的概念,选择器用于监听多个通道的事件(如连接打开、数据到达)。

8.6.1 Path 和 Files 的使用

Path 的对象用于定位文件系统中的一个文件,就是文件的路径。包含一个由目录或者文件名元素构成的序列。它的对象可以使用 Path 接口的 of()函数得到,也可以从 Paths 类的一个静态 get()函数获得,用法相同。

```
public static Path get(String first,String... more)
```

该方法可以把一个路径字符串 first 转成 Path 对象,或者把 first 和 more 表示的一个字符串序列联合起来转成路径对象。例如:

```
Path path1 = Paths.get("src", "code0805", "ReplaceText.java");
```

代表的路径就是 src\code0805\ReplaceText.java。

Path 对象也可以通过 File 类的 toPath()方法获得:

```
File f=new File("src\\code0805\\ReplaceText.java");
Path path2=f.toPath();
```

NIO 的 Files 类是一个工具类,提供了多种静态方法来操作文件或者目录,比直接使用 File 类要方便,其主要功能如下。

(1) copy(Path source,OutputStream out):复制一个文件到输出流。

(2) readAllBytes(Path path):读取一个文件的所有字节。

(3) createDirectories(Path dir,FileAttribute<? >... attrs):创建文件。

(4) delete(Path path):删除一个文件或者目录。

(5) move(Path source,Path target,CopyOption... options):移动文件。

(6) list(Path dir):列出子文件或者目录。

(7) readAllLines(Path path, Charset cs):读取文件的所有行。

(8) write(Path path, byte[] bytes, OpenOption... options):把字节数组写入文件。

(9) walk(Path start, FileVisitOption... options):遍历所有子目录和文件。

其他更多功能可以查看 API 文档。下面举例说明这两个类的使用。

【例 8.13】 Path 和 Files 类的使用。

```
package code0806;
import java.nio.file.Files;
import java.nio.file.Path;
import java.nio.file.Paths;
import java.nio.file.StandardCopyOption;
import java.util.stream.Stream;
public class FilesDemo {
    public static void main(String[] args) throws Exception {
        Path p1=Paths.get("");                              //空表示当前路径
        Path p2=p1.toAbsolutePath();                        //转成绝对路径
        System.out.println("p2:"+p2);
        System.out.println("路径上的所有名字元素:");
        //变量路径上的名字元素
        p2.forEach(x->System.out.print(x.getFileName()+","));
```

```
        System.out.println("\n根元素:"+p2.getRoot());          //得到根元素
        System.out.println("父目录:"+p2.getParent());
                                              //得到父路径,除最后一个元素外的部分
        Path p3=p2.resolve("src");                          //p2 下面的子目录 src
        System.out.println("p3:"+p3.toAbsolutePath());
        Stream<Path> stm = Files.walk(p3);                  //遍历所有子目录和文件
    stm. filter (x - > Files. isDirectory (x)). forEach (x - > System. out. print (x.
getFileName()+","));                                        //输出所有子目录
        Path p4=Path.of("src","code0806","FilesDemo.java");    //源文件
        Path p5=Path.of("src","code0806","FilesDemo.txt");     //目标文件
        Files.copy(p4, p5,StandardCopyOption.REPLACE_EXISTING);
                                                            //复制覆盖已存在文件
    }
}
```

程序运行结果：

```
p2:C:\Users\yangrl\eclipse-workspace-test\第8章代码
路径上的所有名字元素:
Users,yangrl,eclipse-workspace-test,第8章代码,
根元素:C:\
父目录:C:\Users\yangrl\eclipse-workspace-test
p3:C:\Users\yangrl\eclipse-workspace-test\第8章代码\src
src,code0801,code0803,code0804,code0805,code0806,sub1,sub2,
```

分析：首先使用不同的方式构建了 Path 对象，然后利用 Files 类遍历了一个目录的所有子目录和文件，输出了所有目录的名字，最后用 Files 完成了文件复制。今后简单的文件操作，如复制、创建、读取等都可以使用此类完成。

8.6.2 使用通道和缓冲区读写文件

使用 NIO 系统，需要获取用于连接 I/O 设备的通道以及用于容纳数据的缓冲区。然后操作缓冲区，对数据进行处理。简单地讲，Channel 负责传输，Buffer 负责存储。

1. 缓冲区

缓冲区是一个用于特定基本数据类型的容器。所有缓冲区都是抽象类 Buffer 的子类。常用的缓冲区有 ByteBuffer、CharBuffer、ShortBuffer、IntBuffer、LongBuffer、FloatBuffer、DoubleBuffer。

所有缓冲区的操作方式是一致的。使用 allocate()方法分配缓冲区，使用 get()系列方法获取缓冲区的数据，使用 put()系列方法存数据到缓冲区。

缓冲区有以下 4 个核心属性。

(1) capacity：容量，表示缓冲区中最大存储数据的容量。一旦声明不能更改。

(2) limit：界限，表示缓冲区中可以操作数据的大小。注意，limit 后的数据不能读写。

(3) position：位置，表示缓冲区中正在操作数据的位置。

(4) mark：一个位置标记。可以通过 reset()方法恢复 position 到 mark 的位置。

【例 8.14】 Buffer 的使用。

```
package code0806;
import java.nio.ByteBuffer;
import java.util.Arrays;
```

```
public class BufferDemo {
    public static void main(String[] args) {
        ByteBuffer bb = ByteBuffer.allocate(256);
        System.out.printf("容量:%d,limit:%d,position:%d\n", bb.capacity(), bb.
limit(), bb.position());
        bb.put("123456".getBytes());                        //6字节
        bb.put("abcdef".getBytes());                        //6字节
        System.out.printf("容量:%d,limit:%d,position:%d\n", bb.capacity(), bb.
limit(), bb.position());
        bb.flip();                        //把 position 设置为 0,limit 设置为上一个 position
        System.out.printf("容量:%d,limit:%d,position:%d\n", bb.capacity(), bb.
limit(), bb.position());
        byte[] b1 = new byte[6];
        byte[] b2 = new byte[6];
        bb.get(b1);                                         //读取 6 字节
        System.out.printf("容量:%d,limit:%d,position:%d\n", bb.capacity(), bb.
limit(), bb.position());
        bb.get(b2);                                         //读取 6 字节
        System.out.printf("容量:%d,limit:%d,position:%d\n", bb.capacity(), bb.
limit(), bb.position());
        ByteBuffer bb2=ByteBuffer.allocate(4);
        bb2.putInt(16909060);                               //把一个整数放入缓冲区,0x01020304
        byte[] b3=new byte[4];
        bb2.flip();
        bb2.get(b3);                                        //得到整数的字节序列
        System.out.println(Arrays.toString(b3));            //输出每个字节的值
    }
}
```

程序运行结果:

```
容量:256,limit:256,position:0
容量:256,limit:256,position:12
容量:256,limit:12,position:0
容量:256,limit:12,position:6
容量:256,limit:12,position:12
[1, 2, 3, 4]
```

分析:当缓冲区放入数据后,position 会增加;flip 后,position 为 0,limit 设置为数据的数量,可以从头开始读取数据。缓冲区 bb2 放入了一个整数,一个整数由 4 字节构成,读取出来放入字节数组。可以查看整数的每个字节的值。

2. 通道

Channel 类似于传统的流,如果想写数据到文件中,必须先把数据都写入缓冲区,然后缓冲区通过通道进行传输,最后把数据写入文件。如需把数据传输到程序中,先使用通道把数据读入缓冲区,再从缓冲区读取数据。

NIO 提供了 FileChannel、SocketChannel、ServerSocketChannel、DatagramChannel 等 Channel 类。本节以 FileChannel 为例说明通道的使用。FileChannel 是一种类型的字节通道,类似于文件字节流,能够将数据从 I/O 设备读入字节缓冲区,或者将字节缓冲区中的数据写入 I/O 设备中;但它是双向的,既可以读,也可以写。FileChannel 包含一个文件当前读写位置 position,可以设置新的位置,也可以获取当前位置。读写数据时,需要一个 ByteBuffer。

可以使用 FileChannel 的静态 open 函数打开一个文件通道:

```
FileChannel open(Path path,OpenOption... options)
```

其中,参数 path 就是文件的路径;options 是打开选项,枚举 StandardOpenOption 定义了一些打开选项,如 APPEND、CREATE、WRITE 等,具体请参考 API 文档。

Files 类的 newByteChannel(Path path, OpenOption... options)函数也可得到 1 字节通道。

【例 8.15】 文件通道 FileChannel 的使用。

```
package code0806;
import java.io.IOException;
import java.nio.ByteBuffer;
import java.nio.channels.FileChannel;
import java.nio.file.Path;
import java.nio.file.StandardOpenOption;
public class FileChannelDemo {
    public static void main(String[] args) throws Exception {
        Path p1=Path.of("src/code0806/FileChannelDemo.java");
        //构建输入通道
        FileChannel  fis=FileChannel.open(p1, StandardOpenOption.READ);
        Path p2=Path.of("src/code0806/FileChannelDemo.txt");
        //构建输出通道
        FileChannel fos=FileChannel.open(p2, StandardOpenOption.CREATE,
StandardOpenOption.WRITE);
        //分配指定大小的缓冲区
        ByteBuffer byteBuffer = ByteBuffer.allocate(1024);
        System.out.println("fis position:"+fis.position());
        //将通道中的数据存入缓冲区中
        while (fis.read(byteBuffer) != -1) {
                                    //fisChannel 中的数据读入 byteBuffer 缓冲区中
            System.out.println("fis position:"+fis.position());
            byteBuffer.flip();                  //重置 position 和 limit
            fos.write(byteBuffer);              //将缓冲区中的数据写入通道
            byteBuffer.clear();                 //清空缓冲区
        }
        System.out.println("fos size:"+fos.size());
        fos.close();
        fis.close();
    }
}
```

程序运行结果:

```
fis position:0
fis position:1024
fis position:1307
fos size:1307
```

分析:构建了一个输入文件通道和输出文件通道。把输入通道的数据读入缓冲区,然后把缓冲区的数据写入输出通道。文件通道读写时,文件位置会发生变化。另外,复制文件也可以直接使用文件通道的 transferTo()函数。其他关于通道的更多操作请参考 API 文档。

8.6.3 实用案例8.4：文件夹的深度复制

【例8.16】 文件夹的深度复制。

```
package code0806;
import java.io.IOException;
import java.nio.file.Files;
import java.nio.file.Path;
import java.nio.file.Paths;
public class CopyDirectories {
    public static void main(String[] args) throws Exception {
        copyMoreDirectory();
    }
    private static void copyMoreDirectory() throws Exception {
        long start = System.currentTimeMillis();
        String source = "src";
        String target = "dest";
        Path basePath = Paths.get(target);
        //复制多级目录
        Files.walk(Paths.get(source)).forEach(path -> {
            int count = path.getNameCount();    //得到路径中名字元素数量
            Path targetPath;
            if (count == 1) {
                targetPath = basePath;          //设置 src 对应的目标目录为 dest
            } else {
                //截取源路径 src 的子路径,然后变成 dest 的子路径,形成新路径
                targetPath = basePath.resolve(path.subpath(1, count));
            }
            //如果源路径是目录,则创建新目录
            if (Files.isDirectory(path)) {
                try {
                    Files.createDirectory(targetPath);
                } catch (IOException e) {
                    e.printStackTrace();
                }
            }
            //如果是普通文件,则复制文件
            if (Files.isRegularFile(path)) {
                try {
                    Files.copy(path, targetPath);
                } catch (IOException e) {
                    e.printStackTrace();
                }
            }
        });
        long end = System.currentTimeMillis();
        System.out.println("计算出复制文件的时间差(毫秒):" + (end - start));
    }
}
```

分析：本案例的所有源路径包含 src，目标路径是 dest，需要把所有源路径中的 src 元素替换成 dest。使用了子路径函数 subpath()，截取了源路径除 src 外的部分，和 dest 路径形成新的路径。Files 的 walk()函数可以遍历一个路径下的所有子目录和文件。当是目录时，创建新目录；当是文件时，复制文件。最后计算出复制时间。

8.7 串 行 化

8.7.1 串行化的概念

对象的寿命通常随着生成该对象的程序的终止而终止。某些时候,需要将对象的状态保存下来,将来需要时可以恢复,或者把对象传输到其他地方。

把对象的这种能记录自己的状态以便将来再生的能力,称为对象的持续性(Persistence)。对象通过写出描述自己状态的数值来记录自己的过程,称为对象的串行化(Serialization)。

串行化的主要任务是写出对象实例变量的数值。如果变量是另一对象的引用,则引用的对象也要串行化。

Java 提供了对象串行化的机制,在 java.io 包中,定义了一些接口和类作为对象串行化的工具。

1. Serializable 接口

只有实现 Serializable 接口的对象才可以被串行化工具存储和恢复。Serializable 接口没有定义任何成员。它只用来标识一个类可以被串行化。如果一个类可以被串行化,它的所有子类都可以被串行化。

2. ObjectOutput 接口

ObjectOutput 继承 DataOutput 接口并且支持对象串行化。它的 writeObject()方法,可以输出一个对象。

```
final void writeObject(Object obj)        //向流写入对象 obj
```

3. ObjectOutputStream 类

ObjectOutputStream 类继承 OutputStream 类和实现 ObjectOutput 接口。它负责向流写入对象。该类的构造函数如下:

```
ObjectOutputStream (OutputStream out)     //参数 out 是串行化的对象将要写入的输出流
```

4. ObjectInput 接口

ObjectInput 接口继承 DataInput 接口。它支持对象反串行化,其 readObject()方法可以反串行化对象。

```
Object readObject()                        //从流读取一个对象
```

5. ObjectInputStream 类

ObjectInputStream 继承 InputStream 类并实现 ObjectInput 接口。ObjectInputStream 负责从流中读取对象。该类的构造函数如下:

```
ObjectInputStream(InputStream in)         //参数 in 是串行化对象将被读取的输入流
```

6. 串行化注意事项

(1) 串行化只能保存对象的非静态成员变量的值。

(2) 不需要保存的变量前面加上 transient 关键字。

8.7.2 实用案例 8.5:串行化学生对象

1. 定义一个可串行化学生类

被串行化的类必须实现 Serializable 接口。

2. 构造对象输入输出流

要串行化一个对象,必须与对象的输入输出联系起来,通过 writeObject()串行化对象,通过 readObject()方法反串行化对象。

【例 8.17】 串行化学生对象。

```
package code0807;
import java.io.FileInputStream;
import java.io.FileOutputStream;
import java.io.ObjectInputStream;
import java.io.ObjectOutputStream;
import java.io.Serializable;
class Student implements Serializable {
    private static final long serialVersionUID = 1L;
    int id;
    String name;
    int age;
    String department;
    public Student(int id, String name, int age, String department) {
        this.id = id;
        this.name = name;
        this.age = age;
        this.department = department;
    }
}
public class SerializableDemo {
    public static void main(String args[]) throws Exception {
        Student stu = new Student(20221064, "zhang shan", 22, "CQUCS");
        FileOutputStream fout = new FileOutputStream("data1.ser");
        ObjectOutputStream oout = new ObjectOutputStream(fout);
        //输出对象
        oout.writeObject(stu);
        oout.close();
        stu = null;
        FileInputStream fin = new FileInputStream("data1.ser");
        ObjectInputStream oin = new ObjectInputStream(fin);
        //读入对象
        stu = (Student) oin.readObject();
        oin.close();
        System.out.println("学生信息:");
        System.out.println("ID:          " + stu.id);
        System.out.println("name:        " + stu.name);
        System.out.println("age:         " + stu.age);
        System.out.println("department:" + stu.department);
    }
}
```

程序运行结果:

```
学生信息:
ID:        20221064
name:       zhang shan
age:       22
department:CQUCS
```

分析:先构造文件流,然后再构造对象流。对象先被串行化到文件中,然后读入内存再生。从输出结果可以看出，通过串行化机制正确地保存和恢复了对象的状态。在写网络通信程序时，可以通过 Socket 传输串行化的对象封装的网络消息。

8.8 输入输出处理实训任务

【任务描述】

编写一个文件分割与合并程序，把一个大的原始文件分割成多个指定大小的小文件，并且能够把分割后的小文件合并成原始文件。

【任务分析】

要求按照指定大小对文件进行分割。首先需要获得被分割文件的大小。其次根据分割后每个文件的大小可以计算出分割后文件的数量。对分割后的文件进行编号，这样便于对文件进行合并。

可以定义两个函数：一个函数进行文件分割，另一个函数进行文件合并。需要使用文件输入输出流来读写文件内容。

设创建的程序文件名为 FileCutMerge.java。可以定义两个命令行参数，如-c4000 filename 表示分割文件，每个文件大小为 4000 字节；-m prefix 表示把当前目录下文件名前缀为 prefix 的文件合并为一个文件。

程序的基本工作流程如下：

(1) 分析命令行参数决定是分割文件，还是合并文件。

(2) 调用具体的函数进行分割和合并。

【任务解决】

完整程序如下：

```
package code0808;
import java.io.File;
import java.io.FileFilter;
import java.io.FileInputStream;
import java.io.FileOutputStream;
public class FileCutMerge {
    //分析命令行参数，决定是分割文件，还是合并文件
    public static void main(String[] args) {
        //创建一个本类的对象
        FileCutMerge tool = new FileCutMerge();
        //判断命令行参数的数量
        if ((args == null) || (args.length != 2)) {
            tool.help();
        } else if (args[0].startsWith("-c")) {
            //被分割文件的文件名，从当前目录开始
            File f1 = new File(args[1]);
            if (!f1.exists()) {
                System.out.println("指定的文件不存在");
            }
            //从命令行参数获得分割后文件的大小
            int fileSize = Integer.parseInt(args[0].substring(2));
            try {
                //分割文件
                tool.cut(f1, fileSize);
```

```
            } catch (Exception e) {
                e.printStackTrace();
            }
        } else if (args[0].equals("-m")) {
            //被合并的小文件名的前缀,就是"-"前面的内容
            String prefixname = args[1];
            File f = new File(".");
            //列出当前目录下需要合并的文件,使用了一个文件过滤器
            File[] names = f.listFiles(new PrefixFilter(prefixname));
            try {
                //合并文件
                tool.merge(names);
            } catch (Exception e) {
                e.printStackTrace();
            }
        } else {
        }
    }
    public void help() {
        System.out.println("错误的命令行格式,正确的是:");
        System.out.println("FileCutMerge -c 文件大小 filename");
        System.out.println("或者");
        System.out.println("FileCutMerge -m filenameprefix");
    }
    public void cut(File file, int size) throws Exception {
        System.out.println("开始分割文件...");
        //获得被分割文件的父目录,把分割后的文件放到该目录下
        File parent = file.getParentFile();
        long fileLength = file.length();          //获得文件大小
        //获得分割后小文件的数目
        int filenum = (int) (fileLength / size);
        if (fileLength % size != 0) {
            filenum += 1;
        }
        String[] smallfilenames = new String[filenum];
        //创建文件输入流读取被分割的文件
        FileInputStream fin = new FileInputStream(file);
        //构造一个字节数组,每次读取一个字节数组的数据
        byte[] buf = new byte[size];
        for (int i = 0; i < filenum; i++) {
            //构造分割后的文件名
            File outfile = new File(parent, file.getName() + "-" + i);
            //构造文件输出流
            FileOutputStream fout = new FileOutputStream(outfile);
            //读取数据
            int count = fin.read(buf);
            //输出数据
            fout.write(buf, 0, count);
            fout.close();
            smallfilenames[i] = outfile.getName();
        }
        fin.close();
        //输出分割后的文件名
        System.out.println("分割后的文件如下:");
```

```
        for (int i = 0; i < smallfilenames.length; i++) {
            System.out.println(smallfilenames[i]);
        }
        System.out.println("文件分割完成。");
    }
    public void merge(File[] files) throws Exception {
        System.out.println("开始合并文件...");
        //获得目标文件名,来源于被合并的文件
        String smallfilename = files[0].getName();
        int pos = smallfilename.indexOf("-");
        String tagetfilename = "new-" + smallfilename.substring(0, pos);
        System.out.println("合并后的文件为:" + tagetfilename);
        File outFile = new File(files[0].getParentFile(), tagetfilename);
        FileOutputStream fout = new FileOutputStream(outFile);
        //合并文件内容,输出到目标文件
        for (int i = 0; i < files.length; i++) {
            FileInputStream fin = new FileInputStream(files[i]);
            int b;
            while ((b = fin.read()) != -1) {
                fout.write(b);
            }
            fin.close();
        }
        fout.close();

        System.out.println("合并文件完成。");
    }
}
//文件过滤器,列出目录下符合条件的文件,此类是根据文件名的前缀进行过滤
class PrefixFilter implements FileFilter
{
    String prefix = "";
    public MyFilesFilter(String prefix) {
        this.prefix = prefix;
    }
    @Override
    public boolean accept(File f) {
        if (f.getName().length() > prefix.length()
                && f.getName().startsWith(prefix)) {
            return true;
        }
        return false;
    }

}
```

假设把文件 FileCutMerge.java 移动到系统当前目录下。例如,分割文件的命令行

```
java.exe FileCutMerge -c500 FileCutMerge.java
```

表示把 FileCutMerge.java 分割为大小为 500 的小文件,其运行结果如下:

```
开始分割文件...
分割后的文件如下:
FileCutMerge.java-0
```

```
FileCutMerge.java-1
FileCutMerge.java-2
FileCutMerge.java-3
FileCutMerge.java-4
FileCutMerge.java-5
FileCutMerge.java-6
FileCutMerge.java-7
文件分割完成。
```

例如,合并文件的命令行:

```
java.exe FileCutMerge -m FileCutMerge.java
```

表示把当前目录下文件名前缀为 FileCutMerge.java 的文件合并为一个大文件。程序运行结果如下:

```
开始合并文件...
合并后的文件为:new-FileCutMerge.java
合并文件完成。
```

1. 简述可以用哪几种方法对文件进行读写。
2. 使用 File 类列出某一个目录下创建日期晚于 2022-8-12 的文件。
3. 使用 File 类创建一个多层目录 D:\java\my Program。
4. 能否将一个对象写入一个随机访问文件?
5. 从字节流到字符流的转化过程中,有什么注意事项?
6. 读取一个 Java 源程序,找出其中使用到的关键字,并统计其个数。
7. 使用一个文件通道对一个文件进行读写操作。

第 9 章 Java 多线程

本章学习目标

线程也被称为轻型进程，是处理器调度的基本单位，灵活使用多线程进行程序设计，可以提高程序执行效率。例如，当使用浏览器下载一个软件时，只有一个线程在做下载工作，当改用支持多线程的专用下载工具后，一个文件的下载任务被分解成多个线程并发工作，显著地提高了下载的速度和效率。

本章将主要介绍线程的基本概念、创建线程的方法、线程的生命周期及调度、线程的同步、线程池的使用等。本章应该重点掌握以下主要内容：

（1）线程的创建。

（2）线程的调度。

（3）线程的同步。

（4）线程池的使用。

9.1 为什么使用多线程

前面写的 Java 程序在一个时间段内只能完成一个任务，程序代码是顺序执行。有时需要一个 Java 程序在一段时间内并发完成多个任务，这就需要使用多线程。例如，对 1～1000 的数求和、求平方和。普通的 Java 程序可以串行执行这两个任务，使用多线程后，可以让一个线程执行求和任务，另一个线程执行求平方和的任务，两个线程并发执行。

【例 9.1】 使用两个线程分别执行数的求和、求平方和运算。

分析：创建两个线程，一个是 SumThread，求和；另一个是 SquareSumThread，求平方和。Java 中创建线程最简单的方法就是继承 Thread 类。问题解决如下：

```
package code0901;
public class SimpleThreadTest {
    public static void main(String[] args) {
        //创建线程对象：
        Thread sum = new SumThread();
        Thread squareSum = new SquareSumThread();
        //启动线程
        sum.start();
        squareSum.start();
```

```
        }
    }
    //继承 Thread 创建线程
    class SumThread extends Thread {
        public void run() {
            int sum = 0;
            for (int i = 1; i <= 1000; i++) {
                sum +=i;
            }
            System.out.println("和:" + sum);
        }
    }
    //继承 Thread 创建线程
    class SquareSumThread extends Thread {
        public void run() {
            int sum = 0;
            for (int i = 1; i <= 1000; i++) {
                sum += i * i;
            }
            System.out.println("平方和:" + sum);
        }
    }
```

程序运行结果：

```
和:500500
平方和:333833500
```

该问题的求解涉及下面几个关键点：

(1) 通过继承类 Thread,创建自己的线程,本例创建了两个线程。

(2) 线程的核心代码在 run()方法中,也被称作线程体。

(3) 创建线程对象后,run()方法不会自动执行,需要调用 start()方法启动线程。

(4) 多线程的执行是并发,独立线程之间没有确定的顺序关系,输出结果的先后顺序不确定,多次反复执行程序会发现,两个线程的输出结果的先后顺序不确定。

9.2 线程的概念

一般每个程序都有一个入口、一个出口以及一个顺序执行的序列,执行中的程序称为进程,在进程执行过程中的任何指定时刻,都只有一个单独的执行点。在多线程情况下,在单个进程内部,可以在同一时刻进行多种运算,有多个执行点。

一个单独的线程和进程相似,也有一个入口、一个出口以及一个顺序执行的序列,从概念上说,一个线程是一个进程内部的一个顺序控制流,必须在进程中运行。在一个进程中可以实现多个线程,有多个处理器的系统,可以并发运行不同的线程,完成不同的功能。

对比线程与进程：

(1) 两者的粒度不同。进程是由操作系统来管理的,而线程则是在一个进程内。

(2) 不同进程的代码、内部数据和状态都是完全独立的,而一个进程内的多线程是共享进程的内存空间和系统资源,有可能互相影响。

(3) 线程本身的数据通常只有寄存器数据,以及一个程序执行时使用的堆栈,所以线程的切换比进程切换的负担要小。

使用多线程具有如下优点：

(1) 多线程编程简单，效率高，线程间能直接共享数据和资源，多进程不能。

(2) 适合于开发服务程序(如 Web 服务、聊天服务等)，程序的吞吐量会得到改善。

(3) 适合于开发有多种交互接口的程序(如聊天程序的客户端、网络下载工具)。

(4) 适合于有人机交互又有计算量的程序(如字处理程序 Word、表格处理工具 Excel)。

9.3 线程的创建

类 java.lang.Thread 是支持多线程编程的核心类，所有的线程都是它的对象，提供了一些方法来控制线程。创建一个线程就是创建 Thread 类或者它的子类的对象。Thread 类有很多重载的构造方法，例如：

```
Thread(Runnable target, String name)
```

参数 target 是线程执行的目标对象，即线程执行的代码，一般是实现了 java.lang.Runnable 接口的对象；name 是线程的名字。

创建线程主要有 4 种方法：①继承 Thread，重写 run()方法实现线程体；②实现接口 Runnable 的 run()方法提供线程体；③实现 Callable 接口的 call()方法提供线程体，这种方式一般在线程池中使用；④使用定时器 Timer 创建线程。

9.3.1 继承 Thread 创建线程

继承 Thread 创建线程的方法比较简单，主要是通过继承 java.lang.Thread 类，并重写 Thread 类的 run()方法来完成线程的创建。Thread 类封装了线程的行为。要创建一个线程，可以创建一个 Thread 类的子类的对象。

虽然 run()方法是线程体，但不能直接调用 run()方法，而是通过调用 start()方法来启动线程。在调用 start()时，start()方法会首先进行与多线程相关的初始化，然后再调用 run()方法。

【例 9.2】 使用继承创建线程。

```
package code0903;
//继承 Thread 类
public class MyThread extends Thread
{   //count 变量用于统计打印的次数并共享变量
    private static int count = 0;
    public MyThread(String name)
    {
        super(name);                        //调用超类 Thread 的构造函数,传入线程名字
    }
    public static void main(String[] args)
    {   //main 方法开始
        MyThread p = new MyThread("t1");    //创建一个线程实例
        p.start();                          //执行线程
        //主线程 main 方法执行一个循环
        for (int i = 0; i < 5; i++) {
            count++;
```

```
            //主线程中打印 count+"main"变量的值,并换行
            System.out.println(count + ": main");
        }
    }
    public void run()
    {   //线程类必须有的 run()方法
        for (int i = 0; i < 5; i++) {
            count++;
            System.out.println(count + ":" + this.getName());
        }
    }
}
```

程序运行结果:

```
1: main
3: main
4: main
5: main
6: main
2:t1
7:t1
8:t1
9:t1
10:t1
```

分析:上面这段程序中,main()方法作为主线程执行,生成新线程 t1 并启动。线程 t1 通过 for 循环输出变量 count 和线程的名称。main 线程和 t1 线程都在操作变量 count,两个线程可以直接共享变量。

9.3.2 实现接口 Runnable 创建线程

Java 中只允许单继承,如果类已继承了其他的类,那么就无法再继承 Thread 类了。为此,Java 中提供了另外一种方法来实现多线程。

该方法通过实现 Runnable 接口的 run()方法创建线程体,然后把 Runnable 对象传递给 Thread 对象。通过 Thread 类的构造方法 public Thread(Runnable target)来实现。

【例 9.3】 使用接口创建线程。

```
package code0903;
//实现 Runnable 接口
public class RunnableThread implements Runnable
{
    int count = 1, number;
    public RunnableThread(int i)
    {
        number = i;
        System.out.println("创建线程 " + number);
    }
    public void run()
    {
        while (true) {
            System.out.println("线程 " + number + ":计数 " + count);
```

```
            if (++count == 6)
                return;
            {
            }
        }
    }
    public static void main(String args[])
    {
        for (int i = 0; i < 5; i++)
            //通过 Thread 类创建线程对象，并启动
            new Thread(new RunnableThread(i + 1)).start();
    }
}
```

程序运行结果：

```
创建线程 1
创建线程 2
创建线程 3
线程 1:计数 1
线程 1:计数 2
线程 1:计数 3
线程 1:计数 4
线程 1:计数 5
创建线程 4
创建线程 5
线程 2:计数 1
线程 2:计数 2
线程 2:计数 3
…
线程 3:计数 3
线程 3:计数 4
线程 3:计数 5
```

分析：由于多线程之间执行顺序的不确定性，每次执行程序的输出结果可能不同。这种创建线程的方式是比较灵活的方式，也是常用的创建线程的方式。其特点是，专门创建一个类实现 Runnable 接口的 run()方法，提供线程的执行代码作为线程体；然后，再创建一个线程对象去执行这个线程体。实现了线程代码和线程对象的分离。

9.3.3 使用 Timer 创建线程

可以使用 Timer 创建需要周期性执行任务的线程。TimerTask 就是 Timer 执行的周期性任务。TimerTask 实现了 Runnable 接口作为线程体。Timer 使用单一后台线程依次调度执行提交给它的所有 TimerTask 对象。这些任务应该在比较短的时间内完成。Timer 使用一个函数调度执行任务：

```
schedule(TimerTask task, Date firstTime, long period)
```

【例 9.4】 创建周期性线程。

```
package code0903;
import java.util.Timer;
import java.util.TimerTask;
```

```
public class TimerThreadDemo {
    static class Blossom extends TimerTask {
        public void run() {
            System.out.println(Thread.currentThread().getName()+":开花。");
        }
    }
    static class Fruit extends TimerTask {
        public void run() {
            System.out.println(Thread.currentThread().getName()+":结果。");
        }
    }
    public static void main(String[] args) throws InterruptedException {
        Timer timer = new Timer();
        timer.schedule(new Blossom(),300,200);  //300ms 后开始执行,每 200ms 执行一次
        timer.schedule(new Fruit(), 400, 500);  //400ms 后执行,每 500ms 执行一次
        Thread.sleep(1000);                     //主线程休息 1000ms
        timer.cancel();                         //终止定时器,丢弃所有正在执行的任务
    }
}
```

程序运行结果:

```
Timer-0:开花。
Timer-0:结果。
Timer-0:开花。
Timer-0:开花。
Timer-0:结果。
Timer-0:开花。
```

分析:创建了两个 TimerTask 对象,一个是开花,另一个是结果。使用 Timer 定时器分别调度两个任务执行。如果不终止定时器,那么会一直周期性地执行下去。最后,在定时器执行 1000ms 后,终止它的执行。

9.3.4 实用案例 9.1:使用线程池创建线程

前面章节讲解的两种创建线程的方法,都需要新建 Thread 对象,这些线程完成任务后就死亡。有时需要使用已经存在的线程对象,反复执行特定的任务。这种情况下,可以使用线程池。

使用线程池的基本思想是,事先创建一些线程对象放入线程池,如果需要执行任务,则从线程池中取一个线程来执行指定的任务,任务完成后,把线程重新放回线程池。可以重复使用线程,避免了重复创建线程对象,节省了资源。

线程池可以使用两种线程体,一种是实现 Runnable 接口的线程体,另外一种是实现 Callable 接口的线程体。Callable 接口的 call()方法可以返回一个值,而 run()方法不能返回值。

线程池 ExecutorService 使用 execute()方法向线程池提供 Runnable 类型的执行体;使用 submit()方法向线程池提交 Runnable 和 Callable 类型的执行体。

submit()方法返回 Future 对象。Future 对象表示线程的异步计算结果,可以用于检测线程是否完成、得到线程计算结果、取消线程执行等。

【例 9.5】 使用固定线程池创建线程，然后让它们执行特定的任务。

```
package code0903;
import java.util.ArrayList;
import java.util.List;
import java.util.Random;
import java.util.concurrent.Callable;
import java.util.concurrent.ExecutorService;
import java.util.concurrent.Executors;
import java.util.concurrent.Future;
import java.util.concurrent.TimeUnit;
public class ThreadPoolDemo {
    public static void main(String[] args) throws Exception {
        //创建线程池包含 2 个线程
        ExecutorService pool = Executors.newFixedThreadPool(2);
        //创建 10 个线程任务,并提交到线程池
        for (int i = 1; i <= 5; i++) {
            //创建线程执行任务
            RunnableTarget t = new RunnableTarget("线程体" + i);
            //把任务提交到线程池
            pool.execute(t);
        }
        //存放 submit 方法的返回值,将来可以检测线程的执行状态
        List<Future<Integer>> futures = new ArrayList<Future<Integer>>();
        for (int i = 0; i <= 4; i++) {
            //把 Callable 任务提交到线程池,返回 Future 对象,泛型表示线程体的返回值类型
            Future<Integer> result = pool.submit(new CallableTarget("线程体 1" + i));
            futures.add(result);
        }
        for (int i = 0; i <= 4; i++) {
            Future<Integer> f = futures.get(i);
            if (futures.get(i).isDone()) {                     //检测线程是否执行结束

                System.out.println("直接得结果:" + f.get()); //得到线程体的计算结果
            } else {
                Integer r = f.get(1, TimeUnit.SECONDS);       //等待 1s 后获得结果
                System.out.println("等待得结果:" + r);
            }
        }
    }
}
class RunnableTarget implements Runnable {
    private String targetName;                                 //线程体名字
    Random r = new Random();
    public RunnableTarget(String targetName) {
        this.targetName = targetName;
    }
    public void run() {
        int sum = 0;
        for (int i = 0; i < 100; i++) {
            sum += r.nextInt(100);                             //求和
        }
         System.out.println(Thread.currentThread().getName() + "执行:" + this.
targetName + " 结果:" + sum);
    }
```

```
    }
    class CallableTarget implements Callable<Integer> {     //泛型表示 call 的返回类型
        private String targetName;                           //线程体名字
        Random r = new Random();
        public CallableTarget(String targetName) {
            this.targetName = targetName;
        }
        @Override
        public Integer call() {                   //线程体可以返回一个值,类型用泛型说明
            int sum = 0;
            for (int i = 0; i < 100; i++) {
                int tmp = r.nextInt(100);
                sum += tmp * tmp;                            //计算平方和
            }
             System.out.println(Thread.currentThread().getName() + "执行:" + this.
    targetName);
            return sum;
        }

    }
```

程序运行结果:

```
pool-1-thread-1 执行:线程体 1 结果:4786
pool-1-thread-2 执行:线程体 2 结果:4502
pool-1-thread-2 执行:线程体 3 结果:4897
pool-1-thread-1 执行:线程体 4 结果:5338
pool-1-thread-2 执行:线程体 5 结果:5229
pool-1-thread-2 执行:线程体 10
pool-1-thread-1 执行:线程体 11
pool-1-thread-1 执行:线程体 12
pool-1-thread-1 执行:线程体 14
等待得结果:331718
直接得结果:361925
直接得结果:277322
pool-1-thread-2 执行:线程体 13
等待得结果:323848
直接得结果:390113
```

分析:创建了包含了两个线程对象的线程池。向线程池提交了 5 个 Runnable 类型的线程体,5 个 Callable 类型的线程体。10 个线程体共享两个线程对象。提交 Callable 线程体后,可以返回一个 Future 对象,以便将来对线程进行控制等。本例用 Future 对象返回了线程体的执行结果,先判断线程是否执行完成,然后再获取结果;如果没有完成,则等待一段时间,再获取结果。多线程并发执行的顺序每次可能不同。线程池的其他更深入用法,请参考 API 文档。

9.3.3 节介绍了定时器 Timer,它使用单一后台线程执行周期任务,不论任务对象有多少。当有多个周期性任务时,Java 建议使用线程池 ScheduledThreadPoolExecutor。其详细操作请参考 API 文档。

9.4 线程的生命周期及调度

9.4.1 线程生命周期

线程是动态的，具有一定的生命周期，经历从创建、执行直到消亡的过程。JVM 中线程的执行状态可以分为新建（new）、运行（runnable）、阻塞（blocked）、等待（waiting）、限时等待（timed waiting）和终止（terminated）共 6 个状态。Thread 类中定义了一个内部枚举 State 来表示线程状态。线程各个状态之间的状态转换过程如图 9-1 所示。

（1）新建状态：新创建的线程在启动之前处于这个状态。

（2）运行状态：调用 start()方法启动线程后，将线程的状态转换为运行状态。线程在 Java 虚拟机中执行任务，或者可能是在等待操作系统的其他资源，如处理器。运行状态的线程不一定正在处理器上执行。运行状态可以转入除新建状态的其他状态。

（3）阻塞状态：线程等待监视锁（monitor lock）进入阻塞状态。调用 Object.wait()方法后，线程进入阻塞状态；在阻塞状态中等待锁，获得锁后进入同步块（synchronized block）或者同步方法（synchronized method）。线程通过 I/O 请求完成任务，I/O 完成之前，线程进入阻塞状态；I/O 完成后，线程返回运行状态。

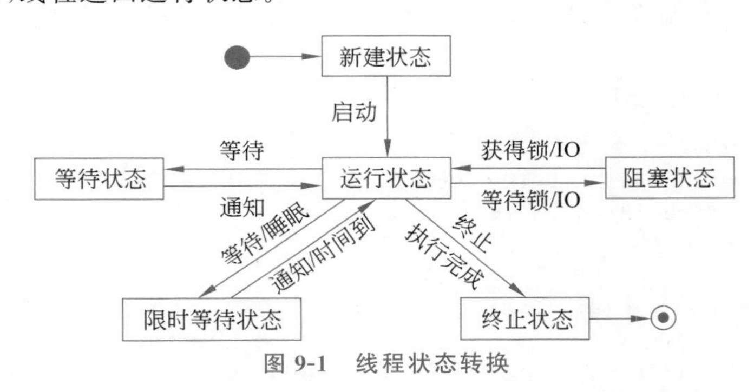

图 9-1 线程状态转换

（4）等待状态：调用无超时参数的 Object.wait()方法后线程进入等待状态。被其他线程通知时返回运行状态，如其他线程调用 Object.notify() 或者 Object.notifyAll()方法。

调用无超时参数的 Thread.join()后，线程进入等待状态，等待其他线程的终止，其他线程终止后，返回运行状态。

（5）限时等待状态：调用 Thread.sleep()方法、有超时参数的 Object.wait()方法、有超时参数的 Thread.join()方法后，线程进入限时等待状态。时间到期后返回运行状态。调用 wait()方法进入限时等待状态后，当被其他线程调用 Object.notify() 或者 Object.notifyAll()方法通知后，也可返回运行状态。拥有监视锁的线程才能调用 wait()和 notify()方法。调用 sleep()方法后，线程不会丢失拥有的锁。

（6）终止状态：当线程体运行结束、产生异常错误或者调用 stop()方法后，线程将终止运行，由 JVM 收回线程占用的资源。

9.4.2 线程调度和优先级

虽然线程是并发运行的，然而事实并非常常如此。当系统中只有一个处理单元时，以某种顺序执行多线程。JVM 采用基于优先级的调度方法执行线程。这种算法根据线程的相对优先级实行调度。在任何时刻,如果有多个线程等待运行,系统选择优先级最高的线程运行。只有当它停止、自动放弃或者由于某种原因成为非运行状态,低优先级的线程才能运行。如果两个线程具有相同的优先级,它们将被交替地运行。JVM 还使用抢先试调度算法,在任何时刻,如果一个比其他线程优先级都高的线程的状态变为可运行状态,JVM 将选择该线程来运行。为了防止低优先级的线程饿死,JVM 酌情分配执行时间给低优先级线程。总之,优先级高的线程分配时间片的数量多于优先级低的线程。

Java 将线程的优先级分为 10 个等级,分别用 1～10 的数字表示。数字越大表明线程的级别越高。相应地,在 Thread 类中定义了表示线程最低、最高和普通优先级的成员变量 MIN_PRIORITY、MAX_PRIORITY 和 NORMAL_PRIORITY,代表的优先级等级分别为 1、10 和 5。当一个线程对象被创建时,其默认的线程优先级是 5。

在创建线程对象之后,可以调用线程的 setPriority()方法改变该线程的运行优先级,同样可以调用 getPriority()方法获取线程的优先级。

在 Java 中比较特殊的线程是被称为守护(Daemon)线程的低级别线程。这个线程具有最低的优先级,用于为系统中的其他对象和线程提供服务。将一个用户线程设置为守护线程的方式是在线程对象启动之前调用 setDaemon()方法。典型的守护线程例子是 JVM 中的系统资源自动回收线程,它始终在低级别的状态中运行,用于实时监控和管理系统中的可回收资源。

【例 9.6】 线程的优先级。

```
package code0904;
public class TestThreadPriority extends Thread{
    public TestThreadPriority(String name) {
        super(name);
    }
    public static void main(String[] args)
    {
        TestThreadPriority t1 = new TestThreadPriority("Thread1");
        t1.setPriority(Thread.MIN_PRIORITY);
        t1.start();
        TestThreadPriority t2 = new TestThreadPriority("Thread2");
        t2.setPriority(Thread.NORM_PRIORITY);
        t2.start();
        TestThreadPriority t3 = new TestThreadPriority("Thread3");
        t3.setPriority(Thread.MAX_PRIORITY);
        t3.start();
    }
    public void run()
    {
        for (int i = 0; i < 5; i++)
            System.out.println(this.getName() + " is running!");
    }
}
```

程序运行结果:

```
Thread1 is running!
Thread2 is running!
```

```
Thread3 is running!
Thread3 is running!
Thread2 is running!
Thread1 is running!
Thread2 is running!
Thread2 is running!
Thread3 is running!
Thread3 is running!
Thread3 is running!
Thread2 is running!
Thread1 is running!
Thread1 is running!
Thread1 is running!
```

分析：由程序的运行结果可以看出，高优先级的线程有优先执行的机会，低优先级的线程也有执行的机会，不一定是高优先级执行完成后，低优先级的才执行。每次执行结果可能不同。

9.4.3 线程的终止

前面的程序，线程体 run()方法执行完成后，线程就终止了。也可以用修改某些变量来指示目标线程应该停止运行。要使线程在完成任务之前可取消，必须采取一定的措施，但应该有一个清晰而安全的机制使线程终止。

Thread 类提供了 stop()方法，任何时候，调用线程的 stop()方法可以终止线程，但这是一个不安全的方法，会破坏线程的状态，已经不推荐使用。

另外一种方法是通过中断线程来请求取消线程的执行，让线程监视并响应中断。中断请求不会强制终止线程，但是会中断线程的睡眠状态，如调用 sleep()和 wait()方法后。

Thread 类提供中断线程的方法有：①interrupt()，向线程发送中断；②isInterrupted()，测试线程是否已经被中断；③interrupted()，测试当前线程是否已经被中断，随后清除线程“中断”状态。

线程的中断状态只能由线程自己清除，当线程侦测到自己被中断时，需要在响应中断之前做某些清除工作。如果被中断的线程正在执行 sleep()或者 wait()方法，就会抛出 InterruptedException 异常。这种抛出异常的中断会清除线程的中断状态。

大体上任何执行阻塞操作的方法都应该通过 interrupt 来取消阻塞操作。

下面的程序，主线程在等待计算线程 2000ms 后，中断计算线程，计算线程由于正在执行 sleep()方法，就会抛出 InterruptedException 异常，终止休眠状态，然后进入异常处理，在 catch 中可以做一些清理工作(如果需要)，然后结束线程执行。

【例 9.7】 中断线程的执行。

```
package code0904;
public class InterruptTest extends Thread {
    static int result = 0;
    public InterruptTest(String name) {
        super(name);
    }
    public static void main(String[] args) {
        System.out.println("主线程执行");
```

```
        Thread t = new InterruptTest("计算线程");
        t.start();
        System.out.println("result:" + result);
        try {
            long start = System.nanoTime();     //获得开始时间
            t.join(2000);                       //等待线程 t 执行 2000ms
            long end = System.nanoTime();       //获得结束时间
            t.interrupt();                      //中断线程 t 的执行
            System.out.println((end - start) / 1000000 + "毫秒后:" + result);
        } catch (InterruptedException e) {
            e.printStackTrace();
        }
    }
    @Override
    public void run() {
        System.out.println(this.getName() + "开始计算...");
        try {
            Thread.sleep(4000);
        } catch (InterruptedException e) {
            System.out.println(this.getName()+"被中断,结束");
            return;
        }
        result = (int) (Math.random() * 10000);
        System.out.println(this.getName() + "结束计算");
    }
}
```

程序运行结果：

```
主线程执行
result:0
计算线程开始计算...
1999 毫秒后:0
计算线程被中断,结束
```

分析：从运行结果可以看出，计算线程被中断后，run()方法中的最后两行语句没有执行。没有产生计算结果。程序中的 join()方法，也是一个控制线程执行的方法，表示一个线程等待另一个线程执行一段时间之后执行，或者等待另一个线程执行结束之后执行。

main 作为主线程，在 main 中调用 t.join()，表示 main 线程等待 t 线程结束之后再执行。t.join(2000)，表示 main 线程等待 t 线程 2000ms 之后再执行。

如果一个线程长时间没有调用能够抛出 InterruptedException 异常的方法，那么线程就必须定期的调用 Thread.interrupted()方法，如果接收到中断，就返回 true，然后退出线程。

9.4.4 实用案例 9.2：周期性检测中断结束线程

一般强行终止线程，有可能破坏线程所操作数据的一致性，应该向线程发送中断信号，线程收到中断信号后，自行执行结束操作。

【例 9.8】 周期性的调用检测中断的方法结束线程。

```
package code0904;
public class InterruptToTerminateThread extends Thread {
    static int result = 0;
```

```
    public InterruptToTerminateThread(String name) {
        super(name);
    }
    public static void main(String[] args) {
        System.out.println("主线程执行");
        Thread t = new InterruptToTerminateThread("计算线程");
        t.start();
        System.out.println("result:" + result);
        try {
            long start = System.nanoTime();
            t.join(10);
            long end = System.nanoTime();
            t.interrupt();
            System.out.println((end - start) / 1000000 + "毫秒后:" + result);
        } catch (InterruptedException e) {
            e.printStackTrace();
        }
    }
    @Override
    public void run() {
        System.out.println(this.getName() + "开始计算...");
        for (int i = 0; i < 1000000; i++) {
            result++;
            if (Thread.interrupted()) {
                System.out.println(this.getName() + "被中断,即将结束");
                return;
            }
        }
        System.out.println(this.getName() + "结束计算");
    }
}
```

程序运行结果：

```
主线程执行
result:0
计算线程开始计算...
11 毫秒后:182167
计算线程被中断,即将结束
```

分析：计算线程原计划执行 1000000 次循环，主线程等待 11ms 后，中断计算线程。计算线程定期检测中断信号，收到中断后，结束执行。

9.5 多线程互斥与同步

9.5.1 线程的互斥

1. 问题的提出

已知一个银行账号，当从多个渠道同时取钱时，有可能造成账号数据被破坏。

【例 9.9】 错误的银行账号对象。

```
package code0905;
```

```
public class WrongAccount {
    double balance;                                         //账号余额
    public WrongAccount(double balance) {
        super();
        this.balance = balance;
    }
}
class AccountThread extends Thread
{
    WrongAccount account;
    int delay=100;
    public AccountThread(WrongAccount account) {
        super();
        this.account = account;
    }
    @Override
    public void run() {
        //下面执行取款操作
        if(account.balance>=1000)
        {
            try {
                sleep(delay);
                account.balance-=1000;
                System.out.println("取款 1000 成功。");
            } catch (InterruptedException e) {
                e.printStackTrace();
            }
        }else
        {
            System.out.println("取款失败。");
        }
    }
    public static void main(String[] args) throws InterruptedException
    {
        //创建账号对象
        WrongAccount account=new WrongAccount(1005);
        //创建 3 个取款线程
        AccountThread t1=new AccountThread(account);
        AccountThread t2=new AccountThread(account);
        AccountThread t3=new AccountThread(account);
        t1.start();
        t2.start();
        t3.start();
        //主线程等待 3 个线程结束后,输出最后余额
        t1.join();
        t2.join();
        t3.join();
        System.out.println("最终账号余额是:"+account.balance);
    }
}
```

程序第一次执行结果：

```
取款 1000 成功。
取款 1000 成功。
```

```
取款 1000 成功。
最终账号余额是:5.0
```

程序再次执行结果：

```
取款 1000 成功。
取款 1000 成功。
取款 1000 成功。
最终账号余额是:-995.0
```

分析：从输出结果可以看出，当多线程同时访问一个账号对象时输出结果是不确定的，3 个线程都取款成功。正确的情况应该是只有一个线程能够取款成功。说明本程序在多线程并发的时候是有问题的。原因在于，当一个线程对数据进行修改后，另外的线程不知道。多线程并发访问账号对象时，应该进行保护，同一时刻只能有一个线程访问对象，保证对象状态的正确性。

2. 互斥对象

通常把多线程并发访问的资源称为临界资源。对临界资源的访问必须是互斥的，JVM 可以为每个对象设置一个“互斥锁”（又称隐含锁、监视锁，intrinsic lock、monitor lock），保证同一时刻只有一个线程拥有互斥锁。其他线程必须等待拥有锁的线程释放锁后才可以获取。

Java 提供了关键字 synchronized 来实现互斥锁。当定义类、方法或者代码片段中，使用该关键字，就表示和该关键字相关联的对象有互斥锁。下面修改例 9.9 的程序，实现多线程互斥。

【例 9.10】 正确银行账号对象。

```
package code0905;
public class RightAccount {
    double balance;                                    //账号余额
    int delay = 100;
    public RightAccount(double balance) {
        super();
        this.balance = balance;
    }
    //同步方法,执行该方法时,必须获得所在对象的互斥锁
    public synchronized void withdraw(double money) {
        //下面执行取款操作
        if (balance >= 1000) {
            try {
                Thread.sleep(delay);
                balance -= 1000;
                System.out.println("取款 1000 成功。");
            } catch (Exception e) {
                e.printStackTrace();
            }
        } else {
            System.out.println("取款失败。");
        }
    }
}
class AccountThread2 extends Thread {
    RightAccount account;
    public AccountThread2(RightAccount account) {
```

```
        super();
        this.account = account;
    }
    @Override
    public void run() {
        account.withdraw(1000);
    }
    public static void main(String[] args) throws InterruptedException {
        //创建账号对象
        RightAccount account = new RightAccount(1005);
        //创建 3 个取款线程
        AccountThread2 t1 = new AccountThread2(account);
        AccountThread2 t2 = new AccountThread2(account);
        AccountThread2 t3 = new AccountThread2(account);
        t1.start();
        t2.start();
        t3.start();
        //主线程等待 3 个线程结束后,输出最后余额
        t1.join();
        t2.join();
        t3.join();
        System.out.println("最终账号余额是:" + account.balance);
    }
}
```

程序运行结果:

```
取款 1000 成功。
取款失败。
取款失败。
最终账号余额是:5.0
```

分析:从输出结果看,用了互斥锁后可以保证账号状态的正确性。只有一个线程可以取款成功。其他线程因为余额不足不能取款。synchronized 关键字加在方法 withdraw()之前,该方法为同步方法,表示线程执行该方法时,必须获得该方法所在账号对象的互斥锁。一般凡是需要获得互斥锁的代码都可以使用 synchronized。对于本例,也可以把 synchronized 用在账号类的前面,表示该类的所有方法都是同步方法。为了更细粒度地并发控制,还可以使用同步语句,即只在需要同步的几个语句前面使用 synchronized,同步语句必须指定一个提供隐含锁的对象。具体使用方法请查阅文献。

9.5.2 线程的同步

多线程并发工作时,有时需要互斥,有时需要相互协作,按照一定的步骤共同完成任务。

以银行账号为例,有两类线程:存款线程和取款线程。存款线程负责向账号存款,取款线程负责从账号取款。操作的条件是:当账户余额为 0 时,不能取款,取款线程等待;当账户余额达到 1 万元时,不能继续存款,存款线程等待。

这个问题符合生产者-消费者模型:①生产者生产商品保存到厂库;②消费者从厂库取走商品;③仓库容量有限,只有当仓库有剩余空间时,生产者可以把商品加入仓库,否则等待;④只有仓库非空时,消费者才能取走产品,否则等待。

Java 为多线程同步提供了两类方法:wait()和 notify()方法。wait()方法是当一个线程

执行了该方法时，放弃互斥锁，进入互斥锁的等待队列。notify()方法是唤醒互斥锁等待队列中的线程，并进入就绪状态。

【例 9.11】 多线程同步操作银行账号。

```
package code0905;
class SynAccount {
    double balance;                                         //余额
    final double MAX = 10000;                               //最高限额
    public SynAccount(double balance) {
        this.balance = balance;
    }
    //取款同步方法
    public synchronized void withdraw(double money) {
        if (balance < money) {
            try {
                System.out.printf("取款%1$,.2f 失败。余额:%2$,.2f\n", money,balance);
                wait();                                     //进入等待队列
            } catch (InterruptedException e) {
                e.printStackTrace();
            }
        }
        balance -= money;
        System.out.printf("取款%1$,.2f 成功。余额:%2$,.2f\n", money,balance);
        notify();                                           //唤醒等待队列的线程
    }
    //存款同步方法
    public synchronized void deposit(double money) {
        if (balance + money >= MAX) {
            try {
                System.out.printf("存款%1$,.2f 失败。余额:%2$,.2f\n", money,
balance);
                wait();                                     //进入等待队列
            } catch (InterruptedException e) {
                e.printStackTrace();
            }
        }
        balance += money;
        System.out.printf("存款%1$,.2f 成功。余额:%2$,.2f\n", money,balance);
        notify();                                           //唤醒等待队列的线程
    }
}
public class SynThreads {
    //取款线程
    static class Withdrawer extends Thread {
        SynAccount account;
        public Withdrawer(SynAccount account) {
            super();
            this.account = account;
        }
        @Override
        public void run() {
            for (int i = 0; i < 8; i++) {
                account.withdraw(2000);
            }
```

```
            }
        }
        //存款线程
        static class Depositor extends Thread {
            SynAccount account;
            public Depositor(SynAccount account) {
                this.account = account;
            }
            @Override
            public void run() {
                for (int i = 0; i < 8; i++) {
                    account.deposit(2000);
                }
            }
        }
        public static void main(String[] args) {
            SynAccount account = new SynAccount(5000);
            Thread withdraw = new Withdrawer(account);
            Thread deposit = new Depositor(account);
            withdraw.start();
            deposit.start();
        }
    }
```

程序输出结果:

```
取款 2,000.00 成功。余额:3,000.00
取款 2,000.00 成功。余额:1,000.00
取款 2,000.00 失败。余额:1,000.00
存款 2,000.00 成功。余额:3,000.00
存款 2,000.00 成功。余额:5,000.00
存款 2,000.00 成功。余额:7,000.00
存款 2,000.00 成功。余额:9,000.00
存款 2,000.00 失败。余额:9,000.00
取款 2,000.00 成功。余额:7,000.00
取款 2,000.00 成功。余额:5,000.00
取款 2,000.00 成功。余额:3,000.00
取款 2,000.00 成功。余额:1,000.00
取款 2,000.00 失败。余额:1,000.00
存款 2,000.00 成功。余额:3,000.00
存款 2,000.00 成功。余额:5,000.00
存款 2,000.00 成功。余额:7,000.00
存款 2,000.00 成功。余额:9,000.00
取款 2,000.00 成功。余额:7,000.00
取款 2,000.00 成功。余额:5,000.00
```

分析:从输出结果可以看出,存款线程和取款线程实现了同步协作工作。条件不满足时,存款和取款都会失败。

多线程同步时,需要注意以下问题:

(1) wait()和 notify()方法必须位于同步代码块中,也就是在 synchronized 代码块中。执行这些方法的线程必须已经获得了互斥锁。这两个方法属于拥有互斥锁的对象。

(2) wait()和 notify()方法必须配对使用,执行 wait()方法进入等待队列的线程,应该由另一个线程执行 notify()方法唤醒。

(3) 在某些情况下,可以使用 notifyAll()方法代替 notify()方法,唤醒等待队列中的所有线程。

9.5.3 实用案例 9.3:使用显式锁实现多线程互斥

包 java.util.concurrent.lock 中的显式锁 ReentrantLock 被作为隐含锁(同步方法自动锁)的替代,在争用条件下有更好的性能。它有一个与锁相关的计数器,如果拥有锁的某个线程得到锁,那么计数器就加 1。锁必须在 finally 块中释放;否则,如果受保护的代码抛出异常,锁就有可能永远得不到释放。

【例 9.12】 用显式锁实现多线程取款操作。

```
package code0905;
import java.util.concurrent.locks.Lock;
import java.util.concurrent.locks.ReentrantLock;
public class LockAccount {
    double balance;                                   //账号余额
    Lock lock = new ReentrantLock();                  //创建锁对象
    public LockAccount(double balance) {
        super();
        this.balance = balance;
    }
    //使用显示锁
    public void withdraw(double money) {
        lock.lock();                                  //锁定
        try {
            if (balance >= 1000) {
                try {
                    Thread.sleep(100);
                    balance -= 1000;
                    System.out.println("取款 1000 成功。");
                } catch (Exception e) {
                    e.printStackTrace();
                }
            } else {
                System.out.println("取款失败。");
            }
        } finally {
            lock.unlock();                            //释放锁
        }
    }
}
class AccountThread3 extends Thread {
    LockAccount account;
    int delay = 100;
    public AccountThread3(LockAccount account) {
        super();
        this.account = account;
    }
    @Override
    public void run() {
        account.withdraw(1000);
    }
    public static void main(String[] args) throws InterruptedException {
```

```
        //创建账号对象
        LockAccount account = new LockAccount(1005);
        //创建 3 个取款线程
        AccountThread3 t1 = new AccountThread3(account);
        AccountThread3 t2 = new AccountThread3(account);
        AccountThread3 t3 = new AccountThread3(account);
        t1.start();
        t2.start();
        t3.start();
        //主线程等待 3 个线程结束后,输出最后余额
        t1.join();
        t2.join();
        t3.join();
        System.out.println("最终账号余额是:" + account.balance);
    }
}
```

程序运行结果:

```
取款 1000 成功。
取款失败。
取款失败。
最终账号余额是:5.0
```

分析:从输出结果可以看出,使用显式锁确保了在多线程并发操作账号的情况下数据的正确性。

9.6 多线程实训任务

【任务描述】

编写一个 Java 多线程程序,同时用 3 种方法实现对一个整型数组的排序。可以使用的排序算法有冒泡法、双向冒泡法和快速排序法。

【任务分析】

3 种排序算法并发执行,需要创建 3 个线程,每个线程执行一种排序算法。为每种算法创建一个线程目标对象。可以使用工具类 System 计算每个线程的执行时间。整型数组可以用随机数产生,并使用 System 类为线程复制数组数据。

3 种排序算法有共同的执行框架,可以定义一个父类 Sort,定义排序线程的执行框架,分别创建 BSort、QSort 和 BBSort 继承 Sort 类。分别表示冒泡排序执行体、快速排序执行体和双向冒泡排序执行体。

最后,定义一个主程序 MainTest 具体创建线程对象,执行排序算法。

【任务解决】

完整程序如下:

```
package code0906;
//排序算法的目标执行对象,定义排序线程执行体的框架,子类覆盖 sort 方法
public class Sort implements Runnable {
    int[] data;                                     //排序算法需要的数据
    public Sort(int[] data) {
```

```
        super();
        this.data = data;
    }
    //具体的排序子类覆盖此方法,实现排序算法
    public void sort() {
        throw new RuntimeException("please do it in subclass");
    }
    @Override
    public void run() {                            //线程体,执行 sort 方法,并输出排序后的结果
        long start = System.nanoTime();   //开始时间
        sort();                                    //执行排序
        long end = System.nanoTime();      //结束时间
        System.out.println(Thread.currentThread().getName()
                + "执行时间:" + (end - start) + "ns");
    }
}
//双向冒泡排序算法
public class BBSort extends Sort {

    public BBSort(int[] data) {
        super(data);
    }
    public void sort() {
        try {
            sort(data);
        } catch (Exception e) {
            e.printStackTrace();
        }
    }
    //双向冒泡排序程序,可以参考相关文献
    void sort(int a[]) throws Exception {
        int j;
        int limit = a.length;
        int st = -1;
        while (st < limit) {
            st++;
            limit--;
            boolean swapped = false;
            for (j = st; j < limit; j++) {

                if (a[j] > a[j + 1]) {
                    int T = a[j];
                    a[j] = a[j + 1];
                    a[j + 1] = T;
                    swapped = true;
                }
            }
            if (!swapped) {
                return;
            } else
                swapped = false;
            for (j = limit; --j >= st;) {
                if (a[j] > a[j + 1]) {
                    int T = a[j];
```

```
                    a[j] = a[j + 1];
                    a[j + 1] = T;
                    swapped = true;
                }
            }
            if (!swapped) {
                return;
            }
        }
    }
}
//冒泡排序程序
public class BSort extends Sort {
    public BSort(int[] data) {
        super(data);
    }
    public void sort() {
        try {
            sort(data);
        } catch (Exception e) {
            e.printStackTrace();
        }
    }
    void sort(int a[]) throws Exception {
        for (int i = a.length; --i >= 0;) {
            boolean swapped = false;
            for (int j = 0; j < i; j++) {
                if (a[j] > a[j + 1]) {
                    int T = a[j];
                    a[j] = a[j + 1];
                    a[j + 1] = T;
                    swapped = true;
                }
            }
            if (!swapped)
                return;
        }
    }
}
//快速排序算法
public class QSort extends Sort {
    public QSort(int[] data) {
        super(data);
    }
    public void sort() {
        try {
            QuickSort(data, 0, data.length - 1);
        } catch (Exception e) {
            e.printStackTrace();
        }
    }
    private void swap(int a[], int i, int j) {
        int T;
        T = a[i];
```

```
        a[i] = a[j];
        a[j] = T;
    }
    void QuickSort(int a[], int lo0, int hi0) throws Exception {
        int lo = lo0;
        int hi = hi0;
        int mid;
        if (hi0 > lo0) {
            mid = a[(lo0 + hi0) / 2];
            while (lo <= hi) {
                while ((lo < hi0) && (a[lo] < mid))
                    ++lo;
                while ((hi > lo0) && (a[hi] > mid))
                    --hi;
                if (lo <= hi) {
                    swap(a, lo, hi);
                    ++lo;
                    --hi;
                }
            }
            if (lo0 < hi)
                QuickSort(a, lo0, hi);
            if (lo < hi0)
                QuickSort(a, lo, hi0);
        }
    }
}
//主程序创建线程,执行 3 种排序算法
public class MainTest {
    public static void main(String[] args) {
        final int count = 10000;
        int[] data = createData(count);           //创建排序数据
        //创建快速排序线程
        Thread qsort = new Thread(new QSort(data), "快速排序");
        int[] data2 = new int[count];
        //复制需要排序的数组,为下一个排序算法准备数据,
        System.arraycopy(data, 0, data2, 0, count);
        //创建冒泡排序线程
        Thread bsort = new Thread(new BSort(data2), "冒泡排序");
        int[] data3 = new int[count];
        //复制需要排序的数组,为下一个排序算法准备数据,
        System.arraycopy(data, 0, data3, 0, count);
        //创建双向冒泡排序线程
        Thread bbsort = new Thread(new BBSort(data3), "双向冒泡排序");
        qsort.start();
        bsort.start();
        bbsort.start();
    }
    //随机产生排序的数据
    private static int[] createData(int count) {
        int[] data = new int[count];
        for (int i = 0; i < data.length; i++) {
            data[i] = (int) (Math.random() * count);
        }
```

```
            return data;
        }
    }
```

程序运行结果：

```
快速排序执行时间:8485500ns
双向冒泡排序执行时间:163917400ns
冒泡排序执行时间:200972800ns
```

分析：可以看出，在对 10000 个数据排序的情况下，快速排序最快，其次是双向冒泡，最慢是冒泡排序算法。每次程序执行结果可能不同。

习题与思考

1. 将窗口分为上、下两个区，分别运行两个线程，一个线程在上面的区域中显示由右向左游动的字符串，另一个线程在下面的区域显示从左向右游动的字符串。

2. 简述程序、进程和线程之间的关系。什么是多线程程序？

3. 线程有哪几个基本状态？它们之间如何转化？简述线程的生命周期。

4. Runnable 接口中包括哪些抽象方法？Thread 类有哪些主要的成员变量和方法？

5. 如何在 Java 程序中实现多线程？简述创建线程的 4 种方法。

6. 使用非固定线程池创建线程，并执行任务。比较和固定线程池的区别。

7. 如何使用线程池执行周期性任务？

8. 使用多线程同步和互斥方法解决 5 个哲学家问题。

9. 编写程序解决仓库问题：现有一个仓库，一个出口，一个入口。当仓库未满时，可以放入货物，当仓库不空时，可以取出货物。假设有一群人同时取货和放入货物。

10. 请改造例 9.10。把存款余额定义为原子对象，然后写出完整的程序。

11. 有 A、B、C 3 个线程，要求按顺序 A→B→C 执行 10 轮，A 输出“发芽”，B 输出“开花”，C 输出“结果”。

12. 请自学 java.util.concurrent 包中定义的阻塞队列 BlockingQueue。

第10章 GUI程序设计

本章学习目标

图形用户界面(Graphic User Interface,GUI)设计是构建可视化应用的重要基础。Java有很多GUI程序设计技术,例如AWT、Swing和SWT等。本章介绍Swing技术,使读者初步掌握GUI程序设计的基本方法,并将其应用于Java小应用程序Applet的可视化设计。本章重点需要掌握的内容包括:

(1) Swing技术及其常用组件。

(2) 基于Swing的界面布局。

(3) GUI中的事件处理机制。

(4) 高级Swing组件JTree和JTable。

10.1 为什么学习GUI程序设计

对于程序的使用者而言,他们更愿意使用那些界面友好的应用程序。相对于命令行程序不友好的界面,GUI程序能够带给用户更好的用户体验。因此大量的桌面应用程序都会以GUI的形式提供给最终用户。先看图10-1所示的GUI程序,这些GUI程序都来自JDK提供的样例程序,是不是非常漂亮?通过本章的学习,读者也能开发出同样漂亮的软件界面。

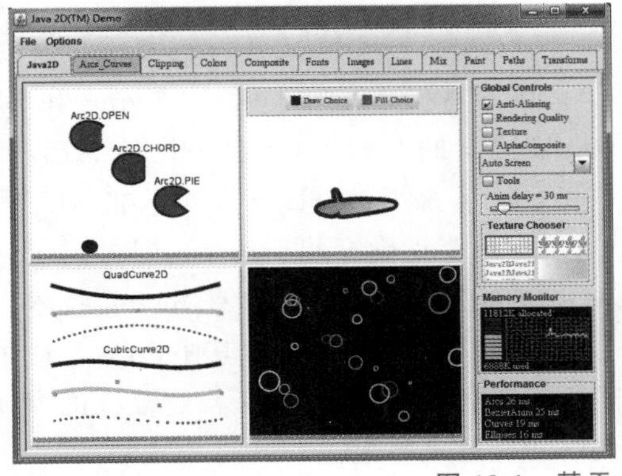

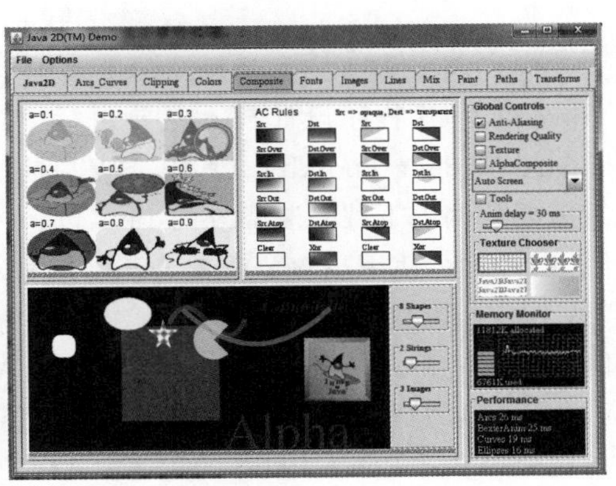

图10-1 基于Swing的GUI程序

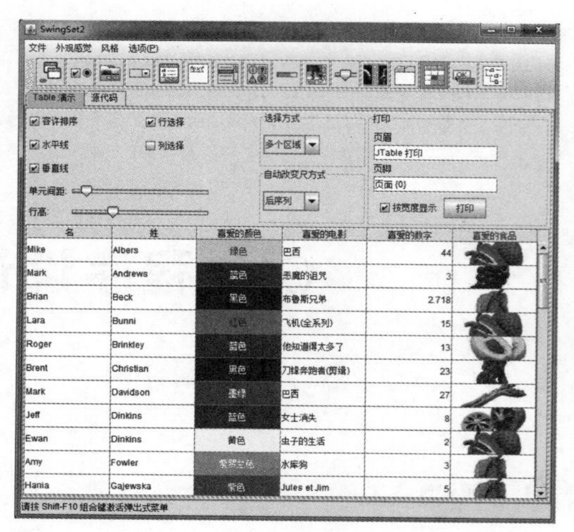

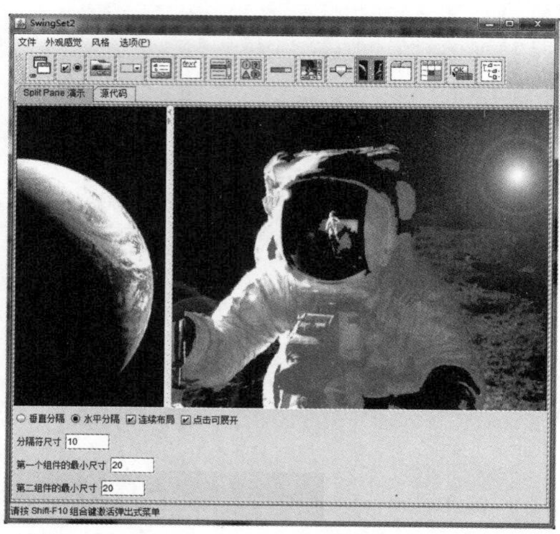

图 10-1 (续)

当然,GUI 程序带给用户的不仅是一种更直观的界面,它也给用户提供了一种更友好的交互方式,用户可以通过鼠标点击、拖动、键盘控制等更灵活的方式进行应用操作。如图 10-2 所示的基于鼠标单击的“运算器”,基于键盘“←→”的“赛车游戏”等。同样,通过 10.4 节中 Java 事件处理机制的学习,读者也能自己开发出类似具有交互功能的应用。

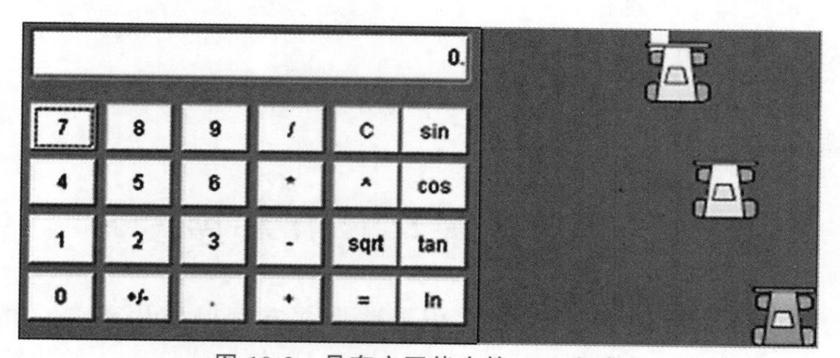

图 10-2 具有交互能力的 GUI 程序

10.2 基于 Swing 的简单界面设计

10.2.1 Swing 简介

Java 的 GUI 程序设计技术主要包括 AWT(Abstract Window Toolkit)、Swing 和 SWT(Standard Widget Toolkit)。其中,SWT 是 IBM 最早提出的开源 GUI 程序设计 API,使用 SWT 需要从网上下载安装额外的 Java 包。AWT 和 Swing 是 Java SE 自带的标准 GUI 程序设计 API,因此,使用 AWT 或者 Swing 编写 GUI 程序,只要正确安装 Java SE 即可。

从 Java 1.0(JDK 1.0)发布开始,AWT 就是 JDK 的一部分,当时还没有 Swing 技术。随着开发人员将 Java 应用于越来越多的平台,AWT 的弱点开始逐渐暴露。其中,AWT 最主要的问题是:AWT 只提供了建立窗口操作应用程序所必需的最少功能,对于构建复杂的窗体程序(如类似 Word、PowerPoint 的程序),AWT 提供的功能远远不足。

Swing 是在 AWT 基础上发展起来的一项技术。Swing 不仅包括了 AWT 所具有的全部组件,而且可以使用树形组件(JTree)、表格(JTable)、选项卡(JTabbedPane)等高级图形组件。同时,Swing 完全由 Java 编写,不再依赖于运行时平台的本地组件(AWT 和 SWT 都存在本地调用),具有良好的可移植性。

由于 Swing 不仅包含了 AWT 的全部功能,而且具有更多高级的特性,随着 Java 技术的发展用 Swing 替代 AWT 已经成为一种趋势。事实上 AWT 和 Swing 技术极为相似,GUI 组件的类名通常只是比 AWT 组件的类名多一个字母 J。例如,按键组件在 Swing 中对应 JButton 类,而在 AWT 中对应 Button 类。因此,一旦掌握了 Swing,要学习 AWT 也是轻而易举的事情。

10.2.2 Swing 的类层次结构

Swing 由很多 Java 类构成。这些 Java 类主要包含在 javax.swing 这个 Java 包中。图 10-3 是 Swing 的类层次结构,图中的 Java 类可以分为 JComponent 及其子类、顶层容器两类。

图 10-3 Swing 的类层次结构

1. JComponent 及其子类

JComponent 及其子类,简称 GUI 组件。GUI 组件中有一部分组件具有图形外观能在图形界面上与用户进行交互,称为可视化组件,例如 JButton、JLabel、JTextField 等。GUI 组件中的另外一些组件没有图形外观,称为非可视化组件。非可视化组件通常需要与可视化组件相结合,共同完成特定的图形功能。例如 JPanel 就是典型的非可视化组件,主要用于界面的布局。

2. 顶层容器

顶层容器(container)包含 JApplet、JDialog、JFrame 和 JWindow 及其子类。通常将 GUI 组件布局在顶层容器上来实现 GUI 程序的界面布局。

JFrame 是最常用的一种顶层容器,它的作用是创建一个顶层的 Windows 窗体。JFrame 的外观就像 Windows 系统下见到的窗体,有标题、边框、菜单等。

JDialog 用于创建对话框。JDialog 创建的对话框相对于 JFrame 创建的窗体没有最大化和最小化按键。在 GUI 编程时可以根据需要选择使用 JDialog 还是 JFrame。

JWindow 创建的窗体没有标题栏,没有最大化、最小化按键。在某些 GUI 应用中,可能需要编写这种不带修饰的窗体,或者用户希望用自己编写的标题栏、最大化、最小化按键来替换 Windows 自带的窗体风格,此时就可以选择创建 JWindow 来实现这些窗体效果。

JApplet 是创建这种程序的顶层容器。Applet 是一种能够嵌入网页中执行的 Java 图形程序。除了顶层容器,Swing 中还有一些轻量级的容器,例如 JPanel。通常可以先将 GUI 组件放置在 JPanel 上,再将 JPanel 加入其他容器或者顶层容器中,从而实现复杂的界面布局。

10.2.3 常见 GUI 组件

Swing 包含了非常多的 GUI 组件,其中常见的组件包括 JButton(按键)、JLabel(标签)、JTextField(单行文本框)、JTextArea(文本输入区)、JTree(树状组件)、JTable(表格)等。

1. 按键

按键是最常用的一个组件,其相应的类是 JButton,以下是其常用的构造方法。

JButton():构造一个没有文字、没有图片的按键。

JButton(Icon icon):构造一个带图片的按键,icon 是按键上的图片。

JButton(String text):构造一个带文字的按键,text 是按键上的文字。

JButton(String text, Icon icon):构造一个带文字、带图片的按键,text 和 icon 分别对应按键所使用的文字和图片。

2. 标签

JLabel 组件用于显示标签。以下是其常用的构造方法。

JLabel(String text):构造一个文字标签,text 是标签对应的文字。

JLabel(Icon image):构造一个图片标签,image 是标签对应的图像。

JLabel(String text, int horizontalAlignment):构造一个文本标签,并通过 horizontalAlignment 指定文本对齐方式,horizontalAlignment 的可取值包括 JLabel.LEFT、JLabel.RIGHT 等。

JLabel(String text, Icon icon, int horizontalAlignment):构造一个带文本和图片的标签,horizontalAlignment 指定对齐方式。

3. 单行文本框

JTextField 用于接受用户的输入信息,但它只能接受一行的用户输入信息,以下是其常用的构造方法。

JTextField():构造一个空的单行文本框。

JTextField(int columns):构造一个空的单行文本框,文本框的宽度(列数)为 columns。

JTextField(String text):构造一个单行文本框,文本框的初始内容为 text。

JTextField(String text, int columns):构造一个单行文本框,文本框的初始内容为 text,

宽度(列数)为 columns。

4. 文本输入区

JTextArea 也称为多行文本框,与 JTextField 不同,JTextArea 可以显示多行多列的文本,以下是其常用的构造方法。

JTextArea():创建一个空的文本输入区。

JTextArea(int rows, int columns):创建一个 rows 行、columns 列的文本输入区。

JTextArea(String text):创建一个文本输入区,文本输入区的初始化文本为 text。

JTextArea(String text, int rows, int columns)创建一个 rows 行、columns 列的文本输入区,文本输入区的初始化文本为 text。

【例 10.1】 使用 Swing JButton 的 GUI 程序,如图 10-4 所示。

```
import java.awt.FlowLayout;
import java.awt.event.ActionEvent;
import java.awt.event.ActionListener;
import javax.swing.JButton;
import javax.swing.JFrame;
import javax.swing.JOptionPane;
public class SwingFrame extends JFrame{
    private JButton button=new JButton("按键");
    public SwingFrame()
    {
        setSize(300,300);                                     //设置窗体大小
        setLocation(400, 400);                                //设置窗体显示位置
        setTitle("ButtonFrame");                              //设置窗体标题栏
        setDefaultCloseOperation(JFrame.EXIT_ON_CLOSE);       //设置窗体默认关闭事件
        setLayout(new FlowLayout());                          //设置布局管理器
        //添加按键事件
        button.addActionListener(new ActionListener(){
            public void actionPerformed(ActionEvent event) {
                JOptionPane.showMessageDialog(null, "点击了按键!");
            }
        });
        add(button);                                          //添加按键
    }
    public static void main(String[] args) {
        SwingFrame frame=new SwingFrame();
        frame.setVisible(true);                               //显示窗体
    }
}
```

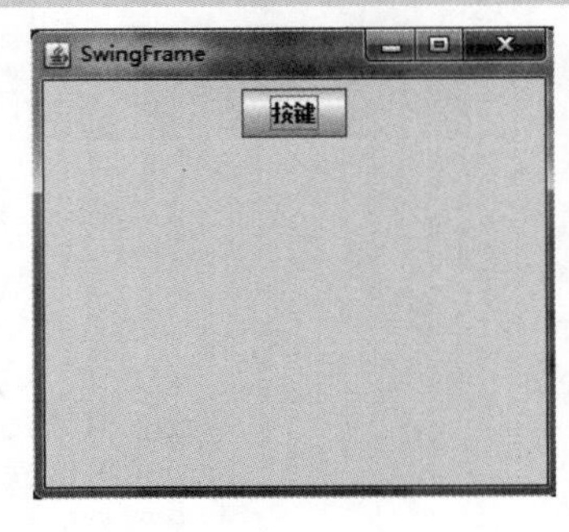

(a)

(b)

图 10-4 使用 Swing JButton 的 GUI 程序

分析：例 10.1 通过从 JFrame 继承，创建了 SwingFrame 类。在 SwingFrame 类的构造函数里，调用 setSize 方法设置窗体大小，setLocation()方法设置窗体显示的位置，setTitle()方法设置窗体的标题。setDefaultCloseOperation()方法用于设置窗体的默认关闭操作，JFrame.EXIT_ON_CLOSE 表明整个应用程序都终止运行。

除此之外，还有 JFrame.DISPOSE_ON_CLOSE、JFrame.DO_NOTHING_ON_CLOSE 和 JFrame.HIDE_ON_CLOSE 等参数。如果不用 setDefaultCloseOperation()方法设置默认关闭操作，则单击窗体的关闭按键时程序不会终止而只会隐藏窗体。button.addActionListener 为按键添加事件响应。setLayout()方法用于设置窗体的布局管理器。最后通过调用 add()方法将一个 JButton 对象添加到窗体上。程序的 main()函数首先创建 SwingFrame 类的实例，然后调用 setVisible()方法显示窗体。

例 10.1 所示代码给出了 GUI 程序的基本流程：

① 根据需要从相应的顶层容器继承(如果创建窗体就继承 JFrame，对话框就继承 JDialog)，新建一个子类。

② 然后设置顶层容器的属性，包括大小、位置、标题、关闭事件等。

③ 设置界面上 GUI 组件的事件响应。

④ 向顶层容器上添加 GUI 组件，并进行布局。

⑤ 创建新建子类的实例，调用 setVisible 方法显示界面。

10.2.4 基于 AWT 的 GUI 程序

看过了 Swing 编写的应用程序，读者一定会好奇 AWT 编写的 GUI 程序又会是怎样的一种形式呢？实际上 AWT 编写的 GUI 程序与 Swing 有非常相似的地方。下面给出一个用 AWT 编写的应用程序，它与前几个程序的不同之处在于：①JFrame 和 JButton 被替代为 Frame 和 Button；②Frame 类没有 setDefaultCloseOperation 方法，因此用 addWindowListener 方法来添加窗口关闭事件；③JOptionPane 类是一个 Swing 类，但由于 AWT 没有类似的类，因此在其中用到了这个类。从中可以看出 Swing 和 AWT 类是可以混用的。可以看出，AWT 程序与 Swing 程序基本类似，但 Swing 在编写 GUI 程序时比 AWT 更为方便。

【例 10.2】 使用 AWT 的 GUI 程序。

```
import java.awt.Button;
import java.awt.FlowLayout;
import java.awt.Frame;
import java.awt.event.ActionEvent;
import java.awt.event.ActionListener;
import java.awt.event.WindowAdapter;
import java.awt.event.WindowEvent;
import javax.swing.JOptionPane;
public class AWTFrame extends Frame{
    private Button button=new Button("按键");
    public AWTFrame()
    {
        setSize(300,300);                          //设置窗体大小
        setLocation(400, 400);                     //设置窗体显示位置
        setTitle("AWTFrame");                      //设置窗体标题栏
```

```
        //设置窗体关闭事件
        this.addWindowListener(new WindowAdapter(){
            public void windowClosing(WindowEvent arg0) {
                System.exit(0);
            }
        });
        setLayout(new FlowLayout());                          //设置布局管理器
        //添加按键事件
        button.addActionListener(new ActionListener(){
            public void actionPerformed(ActionEvent event) {
                JOptionPane.showMessageDialog(null, "点击了按键!");
            }
        });
        add(button);                                          //添加"按键"按钮
    }
    public static void main(String[] args) {
        AWTFrame frame=new AWTFrame();
        frame.setVisible(true);                               //显示窗体
    }
}
```

程序运行结果如图 10-5 所示。

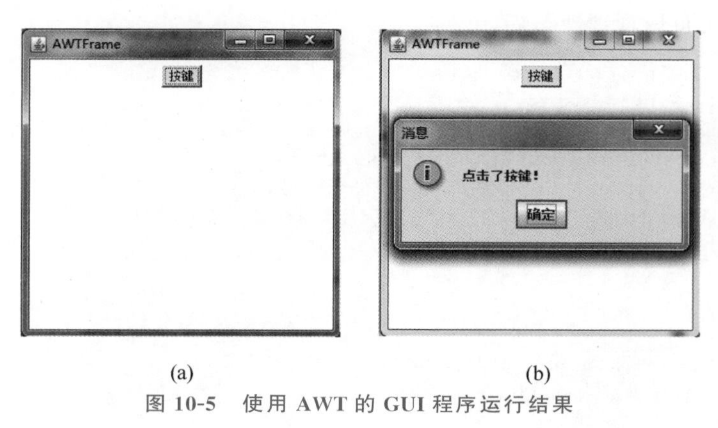

(a) (b)

图 10-5 使用 AWT 的 GUI 程序运行结果

10.3 界面布局

创建 GUI 程序最重要的是界面布局。Swing 采用了两种布局方式：无布局管理器布局和基于布局管理器的布局。其中无布局管理器布局类似于 Visual Basic、Delphi 和 Visual C++ 采用的布局方式，通过指定 GUI 组件在窗体上的绝对位置进行组件布局。

基于布局管理器的布局，首先通过调用容器类(JFrame、JDialog、JPanel 等)的 setLayout()方法设置布局管理器(包括 FlowLayout、BorderLayout、GridLayout 等)。设置好布局管理器之后，容器内的所有组件的布局就由布局管理器负责，包括组件的排列顺序，组件的大小、位置，当窗口移动或调整大小后组件如何变化等。

采用布局管理器比无布局管理方式更具有灵活性。例如，当窗体的大小或者分辨率发生改变时，采用布局管理器方式能够重新布局组件；而采用无布局管理器布局则需要编程者去控制新的组件位置和大小。同时采用布局管理器的 Swing 程序跨平台的布局效果比采用无布局管理器布局更好。

10.3.1 无布局管理器布局

要采用无布局管理器布局,首先要取消 Swing 默认的布局管理器,否则布局方法不会生效。其方法为调用容器的 setLayout()方法,并将布局管理器设置为 null。取消掉默认布局管理器之后,就可以使用 GUI 组件的 setLocation()、setSize()、setBounds() 等布局方法来对 GUI 组件的位置、大小进行设置。表 10-1 列出了 JComponent 及其子类所包含的常用布局方法。

表 10-1 JComponent 及其子类所包含的常用布局方法

函 数	作 用
setLocation(java.awt.Point) setLocation(int, int)	设置组件的坐标位置
setSize(java.awt.Dimension) setSize(int, int)	设置组件的大小
setBounds(java.awt.Rectangle) setBounds(int, int, int, int)	同时设置组件的坐标位置和大小。setBounds(int, int, int, int)的 4 个参数分别是 x、y、width、height,即组件的(x,y)坐标以及组件的 width 和 height

【例 10.3】 无布局管理器布局。

```
import javax.swing.JButton;
import javax.swing.JFrame;
import javax.swing.JTextField;
public class AbsoluteLayoutDemo extends JFrame {
    private JButton button=new JButton("按键");;
    private JTextField textField=new JTextField("文本框");
    public AbsoluteLayoutDemo()
    {
        setSize(500, 200);
        setLocation(400, 400);
        setDefaultCloseOperation(JFrame.EXIT_ON_CLOSE);
        //设置布局管理为 null
        setLayout(null);
        //设置输入框的位置为(20,20),宽 200,高 100
        textField.setBounds(20,20,200,100);
        add(textField);
        //设置按键的位置为(300,50),宽 100,高 20
        button.setLocation(300, 50);
        button.setSize(100, 20);
        add(button);
    }
    public static void main(String[] args) {
        AbsoluteLayoutDemo frame=new AbsoluteLayoutDemo();
        frame.setVisible(true);
    }
}
```

程序运行结果如图 10-6 所示。

无布局管理器布局在对组件的大小和位置的控制上较为灵活,但这种布局方式会导致平台相关,在不同的平台上可能产生不同的显示效果,并且在窗体发生变化时有可能需要进行重新布局。基于布局管理器的布局,就可以很好地解决以上两个问题。

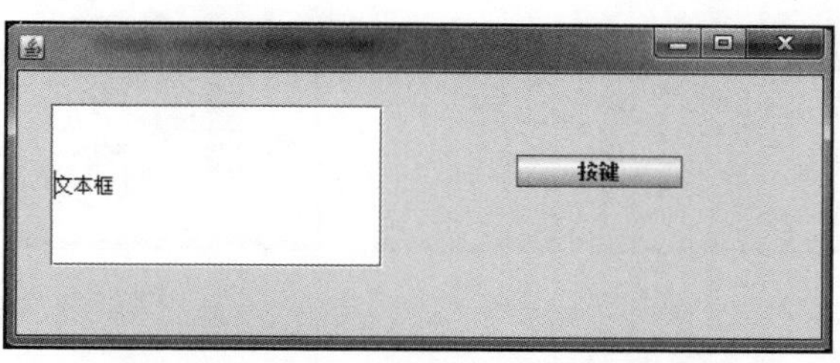

图 10-6　无布局管理器布局的运行结果

10.3.2　FlowLayout

界面采用FlowLayout布局，其GUI组件的放置规律是从左到右、从上到下进行放置。如果界面足够宽，第一个组件先添加到容器中第一行的最左边，后续的组件依次添加到上一个组件的右边，如果当前行已放置不下该组件，则放置到下一行的最左边。

当容器的大小发生变化时，用FlowLayout管理的组件会发生变化，其变化规律是：组件的大小不变，但是相对位置会发生变化。

【例10.4】　使用FlowLayout布局。

```
import java.awt.FlowLayout;
import javax.swing.JButton;
import javax.swing.JFrame;
public class FlowLayoutDemo extends JFrame {
    private JButton button1 = new JButton("First Button");
    private JButton button2 = new JButton("Second Button");
    private JButton button3 = new JButton("Third Button");
    private JButton button4 = new JButton("Fourth Button");
    public FlowLayoutDemo() {
        setSize(300, 150);
        setLocation(400, 400);
        setDefaultCloseOperation(JFrame.EXIT_ON_CLOSE);
        //设置布局方式为FlowLayout
        setLayout(new FlowLayout());
        //添加按键,注意设置布局方式后任何对组件进行设置的方法都会失效
        add(button1);
        add(button2);
        add(button3);
        add(button4);
    }
    public static void main(String arg[]) {
        FlowLayoutDemo frame = new FlowLayoutDemo();
        frame.setVisible(true);
    }
}
```

FlowLayout布局的运行结果如图10-7所示。图10-7(a)是程序默认运行的情况，图10-7(b)是改变窗口大小后的界面情况。可以看出，当窗体的大小发生改变时，程序可以自行调整界面元素的位置布局，而不需要额外编写程序代码。

10.3.3　BorderLayout

BorderLayout布局把界面分成5个区域：North、South、East、West和Center，每个区域

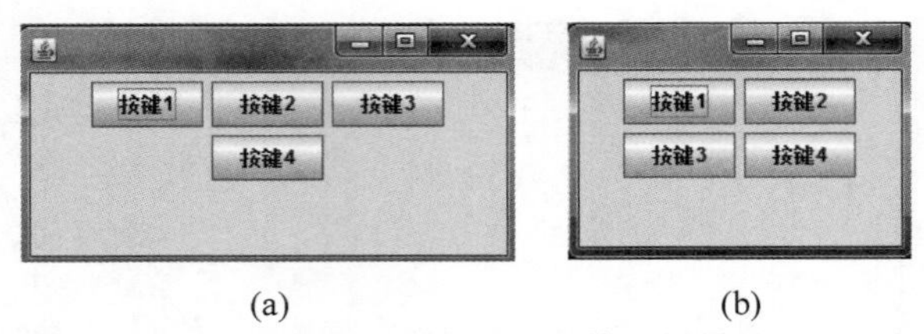

(a) (b)

图 10-7 FlowLayout 布局的运行结果

只能放置一个组件。当界面的大小发生变化时,其变化规律为:组件的相对位置不变,大小发生变化。例如,如果容器变高了,则 North、South 区域不变,West、Center、East 区域变高;如果容器变宽了,则 West、East 区域不变,North、Center、South 区域变宽。由于不一定所有的区域都有组件,如果四周的区域(West、East、North、South 区域)没有组件,则由 Center 区域去补充。

【例 10.5】 使用 BorderLayout 布局。

```
import java.awt.BorderLayout;
import javax.swing.JButton;
import javax.swing.JFrame;
public class BorderLayoutDemo extends JFrame {
    private JButton north = new JButton("北");
    private JButton south = new JButton("南");
    private JButton east = new JButton("东");
    private JButton west = new JButton("西");
    private JButton center = new JButton("中");
    public BorderLayoutDemo() {
        setSize(300, 300);
        setLocationRelativeTo(null);
        setDefaultCloseOperation(JFrame.EXIT_ON_CLOSE);
        //设置布局方式为 BorderLayout
        setLayout(new BorderLayout());
        //添加按键,注意设置布局方式之后任何对
        //组件进行设置的方法,如 setSize、setLocation 等都会失效
        add(north,BorderLayout.NORTH);
        add(south,BorderLayout.SOUTH);
        add(east,BorderLayout.EAST);
        add(west,BorderLayout.WEST);
        add(center,BorderLayout.CENTER);
    }
    public static void main(String arg[]) {
        BorderLayoutDemo frame = new BorderLayoutDemo();
        frame.setVisible(true);
    }
}
```

程序运行结果如图 10-8 所示。图 10-8(a)是程序默认的界面状态,图 10-8(b)是界面大小发生变化之后的状态。

10.3.4 GridLayout

GridLayout 布局管理器将整个界面划分成 N 行 M 列的网格,每个网格的大小相同。布局时,按照组件加入的顺序优先考虑按行布局,当一行布局满之后再布局下一行(每行只能布局 M 个组件)。只有当行列不能满足指定的数值时(N×M 小于组件个数),才按行扩展。例

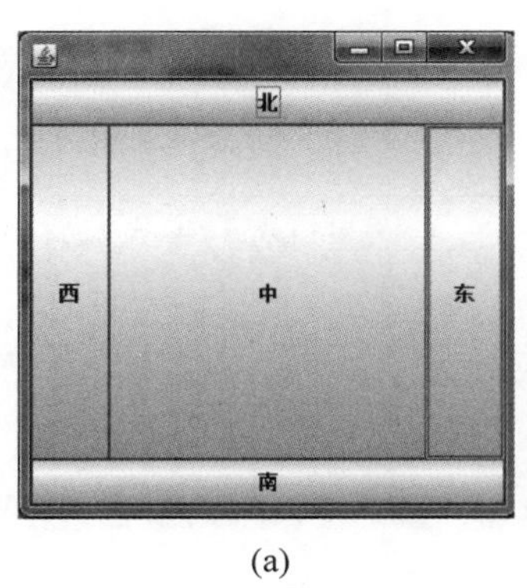

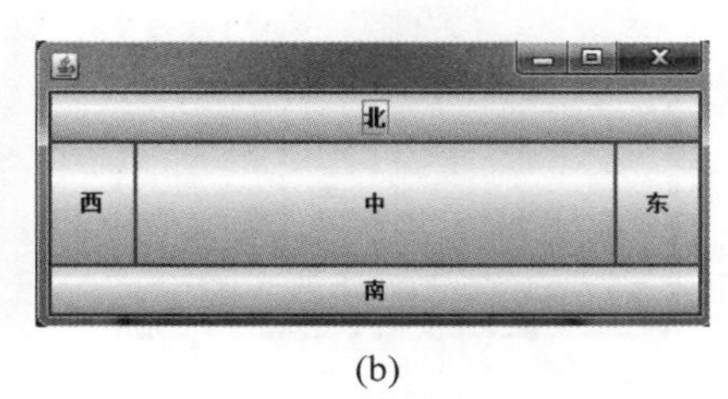

(a) (b)

图 10-8 BorderLayout 布局的运行结果

如,5 个按钮,指定分 2 行 2 列显示,由于 2×2 只能满足 4 个按键,因此自动扩展为 3 列。

【例 10.6】 使用 GridLayout 布局。

```
import java.awt.GridLayout;
import javax.swing.JButton;
import javax.swing.JFrame;
public class GridlayoutDemo extends JFrame {
    private JButton button1 = new JButton("按键 1");
    private JButton button2 = new JButton("按键 2");
    private JButton button3 = new JButton("按键 3");
    private JButton button4 = new JButton("按键 4");
    public GridlayoutDemo() {
        setSize(300, 300);
        setLocation(400, 400);
        setDefaultCloseOperation(JFrame.EXIT_ON_CLOSE);
        //设置布局方式为 GridLayout,2 行、2 列
        setLayout(new GridLayout(2,2));
        //添加组件时不需要设置组件所在行、列
        add(button1);
        add(button2);
        add(button3);
        add(button4);
    }
    public static void main(String arg[]) {
        GridlayoutDemo frame = new GridlayoutDemo();
        frame.setVisible(true);
    }
}
```

程序运行结果如图 10-9 所示。图 10-9(a)是程序默认的界面状态,图 10-9(b)是界面大小发生变化之后的状态。

10.3.5 利用可视化工具进行布局

在 Swing 发展的早期,对界面的布局只能用编写代码的方式完成,因而是一件非常复杂的事情。但随着 Java 开发工具的发展,目前多数集成开发工具都提供了相当好用的可视化布局工具。例如,Eclipse 可以通过安装 WindowBuilder 插件来获得可视化布局工具。下面简单介绍如何用 WindowBuilder 来进行可视化布局。

首先从 Eclipse Marketplace(在 Help 菜单项进入)中安装 WindowBuilder 插件,如图 10-10

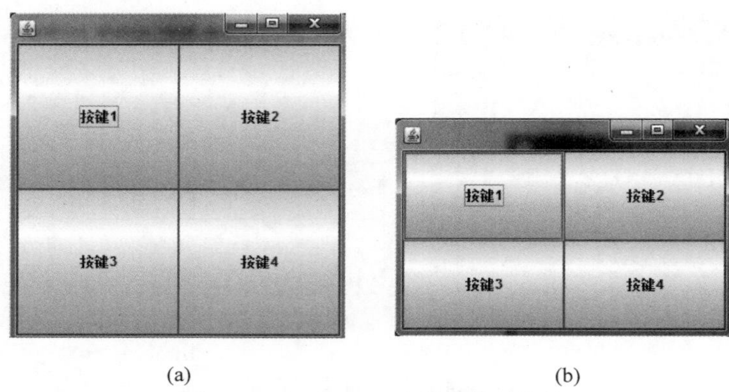

(a) (b)

图 10-9 GridLayout 布局的运行结果

所示。

图 10-10 安装 WindowBuilder 插件

插件安装完毕之后,新建一个 Java 工程(本书为“GUI 程序设计”)。选中该工程,右击,选中菜单项 New,选择 Other。在弹出的对话框中选择 WindowBuilder→Swing Designer→Application Window,如图 10-11(a)所示,最后单击 Next 按钮。在弹出的对话框中输入包名和类名(本书建立的类为 NewWindow),单击 Finish 按钮,新建一个窗体程序,如图 10-11(b)所示。

新建窗体之后,就可以在 Design 视图(图 10-12 所示界面)对窗体进行可视化编辑。默认该界面的右侧是可视化编辑区,左上部是界面结构视图,左下部可以编辑当前选中的 GUI 组件的属性,中间是 GUI 组件,通过拖拽可以将组件拖放到右侧编辑区。

下面演示如何进行界面布局。首先选中窗体,右击,选择 Set Layout 为窗体设置一个布局管理器(图 10-13(a))。可以看到 Swing 除了 FlowLayout、BorderLayout 和 GridLayout 外

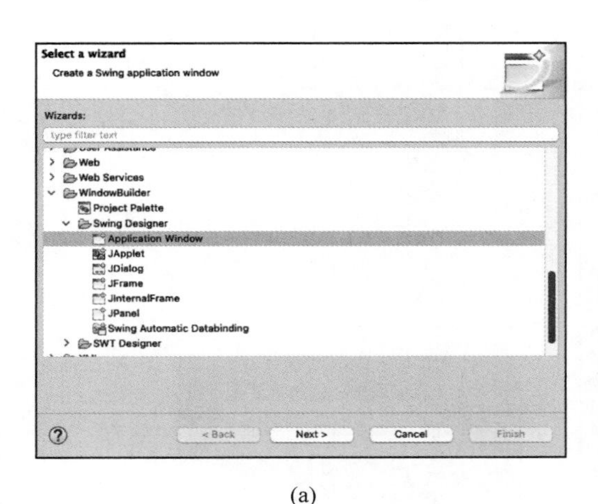

(a)

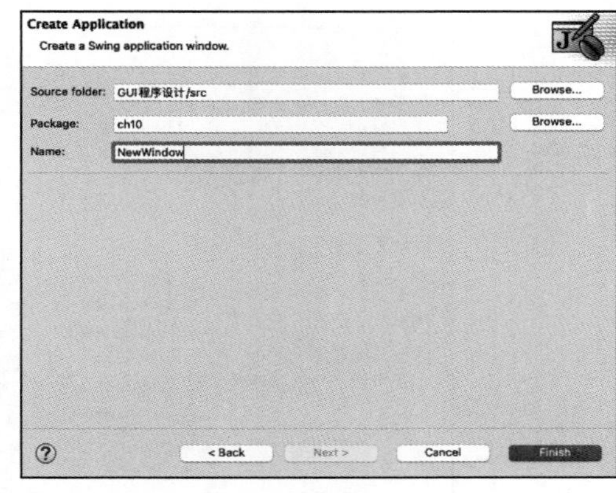

(b)

图 10-11　新建窗体

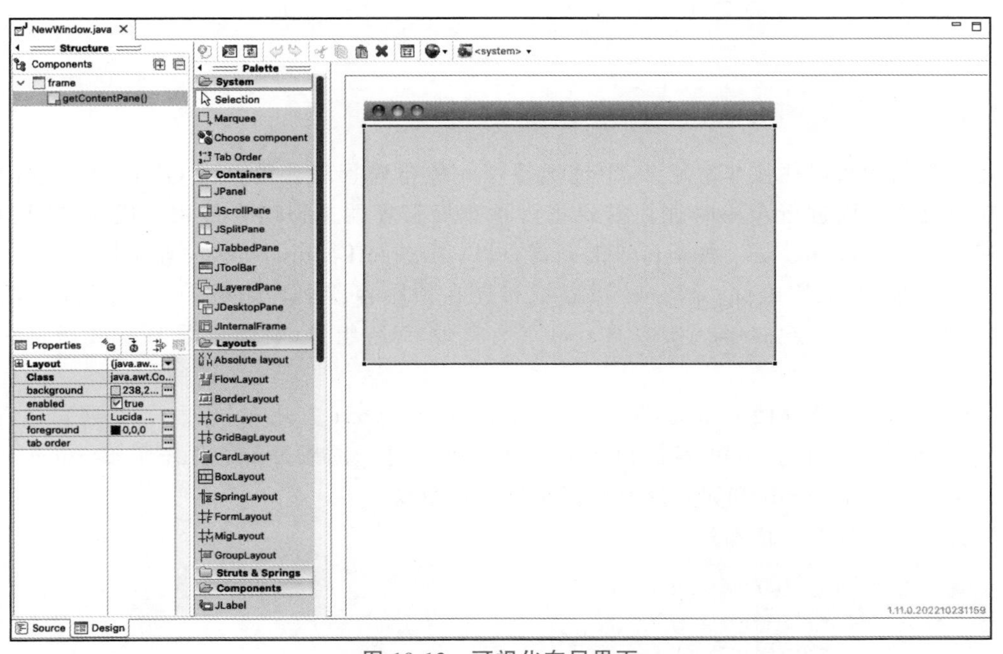

图 10-12　可视化布局界面

还有许多其他布局管理器。这里选择 GroupLayout。GoupLayout 以 Group(组)为单位来管理布局,也就是把多个组件(如 JLable、JButton)按区域划分到不同的 Group(组),再根据各个 Group(组)相对于水平轴(Horizontal)和垂直轴(Vertical)的排列方式来管理。之所以选择这种布局管理器,是因为这种布局管理器就是为了便于可视化工具进行布局而发明的。

接下来可以从工具栏上选择希望布局的组件,将其拖到界面上,并调整大小设置属性(图 10-13(b))。在界面布局的同时,WindowBuilder 会生成对应的 Java 代码(切换到 Source 视图可以查看代码)。

读者如果使用 WindowBuilder 工具,就会发现利用可视化工具进行界面布局是一件相当容易的事情。但本书为什么还要向读者介绍 FlowLayout、BorderLayout 和 GridLayout 这些布局管理器呢?其原因是希望读者知其然也知其所以然。工具毕竟只是一种辅助,某些独特的界面效果,仍然需要通过编写代码来实现。

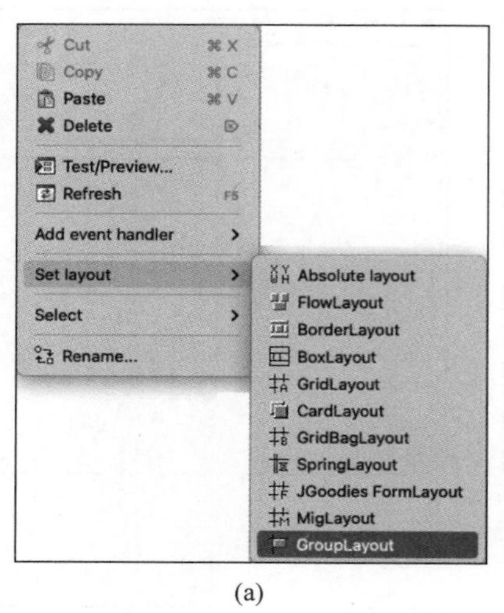

(a)

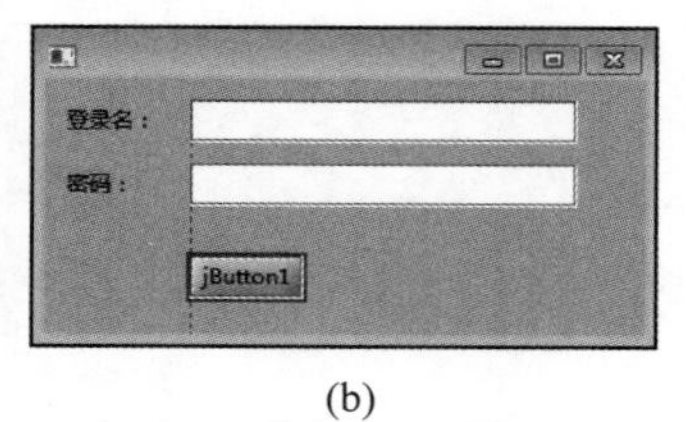

(b)

图 10-13 布局界面

10.3.6 实用案例 10.1:布局复杂界面

复杂界面的布局往往非常复杂,单纯地使用一种布局管理器很难对 GUI 组件进行布局,因此在对复杂界面进行布局时往往需要将多种布局管理器进行组合使用。通常使用 JPanel 类来辅助布局。JPanel 是一种不可见的容器,可以通过 setLayout 方法设置布局方式,也可以用 add 方法添加 GUI 组件。JPanel 可以被布局在顶层容器上。如果 JPanel 上没有任何的 GUI 组件,则显示为空白区域;如果 JPanel 上有其他 GUI 组件,就会依据 JPanel 指定的布局方式进行布局。

下面是一个复杂界面的布局,顶层容器采用 GridLayout 方式布局(2 行、2 列),包含了 4 个 JPanel 容器。4 个 JPanel 容器以从左到右、从上到下的顺序,分别采用 BorderLayout、FlowLayout、GridLayout 和无布局管理器布局进行布局。

【例 10.7】 复杂界面布局。

```
import java.awt.BorderLayout;
import java.awt.FlowLayout;
import java.awt.GridLayout;
import javax.swing.JButton;
import javax.swing.JFrame;
import javax.swing.JPanel;
import javax.swing.JTextField;
public class ComplexLayoutDemo extends JFrame {
    private JPanel panel1=new JPanel();
    private JPanel panel2=new JPanel();
    private JPanel panel3=new JPanel();
    private JPanel panel4=new JPanel();
    public ComplexLayoutDemo()
    {
        setSize(500, 500);
        setLocation(400, 400);
        setDefaultCloseOperation(JFrame.EXIT_ON_CLOSE);
        //对 panel1 进行布局
        layoutPanel1();
```

```
        //对 panel2 进行布局
        layoutPanel2();
        //对 panel3 进行布局
        layoutPanel3();
        //对 panel4 进行布局
        layoutPanel4();
        //对顶层容器进行布局,采用 GridLayout,2 行 2 列
        setLayout(new GridLayout(2,2));
        add(panel1);
        add(panel2);
        add(panel3);
        add(panel4);
    }
    private void layoutPanel1() {
        JButton north = new JButton("北");
        JButton south = new JButton("南");
        JButton east = new JButton("东");
        JButton west = new JButton("西");
        JButton center = new JButton("中");
        //panel1 采用 BorderLayout 布局
        panel1.setLayout(new BorderLayout());
        panel1.add(north,BorderLayout.NORTH);
        panel1.add(south,BorderLayout.SOUTH);
        panel1.add(east,BorderLayout.EAST);
        panel1.add(west,BorderLayout.WEST);
        panel1.add(center,BorderLayout.CENTER);
    }
    private void layoutPanel2() {
        JButton button1 = new JButton("按键 1");
        JButton button2 = new JButton("按键 2");
        JButton button3 = new JButton("按键 3");
        JButton button4 = new JButton("按键 4");
        //panel2 采用 FlowLayout 布局
        panel2.setLayout(new FlowLayout());
        panel2.add(button1);
        panel2.add(button2);
        panel2.add(button3);
        panel2.add(button4);
    }
    private void layoutPanel3() {
        JButton button1 = new JButton("按键 1");
        JButton button2 = new JButton("按键 2");
        JButton button3 = new JButton("按键 3");
        JButton button4 = new JButton("按键 4");
        //panel3 采用 GridLayout 布局,2 行 2 列
        panel3.setLayout(new GridLayout(2,2));
        panel3.add(button1);
        panel3.add(button2);
        panel3.add(button3);
        panel3.add(button4);
    }
    private void layoutPanel4() {
        JButton button=new JButton("按键");;
        JTextField textField=new JTextField("文本框");
```

```
        //panel2 采用无布局管理器布局
        panel4.setLayout(null);
        button.setLocation(20, 20);
        button.setSize(100, 20);
        textField.setBounds(20,50,200,100);
        panel4.add(button);
        panel4.add(textField);
    }
    public static void main(String[] args) {
        ComplexLayoutDemo frame=new ComplexLayoutDemo();
        frame.setVisible(true);
    }
}
```

程序运行结果如图 10-14 所示。

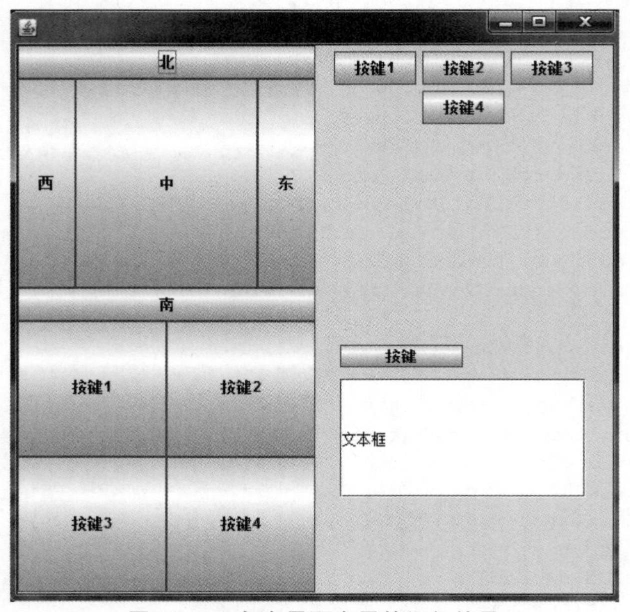

图 10-14　复杂界面布局的运行结果

10.4　响应用户事件

10.4.1　事件处理的基本过程

响应用户事件是 GUI 程序设计的重要内容。用户事件包括鼠标的点击、GUI 组件值的改变、焦点的获取或者丢失、键盘输入等。不同的 GUI 组件可以响应不同类型的事件。当用户事件发生时,该事件会涉及以下三方面的对象。

(1) 事件源(Source):产生事件的对象。这类对象是 GUI 组件,包括 JButton、JTextField 等。

(2) 事件(Event):发生在用户界面上的,由用户交互行为所产生的一种效果。这类对象由 java.awt.AWTEvent 及其子类构成,例如 KeyEvent、MouseEvent 等。

(3) 事件监听器(Listener):接受事件并对其进行处理的对象。这类对象由实现了 java.

util. EventListener 接口的类构成，例如 ActiontListener、MouseListener 等。

举例来说，当用户用鼠标点击一个按键时，鼠标点击动作会触发事件，按键就是事件源，而对该事件进行处理的对象就是事件监听器。通常一个事件源可以触发多种类型的事件，每一个事件可以由一个或者多个事件监听器进行处理。Swing 中每个 GUI 组件(JComponent 及其子类)都有若干名为 addXXXListener 的方法，例如 JButton 类有 addActionListener()、addKeyListener()等 addXXXListener 方法。这类方法被用于注册特定事件的监听器。其中 addActionListener 用于注册按键点击的事件监听器，addKeyListener 用于注册键盘输入的事件监听器。下面以例 10.1 中 JButton 组件的点击事件为例，说明如何编写事件处理程序。例 10.1 中涉及事件处理的代码如下：

```
button.addActionListener(new ActionListener(){
    public void actionPerformed(ActionEvent event) {
        JOptionPane.showMessageDialog(null, "点击了按键!");
    }

});
```

该事件处理过程，首先调用 JButton 的 addActionListener 方法，添加一个事件监听器。负责处理按键点击事件的监听器是实现了 ActionListener 接口的子类。例 10.1 利用匿名类实现了该接口。ActionListener 接口包含一个唯一的方法，即

```
public void actionPerformed(ActionEvent e)
```

程序通过实现该方法来处理按键点击事件。actionPerformed()方法的参数 e 的类型为 ActionEvent。ActionEvent 是对应点击事件的类。通过调用对象 e 的方法可以获取事件的相关属性。例如，调用 getSource()方法将返回事件发生的对象。actionPerformed()方法的实现可以简单，也可以很复杂，具体的情况要根据程序所需要实现的事件处理来确定。例 10.8 给出了另一个按键事件处理程序。

【例 10.8】 一个简单的按键处理程序。

```
import java.awt.FlowLayout;
import java.awt.event.ActionEvent;
import java.awt.event.ActionListener;
import javax.swing.JButton;
import javax.swing.JFrame;
public class EventDemo extends JFrame{
    JButton button=new JButton("点我!");
    public EventDemo() {
        setSize(300,300);
        setLocationRelativeTo(null);
        setDefaultCloseOperation(JFrame.EXIT_ON_CLOSE);
        //设置按键事件,使用了匿名类
        button.addActionListener(new ActionListener(){
            public void actionPerformed(ActionEvent e) {
                //获取被点击的按键
                JButton clickedButton=(JButton) e.getSource();
                //改变被点击按键的标题
                clickedButton.setText("我被点了!");
            }
        });
        setLayout(new FlowLayout());
        add(button);
```

```
        }
        public static void main(String[] args) {
            EventDemo frame=new EventDemo();
            frame.setVisible(true);
        }
    }
```

在例 10.8 中匿名类实现了 ActionListener 接口,并在 actionPerformed 中通过 ActionEvent 的 getSource()方法获取被点击的按键,然后调用按键的 setText 方法替换原有按键的标题。图 10-15(a)是点击按键前的界面,图 10-15(b)是点击按键之后的界面。事件监听器不仅可以实现为匿名类,还可以作为内部类、外部类,也可以作为主类。这几种方式实现的监听器在对事件的处理上并没有本质的区别。

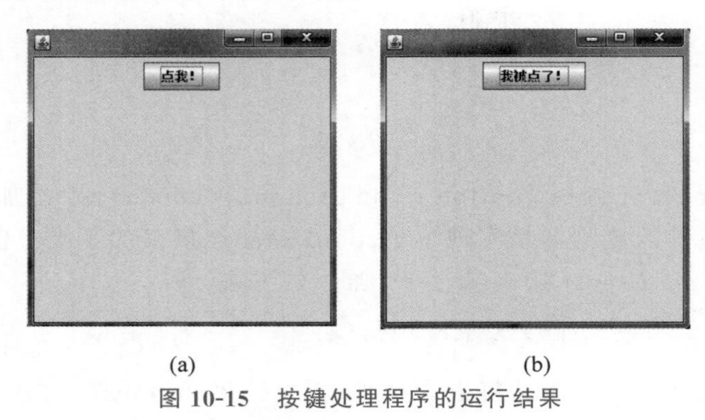

图 10-15 按键处理程序的运行结果

10.4.2 常用事件与事件监听器类

Swing 事件类的基类是 java.awt.AWTEvent,它是 java.util.EventObject 的子类。基类 EventObject 定义了方法 getSource(),该方法返回产生或触发事件的对象。AWTEvent 定义了方法 getID(),该方法的返回值用来区别用同一个事件类所代表的不同类型的事件。AWTEvent 之下常见的事件类及其结构关系,如图 10-16 所示。不同的事件类,除了拥有父类的方法,还定义一些与该事件相关的方法。例如,MouseEvent 有方法 getX()、getY()和 getClickCount(),用于返回鼠标事件产生的坐标位置和鼠标点击的次数。

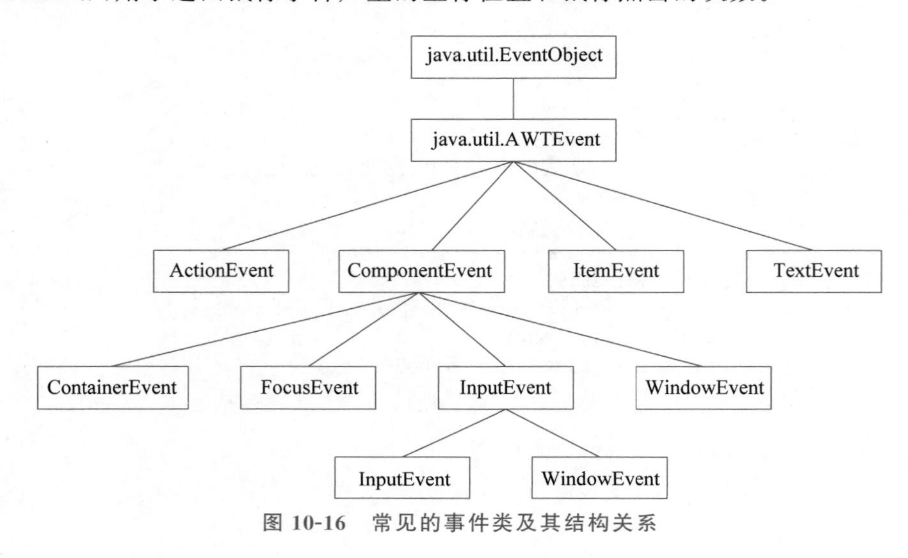

图 10-16 常见的事件类及其结构关系

每个事件类都有对应的事件监听器接口，接口中定义了事件发生时可调用的方法。在处理用户事件时，需要根据事件类型向GUI组件注册对应的事件监听器（调用对应的addXXXListener方法），也就是要实现相应的事件监听接口。表10-2给出了图10-16所示事件类对应的事件监听器接口以及事件产生的原因。

表10-2 常见的事件监听器接口以及事件产生的原因

事件类/监听器接口	接口中声明的方法	事件产生原因
ComponentEvent/ ComponentListener	componentMoved(ComponentEvent e)	移动组件时
	componentHidden(ComponentEvent e)	隐藏组件时
	componentResized(ComponentEvent e)	改变组件大小时
	componentShown(ComponentEvent e)	显示组件时
ContainerEvent ContainerListener	componentAdded(ComponentEvent e)	添加组件时
	ComponentRemoved(ComponentEvent e)	删除组件时
WindowEvent WindowListener	windowOpened(WindowEvent e)	打开窗口时
	windowActivated(WindowEvent e)	激活窗口时
	windowDactivated(WindowEvent e)	窗口失去焦点时
	windowClosing(WindowEvent e)	关闭窗口时
	windowClosed(WindowEvent e)	关闭窗口后
	windowIconified(WindowEvent e)	窗口最小化时
	windowDeiconified(WindowEvent e)	当窗口从最小恢复到正常大小时
ActionEvent/ ActionListener	actionPerformed(ActionEvent e)	单击按钮，文本框中单击回车键，双击列表框选择菜单项等
TextEvent/ TextListener	textValueChanged(TextEvent e)	文本框中修改内容
ItemEvent/ ItemListener	itemStateChanged(ItemEvent e)	选择复选框、选择框，单击列表框，选中带复选框的菜单项
MouseEvent/ MouseMotionListener	mouseDragged(MouseEvent e)	鼠标拖动时
	mouseMoved(MouseEvent e)	鼠标移动时
MouseEvent/ MouseListener	mouseClicked(MouseEvent e)	单击鼠标时
	mouseEntered(MouseEvent e)	鼠标进入时
	mouseExited(MouseEvent e)	鼠标离开时
	mousePressed(MouseEvent e)	按下鼠标时
	mouseReleased(MouseEvent e)	放开鼠标时
KeyEvent/ KeyListener	keyPresssed(KeyEvent e)	按下键盘时
	keyReleased(KeyEvent e)	释放键盘时
	keyTyped(KeyEvent e)	按下键盘按键时

续表

事件类/监听器接口	接口中声明的方法	事件产生原因
FocusEvent/ FocusListener	focusGained(FocusEvent e)	获得焦点时
	focusLost(FocusEvent e)	失去焦点时

10.4.3 键盘与鼠标事件

键盘事件和鼠标事件是 GUI 程序中最常见的两类事件。参见表 10-2,与键盘事件相关的监听器是 KeyListener,与鼠标事件相关的监听器包括 MouseListener、MouseMotionListener 和 MouseWheelListener。为了处理相应的事件,需要调用相应的 addXXXListener 方法,添加相应的事件监听器。

在处理事件时,我们发现事件监听器接口通常有好几个方法,例如 KeyListener 有 3 个方法,如果只关心"按下键盘",而不关心其他两种情况,也必须实现 KeyListener 的 3 个方法。为此 Java 提供了事件适配器类来简化这个问题。java.awt.event 包中定义的事件适配器类包括:①ComponentAdapter 类,实现了 ComponentListener;②ContainerAdapter 类,实现了 ContainerListener;③FocusAdapter 类,实现了 FocusListener;④KeyAdapter 类,实现了 KeyListener;⑤MouseAdapter 类,实现了 MouseListener 和 MouseMotionListener;⑥WindowAdapter 类,实现了 WindowListener。

使用适配器的过程与实际监听器类似,但只需重写事件处理所关心的方法,其他方法不用实现,这样就简化了程序代码。同时适配器是一个类,而不是接口,但使用适配器时仍然必须继承对应的适配器类。键盘和鼠标事件所对应的适配器类分别为 KeyAdapter 和 MouseAdapter。例 10.9 和例 10.10 简要演示了键盘和鼠标事件的处理过程。出于简化程序的目的,示例使用了 KeyAdapter 和 MouseAdapter。

【例 10.9】 键盘事件处理程序。

```
import java.awt.BorderLayout;
import java.awt.event.KeyAdapter;
import java.awt.event.KeyEvent;
import javax.swing.JFrame;
import javax.swing.JLabel;
public class KeyEventDemo extends JFrame{
    JLabel label=new JLabel("按下了按键:");
    public KeyEventDemo() {
        setSize(300,300);
        setLocation(400, 400);
        setDefaultCloseOperation(JFrame.EXIT_ON_CLOSE);
        this.addKeyListener(new KeyAdapter(){
            public void keyPressed(KeyEvent event) {
                switch(event.getKeyCode())
                {
                    case KeyEvent.VK_UP:
                        label.setText("按下了按键:UP");
                        break;
                    case KeyEvent.VK_DOWN:
                        label.setText("按下了按键:DOWN");
```

```
                        break;
                    case KeyEvent.VK_LEFT:
                        label.setText("按下了按键:LEFT");
                        break;
                    case KeyEvent.VK_RIGHT:
                        label.setText("按下了按键:RIGHT");
                        break;
                    default:
                        label.setText("按下了按键:"+event.getKeyChar());
                }
            }
        });
        setLayout(new BorderLayout());
        add(label,BorderLayout.CENTER);
    }
    public static void main(String[] args) {
        KeyEventDemo frame=new KeyEventDemo();
        frame.setVisible(true);
    }
}
```

例 10.9 只重写了 KeyAdapter 的 keyPresssed()方法,用于处理按下键盘按键的事件。在事件处理过程中使用了 KeyEvent。KeyEvent 是对键盘事件的封装,在 KeyEvent 中有两个常用的方法。

(1) getKeyChar():获取触发事件按键对应的字符。例如当按下按键“d”时,所获得的字符就是“d”。

(2) getKeyCode():获取触发事件按键对应的键值。所谓键值,在 KeyEvent 中有若干常量与之对应。例如 KeyEvent.VK_UP 对应键盘方向键“上”的键值,KeyEvent.VK_DOWN 对应键盘方向键“下”的键值,按键“d”的键值是 KeyEvent.VK_D。

运行例 10.10 将显示图 10-17 所示窗体。图 10-17(a)是按下键盘按键“a”所显示的结果,图 10-17(b)是按下键盘方向键“下”所显示的结果。

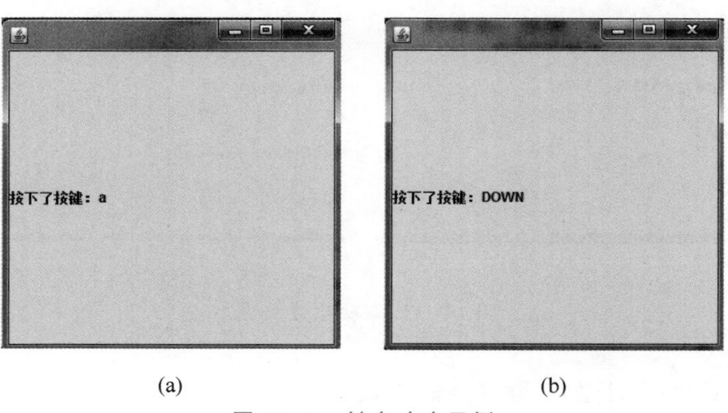

(a)　　　　(b)

图 10-17　键盘响应示例

【例 10.10】　鼠标事件处理程序。

```
import java.awt.BorderLayout;
import java.awt.event.MouseAdapter;
import java.awt.event.MouseEvent;
import javax.swing.JFrame;
```

```
    import javax.swing.JLabel;
    public class MouseEventDemo extends JFrame{
        JLabel label=new JLabel("");
        public MouseEventDemo() {
            setSize(300,300);
            setLocation(400, 400);
            setDefaultCloseOperation(JFrame.EXIT_ON_CLOSE);
            this.addMouseListener(new MouseAdapter(){
                public void mouseClicked(MouseEvent event) {
                    label.setText("鼠标在"+event.getX()+","+event.getY()+"进行了点击");
                }
            });
            this.addMouseMotionListener(new MouseAdapter(){
                public void mouseMoved(MouseEvent event) {
                    label.setText("鼠标移动到了"+event.getX()+","+event.getY());
                }
            });
            setLayout(new BorderLayout());
            add(label,BorderLayout.CENTER);
        }
        public static void main(String[] args) {
            MouseEventDemo frame=new MouseEventDemo();
            frame.setVisible(true);
        }
    }
```

例 10.10 使用 MouseAdapter 为窗体添加了 MouseListener 和 MouseMotionListener 两种类型的监听器。在事件处理中用 MouseEvent 的 getX()和 getY()方法获得事件发生时鼠标所在的位置。程序运行结果如图 10-18 所示。图 10-18(a)是鼠标点击时的事件响应,图 10-18(b)是鼠标移动时的事件响应。

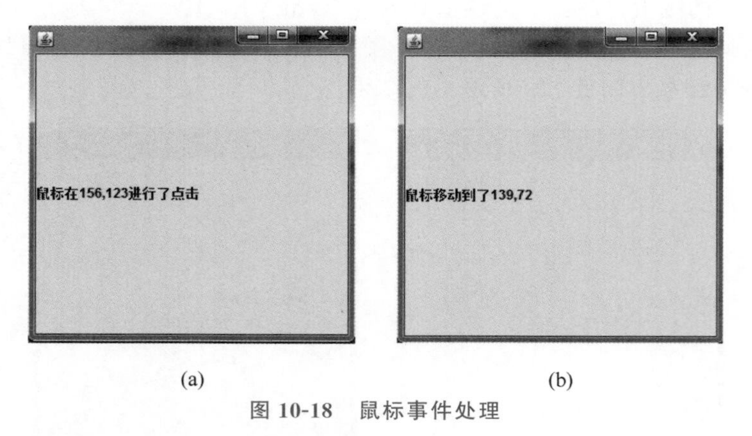

图 10-18　鼠标事件处理

10.4.4　实用案例 10.2:用鼠标绘图

下面的例子使用鼠标事件和窗体事件实现用鼠标绘图。当鼠标在窗体上移动时,同时绘制出图像。本例使用了 BufferedImage 和 Graphics 类,这两个类主要用于图形绘制。

【例 10.11】　用鼠标绘图。

```
    import java.awt.Color;
    import java.awt.Graphics;
```

```
import java.awt.event.ComponentAdapter;
import java.awt.event.ComponentEvent;
import java.awt.event.MouseAdapter;
import java.awt.event.MouseEvent;
import java.awt.image.BufferedImage;

import javax.swing.JFrame;

public class PaintFrame extends JFrame {
    //记录绘图内容
    private BufferedImage canvas;
    //记录上一次鼠标的 x 位置
    private int lastX = -1;
    //记录上一次鼠标的 y 位置
    private int lastY = -1;

    public PaintFrame() {
        //窗口大小改变时间
        this.addComponentListener(new ComponentAdapter() {
            public void componentResized(ComponentEvent e) {
                BufferedImage newCanvas = new BufferedImage(getWidth(),
getHeight(), BufferedImage.TYPE_INT_RGB);
                if (canvas != null) {
                    Graphics g = newCanvas.getGraphics();
                    g.drawImage(canvas, 0, 0, null);
                }
                canvas = newCanvas;
            }
        });
        //鼠标移动事件
        this.addMouseMotionListener(new MouseAdapter() {
            public void mouseMoved(MouseEvent e) {
                int x = e.getX();
                int y = e.getY();
                if (lastX == -1 && lastY == -1) {
                    lastX = x;
                    lastY = y;

                } else {
                    Graphics g = canvas.getGraphics();
                    g.setColor(Color.RED);
                    g.drawLine(lastX, lastY, x, y);
                    lastX = x;
                    lastY = y;
                    //重绘触发 paint
                    repaint();

                }
            }

        });

    }
    //在窗口上绘制图片
```

```
    public void paint(Graphics g) {
        g.drawImage(canvas, 0, 0, null);
    }

    public static void main(String[] args) {
        PaintFrame frame = new PaintFrame();
        frame.setDefaultCloseOperation(JFrame.EXIT_ON_CLOSE);
        frame.setSize(500, 500);
        frame.setVisible(true);
    }
}
```

实用案例 10.2 程序运行结果如图 10-19 所示。

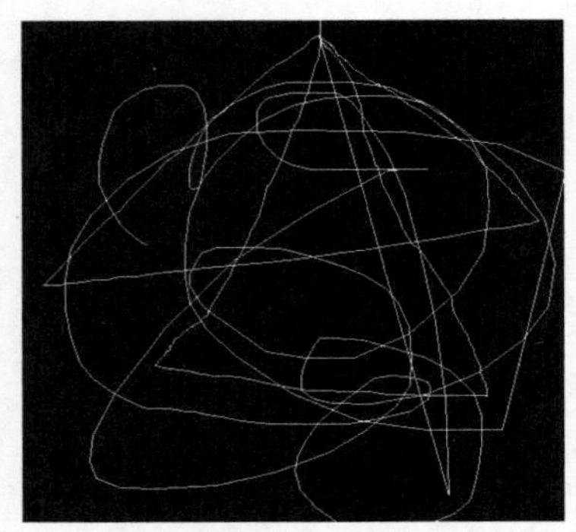

图 10-19　实用案例 10.2 程序运行结果

10.5　高级组件 JTree 和 JTable

10.5.1　JTree 组件

JTree 是 Swing 中的高级界面组件,用于将数据显示为树形结构。以下是其常用的构造方法。

(1) JTree():返回一棵默认的示例树 JTree。

(2) JTree(Hashtable<?,? > value):从 Hashtable 创建 JTree,此方法构造的 JTree 不显示根节点。

(3) JTree(Object[] value):用 Object 数组的每个元素作为 JTree 的子节点,此方法构造的 JTree 不显示根节点。

(4) JTree(TreeModel newModel):使用指定的 TreeModel 数据模型创建 JTree,此方法构造的 JTree 显示根节点。

(5) JTree(TreeNode root):指定 root 作为 JTree 的根节点,并用 root 的子节点作为 JTree 的子节点。

(6) JTree(Vector<? > value):用 Vector 的元素作为 JTree 的子节点,此方法构造的 JTree 不显示根节点。

【例 10.12】　建立一棵默认的树。

```
import javax.swing.JFrame;
```

```
import javax.swing.JScrollPane;
import javax.swing.JTree;
public class SimpleTree extends JFrame{
    public SimpleTree() {
        JTree tree = new JTree();
        JScrollPane scrollPane = new JScrollPane();
        scrollPane.setViewportView(tree);
        add(scrollPane);
    }
    public static void main(String[] args) {
        SimpleTree frame=new SimpleTree();
        frame.setTitle("SimpleTree");
        frame.setSize(300, 300);
        frame.setVisible(true);
    }
}
```

例 10.12 创建了一棵默认的树，如图 10-20 所示。但通常在使用树时，希望树节点的名称、结构都是自定义的。系统默认的树显然不能够满足需求。例 10.13 通过 HashTable 构造 JTree。

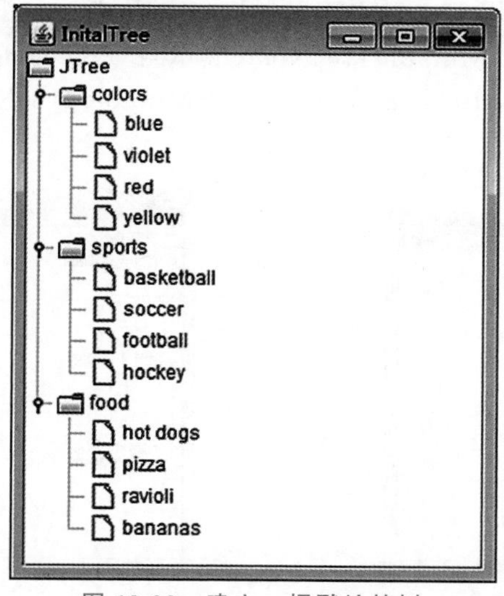

图 10-20　建立一棵默认的树

【例 10.13】 使用 HashTable 构造树。

```
import java.util.Hashtable;
import javax.swing.JFrame;
import javax.swing.JScrollPane;
import javax.swing.JTree;
public class HashConstructTree extends JFrame{
  public HashConstructTree(){
    String[] s1={"A","B","C"};
    String[] s2={"a","b","c"};
    Hashtable<String,String[]> hashtable=new Hashtable<String,String[]>();
    hashtable.put("大写字母",s1);
    hashtable.put("小写字母",s2);
    JTree tree=new JTree(hashtable);
```

```
        JScrollPane scrollPane=new JScrollPane();
        scrollPane.setViewportView(tree);
          add(scrollPane);
    }
    public static void main(String[] args){
        HashConstructTree frame=new HashConstructTree();
        frame.setTitle("HashConstructTree");
        frame.setSize(300, 300);
        frame.setVisible(true);
        frame.setDefaultCloseOperation(JFrame.EXIT_ON_CLOSE);
    }
}
```

例 10.13 创建的树如图 10-21 所示,在实际使用中 HashTable 可嵌套使用构造复杂结构的树。

使用 TreeNode 构造树,是所有方法中最为常用的一种。在这种方法中,JTree 上的每一个节点就是一个实现了 TreeNode 的实例。TreeNode 是一个接口,里面定义了若干有关节点的方法,例如判断是否为树叶节点、有几个子节点、获得父节点等。在实际的应用上,一般不会直接实现 TreeNode,而是采用 DefaultMutableTreeNode 类。DefaultMutableTreeNode 类实现了 MutableTreeNode 接口,MutableTreeNode 继承自 TreeNode,添加了新增节点、删除节点、设置节点等处理方法。例 10.14 演示了 DefaultMutableTreeNode 的用法,其运行效果如图 10-22 所示。

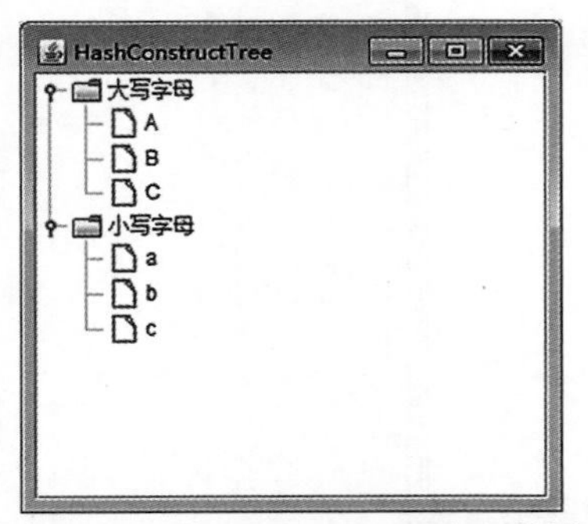

图 10-21 使用 HashTable 建立的树

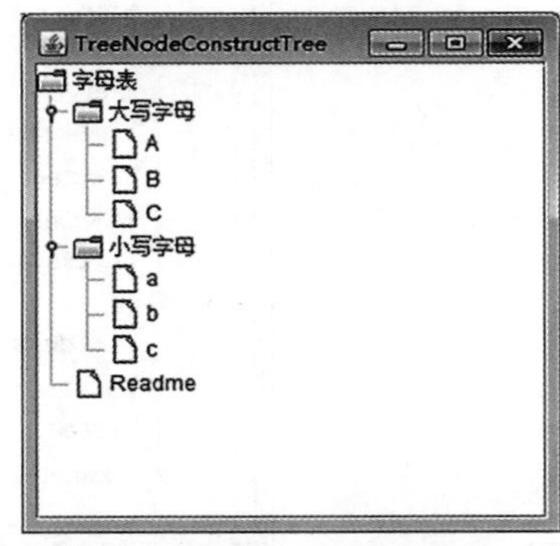

图 10-22 使用 DefaultMutableTreeNode 构造 JTree

【例 10.14】 使用 DefaultMutableTreeNode 构造 JTree。

```
import javax.swing.JFrame;
import javax.swing.JScrollPane;
import javax.swing.JTree;
import javax.swing.tree.DefaultMutableTreeNode;
public class TreeNodeConstructTree extends JFrame{
    public TreeNodeConstructTree() {
        DefaultMutableTreeNode root = new DefaultMutableTreeNode("字母表");
        DefaultMutableTreeNode node1 = new DefaultMutableTreeNode("大写字母");
        DefaultMutableTreeNode node2 = new DefaultMutableTreeNode("小写字母");
        DefaultMutableTreeNode node3 = new DefaultMutableTreeNode("Readme");
        root.add(node1);
```

```
        root.add(node2);
        root.add(node3);
        node1.add(new DefaultMutableTreeNode("A"));
        node1.add(new DefaultMutableTreeNode("B"));
        node1.add(new DefaultMutableTreeNode("C"));
        node2.add(new DefaultMutableTreeNode("a"));
        node2.add(new DefaultMutableTreeNode("b"));
        node2.add(new DefaultMutableTreeNode("c"));
        JTree tree = new JTree(root);
        JScrollPane scrollPane = new JScrollPane();
        scrollPane.setViewportView(tree);
        add(scrollPane);
    }
    public static void main(String[] args) {
        TreeNodeConstructTree frame=new TreeNodeConstructTree();
        frame.setTitle("TreeNodeConstructTree");
        frame.setSize(300, 300);
        frame.setVisible(true);
        frame.setDefaultCloseOperation(JFrame.EXIT_ON_CLOSE);
    }
}
```

10.5.2 JTable组件

JTable表格组件以表格的形式展示数据。以下是JTable常用的构造方法。

(1) JTable()：使用默认模型创建JTable。

(2) JTable(int numRows,int numColumns)：创建具有numRows行、numColumns列的JTable。

(3) JTable(Object[][] rowData,Object[][] columnNames)：建立一个显示二维数组数据的表格，其中rowData是表格的数据，columnNames是表格的列名称。

(4) JTable(TableModel dm)：使用TableModel创建JTable。

下面演示使用JTable(Object[][] rowData,Object[][] columnNames)构造的JTable。

【例10.15】 使用二维数组构造JTable。

```
import javax.swing.JFrame;
import javax.swing.JScrollPane;
import javax.swing.JTable;

public class SimpleTable extends JFrame {
    public SimpleTable() {
        String[] columnNames = { "First Name", "Last Name", "Sport",
                "#of Years", "Vegetarian" };                //表头名称
        Object[][] data = {
                { "Kathy", "Smith", "Snowboarding", new Integer(5),
                        new Boolean(false) },
                { "John", "Doe", "Rowing", new Integer(3), new Boolean(true) },
                { "Sue", "Black", "Knitting", new Integer(2),
                        new Boolean(false) },
                { "Jane", "White", "Speed reading", new Integer(20),
                        new Boolean(true) },
                { "Joe", "Brown", "Pool", new Integer(10), new Boolean(false) } };
                                                            //表数据
```

```
        JTable table = new JTable(data, columnNames);
        JScrollPane scrollPane = new JScrollPane(table);
        add(scrollPane);
    }

    public static void main(String[] args) {
        SimpleTable frame = new SimpleTable();
        frame.setSize(300, 300);
        frame.setVisible(true);
        frame.setTitle("Simple Table");
        frame.setDefaultCloseOperation(JFrame.EXIT_ON_CLOSE);
    }
}
```

程序运行结果如图 10-23 所示。

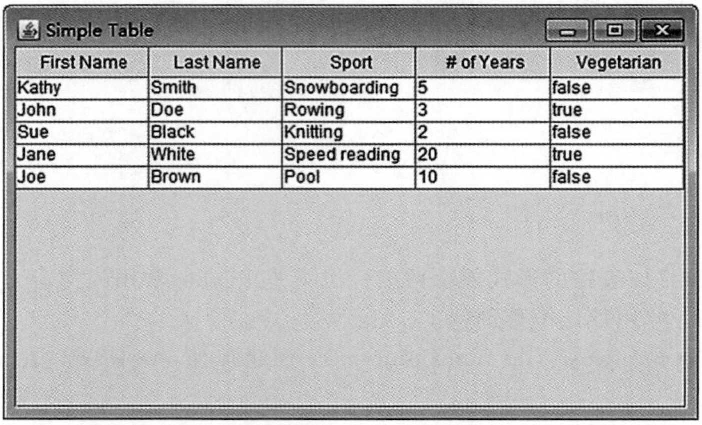

Simple Table

First Name	Last Name	Sport	# of Years	Vegetarian
Kathy	Smith	Snowboarding	5	false
John	Doe	Rowing	3	true
Sue	Black	Knitting	2	false
Jane	White	Speed reading	20	true
Joe	Brown	Pool	10	false

图 10-23 使用二维数组构造 JTable

上例中数组 data 包含了构成表格的数据,columnNames 对应表格的表头部分。

由于 Java Swing 采用了 MVC 的设计模式,所以 JTable 只用于视图展示,并不存储数据,真正用来存储和维护数据的是 TableModel 这个接口的实现类。在所有的构造方法中,TableModel 构造 JTable 是较为常用的一种。

TableModel 是一个接口,在这个接口里面定义了若干的方法,包括了存取表格单元的内容、计算表格的列数等基本存取操作,让设计者可以简单地利用 TableModel 制作想要的表格。Swing 提供了 AbstractTableModel 抽象类。该类实现了 TableModel 的方法,用户可以很有弹性地构造自己的表格模式。例 10.16 演示如何使用 AbstractTableModel 构造 JTable。

【例 10.16】 使用 AbstractTableModel 构造 JTable。

```
import javax.swing.JFrame;
import javax.swing.JScrollPane;
import javax.swing.JTable;

public class AbstractTableModelDemo extends JFrame{
public AbstractTableModelDemo() {
    MyTable myTable=new MyTable();
    JTable t=new JTable(myTable);
    JScrollPane s = new JScrollPane(t);
    add(s);
```

```
    }
    public static void main(String args[]) {
        AbstractTableModelDemo frame=new AbstractTableModelDemo();
        frame.setTitle("JTable");
        frame.setSize(300, 300);
        frame.setVisible(true);
        frame.setDefaultCloseOperation(JFrame.EXIT_ON_CLOSE);
        }
    }
    import javax.swing.table.AbstractTableModel;
    public class MyTable extends AbstractTableModel{
        String[] columnNames = {"First Name", "Last Name",  "Sport",  "#of Years",
    "Vegetarian"};                                              //表头名称
        Object[][] data = {                                     //表数据
            {"Kathy", "Smith", "Snowboarding", new Integer(5), new Boolean(false)},
            {"John", "Doe",  "Rowing", new Integer(3), new Boolean(true)},
            {"Sue", "Black", "Knitting", new Integer(2), new Boolean(false)},
            {"Jane", "White", "Speed reading", new Integer(20), new Boolean(true)},
            {"Joe", "Brown", "Pool", new Integer(10), new Boolean(false)}
        };
        public Class getColumnClass(int c) {
            return getValueAt(0, c).getClass();
        }
        public int getRowCount() {
            return data.length;
        }
        public int getColumnCount() {
            return columnNames.length;
        }
        public Object getValueAt(int rowIndex, int columnIndex) {
            return data[rowIndex][columnIndex];
        }

    }
```

程序运行结果如图 10-24 所示。

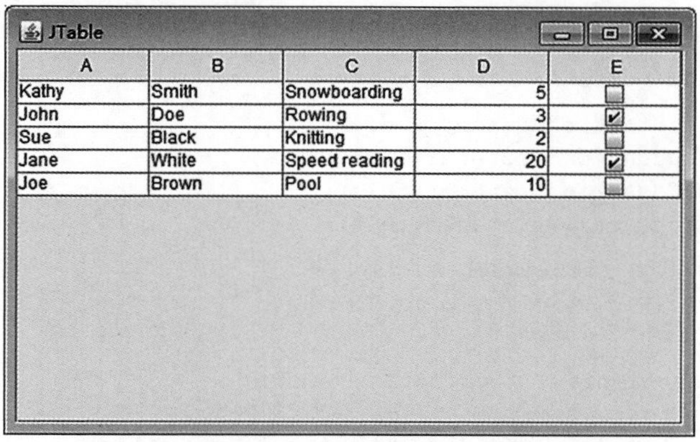

JTable

A	B	C	D	E
Kathy	Smith	Snowboarding	5	☐
John	Doe	Rowing	3	☑
Sue	Black	Knitting	2	☐
Jane	White	Speed reading	20	☑
Joe	Brown	Pool	10	☐

图 10-24　使用 AbstractTableModel 构造 JTable

10.5.3 实用案例 10.3:动态表格

本例使用 DefaultTableModel(AbstractTableModel 的子类)实现对表格的增加行列、删除行列的操作。

【例 10.17】 动态表格。

```
import java.awt.BorderLayout;
import java.awt.event.ActionEvent;
import java.awt.event.ActionListener;
import java.util.Vector;
import javax.swing.JButton;
import javax.swing.JFrame;
import javax.swing.JPanel;
import javax.swing.JScrollPane;
import javax.swing.JTable;
import javax.swing.table.DefaultTableModel;
import javax.swing.table.TableColumn;
import javax.swing.table.TableColumnModel;

public class AddAndRemoveCells extends JFrame implements ActionListener {
    private JTable table ;
    private DefaultTableModel defaultModel;

    public AddAndRemoveCells() {
        String[] name = { "First Name", "Last Name", "Sport" };      //表头名称
        Object[][] data = {                                          //表数据
                { "Kathy", "Smith", "Snowboarding" }, { "John", "Doe", "Rowing" },
{ "Sue", "Black", "Knitting" },
                { "Jane", "White", "Speed reading" }, { "Joe", "Brown", "Pool" } };
        defaultModel = new DefaultTableModel(data, name);
        table = new JTable(defaultModel);
        JScrollPane scrollPan = new JScrollPane(table);

        JButton button1 = new JButton("增加行");
        JButton button2 = new JButton("增加列");
        JButton button3 = new JButton("删除行");
        JButton button4 = new JButton("删除列");

        JPanel panel = new JPanel();
        panel.add(button1);
        panel.add(button2);
        panel.add(button3);
        panel.add(button4);

        button1.addActionListener(this);
        button2.addActionListener(this);
        button3.addActionListener(this);
        button4.addActionListener(this);

        add(panel, BorderLayout.NORTH);
        add(scrollPan, BorderLayout.CENTER);
    }

    public void actionPerformed(ActionEvent e) {
        if (e.getActionCommand().equals("增加列"))
            defaultModel.addColumn("新增列");
        if (e.getActionCommand().equals("增加行"))
            defaultModel.addRow(new Vector());
```

```
        if (e.getActionCommand().equals("删除列")) {
            int columncount = defaultModel.getColumnCount() - 1;
            if (columncount >= 0) {
                //若 columncount<0 代表已经没有任何列了
                TableColumnModel columnModel = table.getColumnModel();
                TableColumn tableColumn = columnModel.getColumn(columncount);
                columnModel.removeColumn(tableColumn);
                defaultModel.setColumnCount(columncount);
            }
        }
        if (e.getActionCommand().equals("删除行")) {
            //getRowCount 返回行数,rowcount<0 代表已经没有任何行了
            int rowcount = defaultModel.getRowCount() - 1;
            if (rowcount >= 0) {
                defaultModel.removeRow(rowcount);
                defaultModel.setRowCount(rowcount);
            }
        }
        table.revalidate();
    }
    public static void main(String args[]) {
        AddAndRemoveCells frame = new AddAndRemoveCells();
        frame.setSize(500, 300);
        frame.setVisible(true);
        frame.setDefaultCloseOperation(JFrame.EXIT_ON_CLOSE);
    }
}
```

程序运行结果如图 10-25 所示。

增加行　增加列　删除行　删除列

First Name	Last Name	Sport	新增列
Kathy	Smith	Snowboarding	
John	Doe	Rowing	
Sue	Black	Knitting	
Jane	White	Speed reading	
Joe	Brown	Pool	

图 10-25　实用案例 10.3 程序运行结果

10.6　GUI 程序设计实训任务

【任务描述】

编写一个 Swing,模拟实现一个可视化的简单计算器,至少提供包括加法、减法、乘法、除法等基本操作,希望能支持包括正负号、平方根、清零等其他功能。

【任务分析】

该任务的实现需要解决好以下三个问题：首先要对计算器的按钮、显示栏等进行规划、布局;然后对需要进行事件处理和响应的组件设置事件监听器,以获取用户的行为;最后在方法 actionPerformed()中逐一判断、分析用户的操作,并根据用户操作(如单击“+”)的不同进行

事件的响应(如执行加法),并显示运算结果。

【任务解决】

【例 10.18】 可视化的简单计算器。

完整示例代码如下：

```
import java.awt.BorderLayout;
import java.awt.Color;
import java.awt.GridLayout;
import java.awt.event.ActionEvent;
import java.awt.event.ActionListener;
import javax.swing.JButton;
import javax.swing.JFrame;
import javax.swing.JPanel;
import javax.swing.JTextField;
public class Calculator extends JFrame implements ActionListener {
    private JPanel centerPanel = new JPanel();
    private String x = "";
    private String y = "";
    private String fh = "";
    private double answer;
    private JTextField tfAnswer;

    public Calculator() {
        this.setBackground(Color.lightGray);
        setLayout(new BorderLayout());
        tfAnswer = new JTextField();
        tfAnswer.setHorizontalAlignment(JTextField.RIGHT);
        tfAnswer.setText("0.");
        add(tfAnswer, BorderLayout.NORTH);

        centerPanel.setLayout(new GridLayout(5, 5));
        addButton("7", Color.blue);
        addButton("8", Color.blue);
        addButton("9", Color.blue);
        addButton("/", Color.red);
        addButton("C", Color.red);
        addButton("4", Color.blue);
        addButton("5", Color.blue);
        addButton("6", Color.blue);
        addButton("*", Color.red);
        addButton("^", Color.blue);
        addButton("1", Color.blue);
        addButton("2", Color.blue);
        addButton("3", Color.blue);
        addButton("-", Color.red);
        addButton("sqrt", Color.blue);
        addButton("0", Color.blue);
        addButton("+/-", Color.blue);
        addButton(".", Color.blue);
        addButton("+", Color.red);
        addButton("=", Color.red);
        addButton("sin", Color.red);
        addButton("cos", Color.red);
        addButton("tan", Color.red);
        addButton("ln", Color.red);
```

```
        add(centerPanel, BorderLayout.CENTER);
    }
    public void addButton(String name, Color color) {
        JButton bt = new JButton(name);
        bt.setBackground(Color.white);
        bt.setForeground(color);
        bt.addActionListener(this);
        centerPanel.add(bt);
    }
    //计算功能实现
    public void dengyu(String z) {
        if (z.equals("+"))
            answer = Double.parseDouble(x) + Double.parseDouble(y);
        if (z.equals("-"))
            answer = Double.parseDouble(x) - Double.parseDouble(y);
        if (z.equals("*"))
            answer = Double.parseDouble(x) * Double.parseDouble(y);
        if (z.equals("/"))
            answer = Double.parseDouble(x) / Double.parseDouble(y);
        if (z.equals("^"))
            answer = Math.pow(Double.parseDouble(x), Double.parseDouble(y));
        if (z.equals("sin"))
            answer = Math.sin(Double.parseDouble(x));
        if (z.equals("cos"))
            answer = Math.cos(Double.parseDouble(x));
        if (z.equals("tan"))
            answer = Math.tan(Double.parseDouble(x));
        if (z.equals("ln"))
            answer = Math.log(Double.parseDouble(x));
        x = Double.toString(answer);
        tfAnswer.setText(x);
        y = "";
        fh = "";
    }
    public void actionPerformed(ActionEvent e) throws IndexOutOfBoundsException {
        if (e.getActionCommand().equals("0")
                || e.getActionCommand().equals("1")
                || e.getActionCommand().equals("2")
                || e.getActionCommand().equals("3")
                || e.getActionCommand().equals("4")
                || e.getActionCommand().equals("5")
                || e.getActionCommand().equals("6")
                || e.getActionCommand().equals("7")
                || e.getActionCommand().equals("8")
                || e.getActionCommand().equals("9")) {
            if (fh.equals("")) {
                x = x + e.getActionCommand();
                if (x.startsWith("00")) {
                    x = x.substring(1);
                }
                tfAnswer.setText(x);
            } else {
                y = y + e.getActionCommand();
```

```
            if (y.startsWith("00")) {
                y = y.substring(1);
            }
            tfAnswer.setText(y);
        }
    }
    if (e.getActionCommand().equals(".")) {
        if (fh.equals("")) {
            int i = 0, j = 0;
            for (i = 0; i < x.length(); i++)
                if (x.charAt(i) == '.')
                    j++;
            if (j == 0)
                x = x + ".";
            tfAnswer.setText(x);
        } else {
            int i = 0, j = 0;
            for (i = 0; i < y.length(); i++)
                if (y.charAt(i) == '.')
                    j++;
            if (j == 0)
                y = y + ".";
            tfAnswer.setText(y);
        }
    }
    if (e.getActionCommand().equals("C")) {
        x = "";
        y = "";
        fh = "";
        tfAnswer.setText("0.");
    }
    if (e.getActionCommand().equals("+/-")) {
        if (fh.equals("")) {
            if (x.substring(0, 1).equals("-"))
                x = x.substring(1);
            else
                x = "-" + x;
            tfAnswer.setText(x);
        } else {
            if (y.substring(0, 1).equals("-"))
                y = y.substring(1);
            else
                y = "-" + y;
            tfAnswer.setText(y);
        }
    }
    if (e.getActionCommand().equals("sqrt")) {
        if (fh != "")
            dengyu(fh);
        answer = Math.sqrt(Double.parseDouble(x));
        x = Double.toString(answer);
        tfAnswer.setText(x);
    }
```

```
        if (e.getActionCommand().equals("+")) {
            if (fh != "")
                dengyu(fh);
            fh = "+";
        }
        if (e.getActionCommand().equals("-")) {
            if (fh != "")
                dengyu(fh);
            fh = "-";
        }
        if (e.getActionCommand().equals("*")) {
            if (fh != "")
                dengyu(fh);
            fh = "*";
        }
        if (e.getActionCommand().equals("/")) {
            if (fh != "")
                dengyu(fh);
            fh = "/";
        }
        if (e.getActionCommand().equals("^")) {
            if (fh != "")
                dengyu(fh);
            fh = "^";
        }
        if (e.getActionCommand().equals("sin")) {
            if (fh != "")
                dengyu(fh);
            answer = Math.sin(Double.parseDouble(x));
            x = Double.toString(answer);
            tfAnswer.setText(x);
        }
        if (e.getActionCommand().equals("cos")) {
            if (fh != "")
                dengyu(fh);
            answer = Math.cos(Double.parseDouble(x));
            x = Double.toString(answer);
            tfAnswer.setText(x);
        }
        if (e.getActionCommand().equals("tan")) {
            if (fh != "")
                dengyu(fh);
            answer = Math.tan(Double.parseDouble(x));
            x = Double.toString(answer);
            tfAnswer.setText(x);
        }
        if (e.getActionCommand().equals("ln")) {
            if (fh != "")
                dengyu(fh);
            answer = Math.log(Double.parseDouble(x));
            x = Double.toString(answer);
            tfAnswer.setText(x);
        }
        if (e.getActionCommand().equals("="))
```

```
            dengyu(fh);
    }

    public static void main(String args[]) {
        Calculator c = new Calculator();
        c.setDefaultCloseOperation(JFrame.EXIT_ON_CLOSE);
        c.setSize(400, 400);
        c.setVisible(true);
    }
}
```

程序运行结果如图 10-26 所示。

图 10-26　计算器

1. Java 有哪些 GUI 程序设计技术？各有什么区别？

2. 简述 Swing 的类层次结构。

3. 简述 BorderLayout、FlowLayout 和 GridLayout 布局方式的用途。

4. 为什么要使用布局管理器？无布局管理器的布局与有布局管理器的布局二者有何区别？

5. 在 Swing 中如何处理鼠标事件和键盘事件？

6. 编写一个 GUI 程序，包含一个带图标的 JButton 对象。当用户单击这个按钮时，把窗体的标题修改为“单击按钮”。按下按钮和鼠标经过按钮时，JButton 上的图标有不同的效果。

第三篇　网络篇

第11章 网络通信

本章学习目标

要使得网络中的两台计算机之间能够相互通信，如果从网络协议的底层去实现也许比较困难，但是通过一些编程接口和通信模型来实现就简单得多。Java 提供了一系列的编程接口来实现这些功能，本章将主要介绍这些通信接口，以及基于这些接口的具体实现。通过本章的学习，应该重点掌握以下内容：

(1) URL 类及其 WWW 连接。

(2) 用 ServerSocket 和 Socket 类实现 TCP/IP 客户和服务器。

(3) 用 DatagramPacket 和 DatagramSocket 进行基于 UDP 的网络通信。

(4) 用 MulticastSocket 实现多点广播。

11.1 类 URL 与 URLConnection

通过一个 URL 连接就可以确定资源的位置，如网络文件、网页以及网络应用程序等，其中包含了许多具体的语法元素。例如：

```
http://home.netscape.com:80/home/welcome.html
```

规定使用 http 协议，主机名称为 home.netscape.com，端口号为 80。这个 URL 的其他部分/home/welcome.html，则确定了要在这个站点上所要访问的资源。

为了表示 URL，Java 中定义了一个 URL 类，允许程序人员通过它打开特定的 URL 连接，并对连接所对应的资源进行各种读写操作，整个访问过程就像访问本地文件一样方便快捷。

为指向要访问的 URL 资源，我们必须先通过构造方法初始化一个 URL 对象。

(1) public URL(String spec) 通过一个表示 URL 地址的字符串可以构造一个 URL 对象。

```
URL urlBase=new URL("http://www.cqu.edu.cn/");
```

(2) public URL (URL context,String spec)通过基 URL 和相对 URL 构造 URL 对象。

```
URL urlBase =new URL("http://www.cqu.edu.cn/pages/");
URL cquGames=new URL(urlBase, "cqugame.html");
```

(3) public URL(String protocol, String host, String file)。

```
new URL("http", "www.gamelan.com", "/pages/Gamelan.net.html");
```

(4) public URL(String protocol, String host, int ort, String file)。

```
URL cqu=new URL ("http", "www.cqu.edu.cn",80, "pages/cqunetwork.html");
```

URL 对象一旦生成,其属性是不能被更改的,但是我们可以通过类 URL 所提供的方法来获取这些属性。

public String getProtocol():获取该 URL 的协议名。

public String getHost():获取该 URL 的主机名。

public int getPort():获取该 URL 的端口号,如果没有设置端口,则返回-1。

public String getFile():获取该 URL 的文件名。

public String getRef():获取该 URL 在文件中的相对位置。

下面的例子生成一个 URL 对象,并获取它的各个属性。

【例 11.1】 URL 对象的属性访问。

```
package code1101;
import java.net.*;
public class ParseURL {
    public static void main(String[] args) throws Exception {
        URL cqu = new URL("http://www.cqu.edu.cn/index.html#top");
        System.out.println("protocol=" + cqu.getProtocol());  //协议
        System.out.println("host=" + cqu.getHost());           //主机名
        System.out.println("filename=" + cqu.getFile());       //文件名
        System.out.println("port=" + cqu.getPort());           //端口
        System.out.println("ref=" + cqu.getRef());             //文件内部的一个引用
    }
}
```

程序运行结果:

```
protocol=http
host=www.cqu.edu.cn
filename=index.html
port=80
ref=top
```

获得 URL 对象之后就可以通过 URL 的 openStream()方法读取指定的 WWW 资源,其定义如下:

```
InputStream openStream();
```

方法 openStream()与指定的 URL 建立连接并返回 InputStream 类的对象,以便从这一连接中读取数据。

此外,URL 类中还有一个常用的建立远程对象连接的方法:

```
URLConnection openConnection()
```

返回一个 URLConnection 对象,URLConnection 表示应用程序和 URL 之间的通信连接,通过该对象可读写此 URL 引用的资源,常用的方法包括:

Object getContent():获取此 URL 连接的内容。

String getHeaderField(String name)：返回指定的头字段的值。

int getContentLength()：返回 content-length 头字段的值。

String getContentType()：返回 content-type 头字段的值。

long getLastModified()：返回 last-modified 头字段的值

InputStreamgetInputStream()：返回从此打开的连接读取的输入流。

OutputStreamgetOutputStream()：返回写入到此连接的输出流。

【例 11.2】 通过 URL 和 URLConnection 访问 WWW 资源。

```
package code1101;
import java.util.*;
import java.net.*;
import java.io.*;
public class URLReader {
    public static void main(String[] args) throws Exception {
        try {
            //通过 URL 类建立远程连接,并获取连接内容
            URL obj = new URL("http://www.cqu.edu.cn/");
            BufferedReader in = new BufferedReader(new InputStreamReader(obj.
openStream()));
            String inputLine;
            while ((inputLine = in.readLine()) != null)
                System.out.println(inputLine);
            in.close();
            //通过 URLConnection 获取响应 Header 信息
            URLConnection conn = obj.openConnection();
            conn.connect();
            System.out.println("获取到的响应长度");
            System.out.println(conn.getContentLength());
            System.out.println("获取到的响应类型 ");
            System.out.println(conn.getContentType());
            //定义 bufferedReader 输入流来读取 URL 的响应
            in = new BufferedReader(new InputStreamReader(conn.getInputStream()));
            String line;
            String result = null;
            while ((line = in.readLine()) != null) {
                result += line;
            }
            System.out.println(result);
        } catch (Exception e) {
            e.printStackTrace();
        }
    }
}
```

程序运行结果：

```
<!Doctype Html public "-//W3C//DTD//Html 4.0 Final//EN">
<html>
<head>
…
…
…
```

```
</body>
</html>
获取到的响应长度
17666
获取到的响应类型
text/html; charset=gb2312
…
```

分析：首先生成一个 URL 对象 cqu，指向重庆大学的主页，然后调用 cqu.openStream()方法生成该 URL 的一个输入流，这是一个字节流，在此基础上进一步通过 InputStreamReader 和 BufferedReader 构造一个带有缓冲功能的字符流，并通过该字符流对象读取该 URL 的 html 内容，进而输出到屏幕上。URLConnection 类也可以用来对由 URL 引用的资源进行读写操作，前提是先通过 connect()方法建立连接，然后再获取响应头信息或响应内容。

实用案例 11.1：实现单线程的资源下载器

实现一个简单的基于单线程的资源下载器，如图 11-1 所示。用户可以任意指定待下载资源的链接地址，系统根据该地址判断资源是否存在，如果存在，则将该资源文件下载到本地。

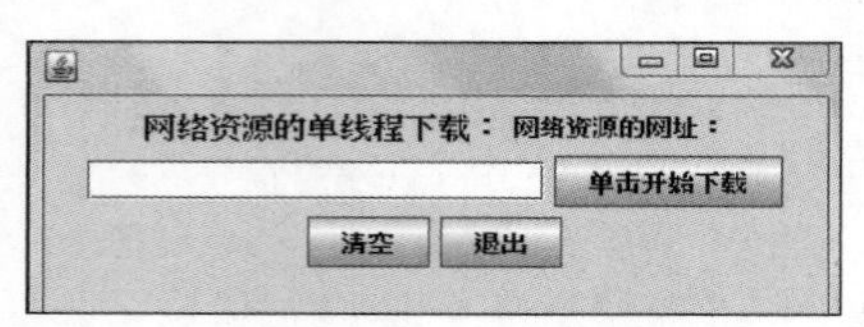

图 11-1　资源下载器基本界面

【例 11.3】 单线程的资源下载器。

```
package code1101;
public class SingleThreadDown extends JFrame implements ActionListener {
    private final JPanel panel = new JPanel();
    private final JLabel label1 = new JLabel("网络资源的单线程下载:");
    private final JLabel label2 = new JLabel("网络资源的网址:");
    JButton StartButton = new JButton("单击开始下载");
    JButton resetButton = new JButton("清空");
    JButton exitButton = new JButton("退出");
    JTextField urlField = new JTextField(20);

    public SingleThreadDown() {
        panel.setLayout(new FlowLayout());
        label1.setFont(new Font("雅黑", Font.BOLD, 15));
        panel.add(label1);
        panel.add(label2);
        panel.add(urlField);
        panel.add(StartButton);
        panel.add(resetButton);
        panel.add(exitButton);
        setContentPane(panel);
        StartButton.addActionListener(this);
        resetButton.addActionListener(this);
        exitButton.addActionListener(this);
        setSize(400, 400);
```

```
        setVisible(true);
        setDefaultCloseOperation(JFrame.EXIT_ON_CLOSE);
    }
    public void download(String address) throws Exception {
        URL url = new URL(address);
        URLConnection urlcon = url.openConnection();
        urlcon.connect();
        InputStream in = urlcon.getInputStream();
        String filePath = url.getFile();
        int pos = filePath.lastIndexOf("/");
        String fileName = filePath.substring(pos + 1);
        FileOutputStream out = new FileOutputStream("D:\\" + fileName);
        byte[] bytes = new byte[1024];
        int len = in.read();
        while (len != -1) {
            out.write(bytes, 0, len);
            len = in.read();
        }
        out.close();
        in.close();
        JOptionPane.showMessageDialog(this, "下载完毕");
    }
    public void actionPerformed(ActionEvent e) {
        if (e.getSource() == StartButton) {
            if ("".equals(urlField.getText())) {
                JOptionPane.showMessageDialog(this, "请输入资源地址");
            }
            String url = urlField.getText();
            try {
                download(url);
            } catch (Exception e1) {
                JOptionPane.showMessageDialog(this, "资源地址有误,请检查,谢谢!");
                e1.printStackTrace();
            }
        } else if (e.getSource() == resetButton) {
            urlField.setText("");
        } else {
            System.exit(0);
        }
    }
    public static void main(String[] args) {
        new SingleThreadDown();
    }
}
```

需要说明的是,为了更加高效地读取远程资源信息,实用的资源下载器(如迅雷等)均为多线程开发,读者可以根据第 9 章的知识对上述单线程资源下载器进行扩展。

11.2 类 InetAddress

Internet 上的主机通常有两种表示地址的方式:域名和 IP 地址。有时需要通过域名来查找它对应的 IP 地址,有时又需要通过 IP 地址来查找主机名。这时可以利用 java.net 包中的

InetAddress 类来完成任务。

InetAddress 类是 IP 地址封装类,它没有提供可用的构造方法,因此只能利用该类的一些静态方法来获取对象实例,然后再通过这些对象实例来对 IP 地址或主机名进行处理。以下是该类常用的一些方法。

pulic static InetAddress getByName(String hostname):在给定主机名的情况下确定主机的 IP 地址。

public static InetAddress getByAddress(byte[] addr):在给定原始 IP 地址的情况下,返回 InetAddress 对象。

public String getHostAddress():获取 IP 地址。

public String getHostName():获取主机名。

publicboolean isReachable(int timeout):测试是否可以达到该地址。

【例 11.4】 根据指定的域名查找 IP 地址。

```
package code1102;
import java.net.*;
public class GetIP {
    public static void main(String[] args) {
        try{
            System.out.println("本机 IP 为:"+InetAddress.getByName(args[0]));
        }catch(UnknownHostException el){
            el.printStackTrace();
        }
    }
}
```

执行该应用,指定域名 java GetIP www.cqu.edu.cn 或者在 Eclipse 中设置参数如图 11-2 所示。则运行结果如下:

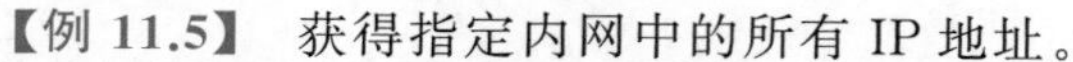

```
本机 IP 为:www.cqu.edu.cn/222.178.10.35
```

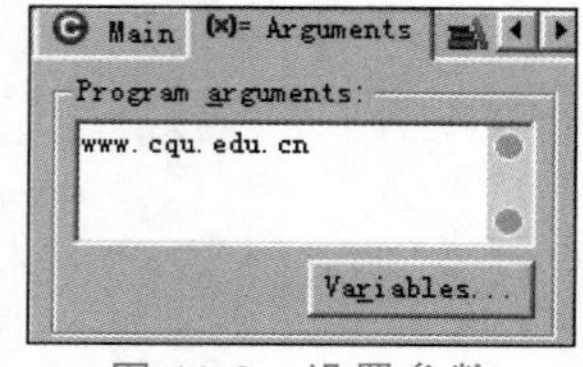

图 11-2 设置参数

实用案例 11.2:获得指定内网中所有活动 IP

可以通过循环生成指定内网中的 IP 地址,再利用 InetAddress 类中 isReachable()方法对待测试的 IP 的可达性进行分析。

【例 11.5】 获得指定内网中的所有 IP 地址。

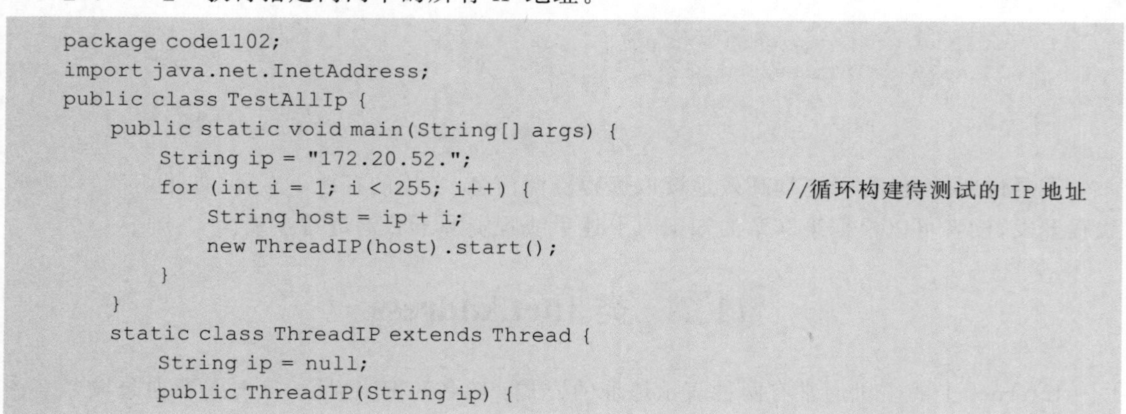

```
package code1102;
import java.net.InetAddress;
public class TestAllIp {
    public static void main(String[] args) {
        String ip = "172.20.52.";
        for (int i = 1; i < 255; i++) {                    //循环构建待测试的 IP 地址
            String host = ip + i;
            new ThreadIP(host).start();
        }
    }
    static class ThreadIP extends Thread {
        String ip = null;
        public ThreadIP(String ip) {
```

```
        super();
        this.ip = ip;
    }
    @Override
    public void run() {
        super.run();
        try {
            InetAddress ia = InetAddress.getByName(ip);
            boolean bool = ia.isReachable(1500);        //判断 IP 是否正在被使用
            if (bool) {
                System.out.println("主机: " + ip + " 可用");
            }
        } catch (Exception e) {
            e.printStackTrace();
        }
    }
  }
}
```

11.3 Socket 通信

Java 中,客户与服务器之间的通信编程一般是基于 Socket 实现的。Socket 是两个实体之间进行通信的有效端点,通过 Socket 可以获得源 IP 地址和源端口、终点 IP 地址和终点端口,并创建一个能被多人使用的分布式应用程序,实现与服务器的双向自由通信。

11.3.1 基于 TCP 的 Socket 通信

流式通信协议 TCP 是一种可靠的、基于连接的协议,发送方和接收方所对应的两个 Socket 之间必须建立连接,以便在 TCP 的基础上进行通信,当一个 Socket(通常是 Server Socker)等待建立连接时,另一个 Socket 可以要求进行连接,一旦这两个 Socket 连接起来就可以进行双向数据传输,双方都可以进行发送或接收操作。在 Java 编程语言中,TCP Socket 连接是用 java.net 包中的类实现的。图 11-3 说明了服务器与客户端基于 TCP 的 Socket 通信过程。

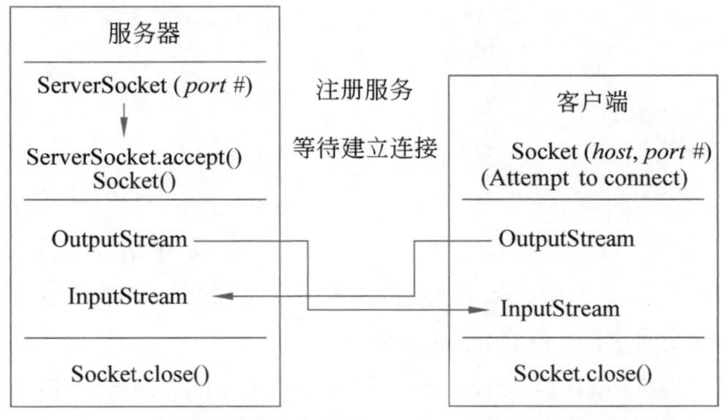

图 11-3 服务器与客户端基于 TCP 的 Socket 通信过程

一个完整的 Socket 通信程序通常包括以下几个基本步骤:

(1) 创建 Socket。首先创建 ServerSocket,在客户端创建 Socket 后,连接服务器,在服务器端创建一个和它对应的一个 Socket。

(2) 打开连接到 Socket 的输入输出流。在客户端和服务器端分别用 Socket 创建输入流和输出流,把客户端的输入流和服务器的输出流连接起来,客户端的输出流和服务器端的输入流连接起来。

(3) 使用 InputStream 和 OutputStream 对 Socket 进行读写操作。

(4) 使用 close()方法关闭 Socket 连接。

通常,程序员主要是针对所要完成的功能在第(3)步进行编程,第(1)步、第(2)步和第(4)步对任何程序几乎是一样的。

Java.net 包中提供了两个类 Socket 和 ServerSocket,分别用来表示双向连接的客户端和服务端。其构造方法如下:

```
Socket(InetAddress address,int port);
Socket(InetAddress address, int  port, Boolean stream);
Socket(String host, int port);
Socket(String host, int port,boolean stream);
ServerSocket(int port);
ServerSocket(int port, int count);
```

其中,address、host 和 port 分别是双向连接另一方的 IP 地址、主机名和端口号,stream 指明 Socket 是流 Socket,还是数据报 Socket, count 则表示服务端所能支持的最大连接数。例如,对于客户端程序,可以通过生成一个 Socket 对象打开 Socket:

```
Socket client;
client=new Socket("Machine name", portNumber);
```

注意:只有给出正确的端口号,才能获得相应的服务。通常,0~1023 的端口号为系统所保留,例如 HTTP 服务的端口号为 80,TELNET 服务的端口号为 23,FTP 服务的端口号为 21,所以在选择端口号时,应选择一个大于 1023 的数以防止发生冲突。

对于服务方,通过生成一个 ServerSocket 对象打开服务器 ServerSocket,然后调用方法 accept()准备接受客户发来的连接请求,生成一个和客户端对应的 Socket。

```
ServerSocket server=null;
try{
    server=new ServerSocket(2000);                //2000 为端口号
    Socket socket=server.accept();
}
catch(IOException e){
    System.out.println("Error:"+e);
}
```

方法 accept()等待客户的请求,直到有一个客户启动并请求连接到相同的端口,然后 accept()返回一个对应于客户的 Socket。这时,客户方和服务方都建立了用于通信的 Socket,接下来,由各个 Socket 分别打开各自的输入输出流。

类 Socket 提供了方法 getInputStream()和 getOutputStream()来得到对应的输入输出流以进行读写操作,这两个方法分别返回 InputStream 和 OutputStream 类对象。为了便于读写数据,可以在返回的输入输出流对象上建立过滤流,如 DataInputStream、DataOutput Stream 类对象,对于文本方式流对象,可以采用 InputStreamReader 和 OutputStreamWriter、PrintWriter 等处理。例如:

```
PrintStream os=new PrintStream(new BufferedOutputStream(socket.getOutputStream()));
DataInputStream is=new DataInputStream(socket.getInputStream());
PrintWriter out=new printWriter(socket.getOutputStream(),true);
BufferedReader in = new BufferedReader ( new InputStreamReader ( socket.
getInputStream()));
```

使用完 Socket 连接后，应将与 Socket 通信相关的所有资源进行关闭。

```
os.close();                                        //关闭输出流
is.close();                                        //关闭输入流
socket.close();                                    //关闭 Socket
```

提示：一定要注意关闭的顺序，与 Socket 相关的输入输出流应该首先全部关闭，然后再关闭 Socket。

11.3.2 实用案例 11.3：单客户端 Socket 通信

【例 11.6】 单客户端 Socket 通信。

服务器程序代码如下：

```
package code1103;
import java.net.*;
import java.io.*;
public class chatServer {
    public static void main(String args[]) {
        ServerSocket s = null;
        Socket s1;
        String sendString = "Hello Net World!";
        OutputStream slout;
        DataOutputStream dos;
        try {
            s = new ServerSocket(5600);                 //使用本地 5600 端口提供服务
        } catch (Exception e) {
            e.printStackTrace();
        }
        while (true) {
            try {
                s1 = s.accept();      //等待客户端的连接,如果连接成功就返回 Socket 对象
                slout = s1.getOutputStream();
                dos = new DataOutputStream(slout);    //建立输出流
                dos.writeUTF(sendString);
                slout.close();
                s1.close();
            } catch (Exception e) {
                e.printStackTrace();
            }
        }
    }
}
```

客户端程序代码如下：

```
package code1103;
import java.net.*;
import java.io.*;
```

```
public class chatClient {
    public static void main(String args[]) throws Exception {
        int c;
        Socket s1;
        InputStream s1In;
        DataInputStream dis;
        s1 = new Socket("localhost",5600);       //创建客户端 Socket,连接服务器
        s1In = s1.getInputStream();
        dis = new DataInputStream(s1In);         //建立输入流
        String st = new String (dis.readUTF());
        System.out.println(st);
        s1In.close();
        s1.close();
    }
}
```

上述程序只能响应一个客户端程序的连接请求,在实际应用中,服务器一般需要同时响应多个客户请求。因此,ServerSocket 对象的 accept()方法每当有一个连接请求发生时,就会产生一个 Socket 对象,所以只要用此方法反复监听客户请求,就可以为每一个客户生成一个专用的 Socket 对象进行通信。

那么如何管理这么多的 Socket 对象呢? 最好的解决办法是将 Socket 对象放置到线程中,这样当每一个 Socket 对象执行任务完成后,只有包含该 Socket 对象的线程会终止,对其他线程没有任何影响。

11.3.3 实用案例 11.4: 多客户端 Socket 通信

【例 11.7】 一对多 Socket 通信。

服务器线程程序代码如下:

```
package code1103;
import java.io.*;
import java.net.*;
public class ServerThread extends Thread{
  private Socket s;
  private DataInputStream is;
  private DataOutputStream os;
  //在构造方法中为每个套接字连接输入和输出流
  public ServerThread(Socket socket) throws IOException{
     super();
     s=socket;
     is=new DataInputStream(s.getInputStream());
     os=new DataOutputStream(s.getOutputStream());
     start();                                   //启动 run()方法
  }
  //在 run()方法中与客户端通信
  public void run() {
     try{
         String str;
         double result,zhi;
         boolean NotEnd=true;
         while(NotEnd){
             str=is.readUTF();
```

```
            if(!str.equals("bye")){
               zhi=Double.parseDouble(str);
               System.out.println("接收到的值为:"+zhi);
               result=zhi*zhi;
               str=Double.toString(result);
               os.writeUTF(str);
               os.flush();
               System.out.println("平方值"+str+"已经发送");
            }else{
                NotEnd=false;
               os.writeUTF("bye");
               os.flush();
            }
         }
         is.close();
         os.close();
         s.close();
      }catch(IOException e){
         e.printStackTrace();
      }
   }
}
```

服务器主程序代码如下：

```
package code1103;
import java.io.*;
import java.net.*;
public class MultiServer {
   public static void main(String[] args) {
      try{
         System.out.println("等待连接");
         ServerSocket serverSocket=new ServerSocket(5500);
         Socket s=null;
         while (true){
            //等待客户端的请求
            s=serverSocket.accept();
            //每次请求都启动一个线程来处理
            new ServerThread(s);
         }
      }catch(IOException e){
         e.printStackTrace();
      }
   }
}
```

客户端程序代码如下：

```
package code1103;
import java.io.*;
import java.net.*;
public class Client {
 public static void main(String[] args) {
  try{
    //连接到本机,端口号 5500
```

```
        Socket s=new Socket("localhost",5500);
        //将数据输入流连接到 Socket 上
        DataInputStream is=new DataInputStream(s.getInputStream());
        //将数据输出流连接到 Socket 上
        DataOutputStream os=new DataOutputStream(s.getOutputStream());
        System.out.println("输入待求平方值,输入 bye 结束。");
        String outStr,inStr;
        boolean NotEnd=true;
        BufferedReader buf=new BufferedReader(new InputStreamReader(System.in));
        //反复读用户的数据并计算
        while(NotEnd){
           outStr=buf.readLine();                    //读入用户的输入
           os.writeUTF(outStr);                      //写到 Socket 中
           os.flush();                               //清空缓冲区,立即发送
           inStr=is.readUTF();                       //从 Socket 中读数据
           if(!inStr.equals("bye"))
              System.out.println("返回结果:"+inStr);
           else
                NotEnd=false;
        }
        is.close();
        os.close();
        s.close();
     }catch(IOException e){
         e.printStackTrace();
     }
   }
 }
```

可以先运行服务器主程序,然后多次运行客户端程序,实现多个客户端与同一服务器同时通信的功能。如图 11-4 所示,其中 11-4(a)为服务器端结果,11-4(b)为客户端 1 的结果,11-4(c)为客户端 2 的结果。

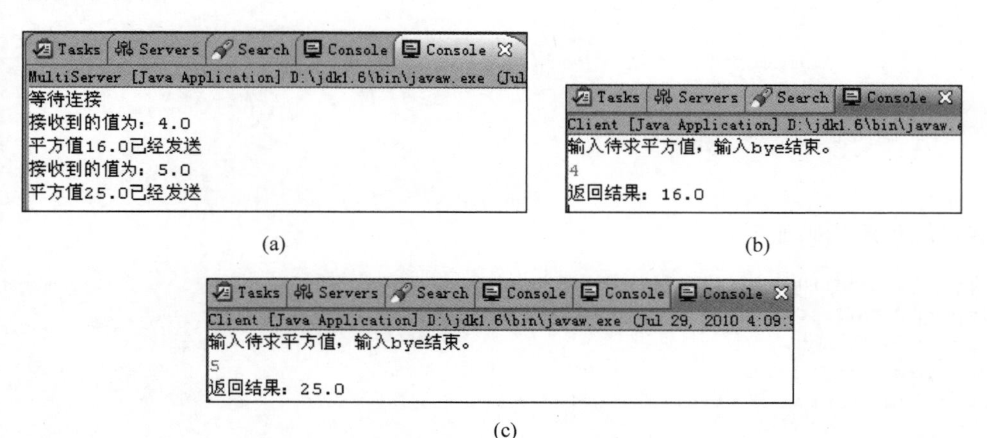

图 11-4　实用案例 11.4 程序运行结果

11.3.4 基于 UDP 的网络通信

与 TCP/IP 不同,数据报通信协议(User Datagram Protocol,UDP)是一种无连接的协议。每个数据报都是一个独立的信息,包括完整的源地址或目的地址,它在网络上以任何可能的路径传往目的地,因此能否到达目的地,到达目的地的时间以及内容的正确性都是不能保证的。

要区分上述两种协议,一种简单的方法就是把它们比作电话呼叫和邮递信件。电话呼叫保证有一个同步通信;消息按给定次序发送和接收。而对于邮递信件,即使能收到所有的消息,它们的顺序也可能不同。

Java.net 中提供了 DatagramSocket 和 DatagramPacket 两个类用来支持数据报通信。DatagramSocket 用于在程序之间建立传送数据报的通信连接,DatagramPacket 则用来表示一个数据报。DatagramSocket 的构造方法如下:

```
DatagramSocket();
DatagramSocket(int port);
```

其中,port 指明 Socket 所使用的端口号。

用数据报方式编写 Client/Server 程序时,无论在客户方还是服务方,首先都要建立一个 DatagramSocket 对象,用来接收或发送数据报,然后使用 DatagramPacket 类对象作为传输数据的载体。DatagramPacket 的构造方法如下。

接收用:

```
DatagramPacket(byte  ibuf[],int  ilength);
```

发送时用:

```
DatagramPacket(byte  ibuf[],int  ilength,InetAddress  iaddr,int iport);
```

其中,ibuf 中存放数据报数据,ilength 为数据报中数据的长度,iaddr 和 iport 指明目的地址。

在接收数据前,应用上面的第一种方法先生成一个 DatagramPacket 对象,给出接收数据的缓冲区及其长度。然后调用 DatagramSocket 的方法 receive()等待数据报的到来,receive()将一直等待,直到收到一个数据报为止。

```
DatagramPacket packet=new DatagramPacket(buf,256);
socket.receive(packet);
```

发送数据前,也要先生成一个 DatagramPacket 对象,这时要使用上面的第二种构造方法,在给出存放发送数据的缓冲区的同时,还要给出完整的目的地址,包括 IP 地址和端口号。发送数据是通过 DatagramSocket 的方法 Send()实现的,Send()根据数据报的目的地址来寻径,以传递数据报。

```
DatagramPacket packet=new DatagramPacket(buf,buf.length,address,port);
socket.send(packet);
```

在构造数据报时,要给出 InetAddress 类参数。类 InetAddress 在包 java.net 中定义,用来表示一个 Internet 地址,可以通过它提供的类方法 getByName()从一个表示主机名的字符串获取该主机的 IP 地址,然后再获取相应的地址信息。

11.3.5 实用案例 11.5：简单的 UDP 通信示例

【例 11.8】 简单的 UDP 通信。

客户端程序代码如下：

```
package code1103;
import java.io.*;
import java.net.*;
import java.util.*;
public class QuoteClient {
    public static void main(String[] args) throws IOException {
        if (args.length != 1) {
            System.out.println("Usage:java QuoteClient <hostname>");
            return;
        }
        DatagramSocket socket = new DatagramSocket();               //创建数据报 Socket
        //发送请求
        byte[] buf = new byte[256];
        InetAddress address = InetAddress.getByName(args[0]);  //获取目标地址
        DatagramPacket packet = new DatagramPacket(buf, buf.length, address, 4445);
        //创建发送数据报
        socket.send(packet);                                        //发送
        //获取响应
        packet = new DatagramPacket(buf, buf.length);               //创建接收数据报
        socket.receive(packet);                                     //接收数据报
        //显示响应
        String received = new String(packet.getData());
        System.out.println("Quote of the Moment:" + received);
        socket.close();                                             //关闭 Socket
    }
}
```

服务端主程序代码如下：

```
package code1103;
public class QuoteServer {
    public static void main(String args[]) throws java.io.IOException {
        new QuoteServerThread().start();                      //启动,创建服务器线程
    }
}
```

服务器线程程序代码如下：

```
//QuoteServerThread.java
package code1103;
import java.io.*;
import java.net.*;
import java.util.*;
public class QuoteServerThread extends Thread {
    protected DatagramSocket socket=null;
    protected BufferedReader in=null;
    protected boolean moreQuotes=true;
    public QuoteServerThread() throws IOException {
        this("QuoteServerThread");
    }
```

```
    public QuoteServerThread(String name) throws IOException {
        super(name);
        socket=new DatagramSocket(4445);        //创建数据报 Socket
        try{ in= new BufferedReader(new FileReader("one.txt"));
        }catch(FileNotFoundException e) {
            System.err.println("Could not open quote file. Serving time instead.");
        }
    }
    public void run() {
        while(moreQuotes) {
            try{
                byte[] buf=new byte[256];
                DatagramPacket packet=new DatagramPacket(buf,buf.length);
                //生成数据报,准备接收
                socket.receive(packet);        //接收
                String dString=null;
                if(in==null)
                    dString=new Date().toString();
                else
                    dString=getNextQuotes();
                buf=dString.getBytes();
                //发送响应给客户端,使用收到的数据报的"address"和"port"。
                InetAddress address=packet.getAddress();
                int port=packet.getPort();
                packet=new DatagramPacket(buf,buf.length,address,port);
                                               //创建发送数据报
                socket.send(packet);           //发送
            }catch(IOException e) {
                e.printStackTrace();
                moreQuotes=false;
            }
        }
        socket.close();
    }
    //获取下一个需要发送的数据内容
    protected String getNextQuotes(){
      String returnValue=null;
      try {
        if((returnValue=in.readLine())==null) {
         in.close();
         moreQuotes=false;
         returnValue="No more quotes.Goodbye.";
        }
      }catch(IOException e) {
         returnValue="IOException occurred in server";
      }
      return returnValue;
    }
}
```

从例子中可以看出,编写数据报通信程序,还是比较简单,其重点是掌握 DatagramPacket 和 DatagramSocket 的使用,包括如何创建接收数据报和接收数据报 Socket,和发送数据报和发送数据报 Socket。创建数据报通信程序,不需要在客户端和服务器端之间建立连接。

11.3.6 基于 MulticastSocket 实现多点广播

DatagramSocket 只允许数据报发送给指定的目标地址,而 MulticastSocket 可以将数据报以广播方式发送到数量不等的多个客户端。

若要使用多点广播时,则需要让一个数据报标有一组目标主机地址,当数据报发出后,整个组的所有主机都能收到该数据报。IP 多点广播实现了将单一信息发送到多个接收者的广播,如图 11-5 所示,其思想是设置一组特殊网络地址作为多点广播地址,每一个多点广播地址都被看作一个组,当客户端需要发送、接收广播信息时,加入该组即可。IP 协议为多点广播提供了这批特殊的 IP 地址,这些 IP 地址的范围是 224.0.0.0~239.255.255.255。

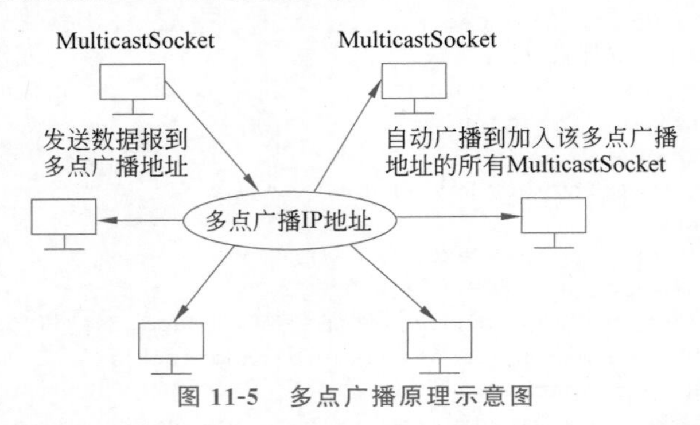

图 11-5 多点广播原理示意图

通过 Java 实现多点广播时,MulticastSocket 类是实现这一功能的关键,当 MulticastSocket 把一个 DatagramPacket 发送到多点广播 IP 地址,该数据报将被自动广播到加入该地址的所有 MulticastSocket。MulticastSocket 类既可以将数据报发送到多点广播地址,也可以接收其他主机的广播信息。

MulticastSocket 类似于 DatagramSocket,要发送一个数据报时,可使用随机端口创建 MulticastSocket,也可以在指定端口来创建 MulticastSocket,下面是对应的主要构造方法。

(1) public MulticastSocket():使用本机地址、随机端口创建 MulticastSocket 对象。

(2) public MulticastSocket(int portNumber):使用本机地址、指定端口创建 MulticastSocket 对象。

(3) public MulticastSocket(SocketAddress bindaddr):使用本机指定 IP 地址、指定端口创建 MulticastSocket 对象。

如果创建仅用于发送数据报的 MulticastSocket 对象,则使用默认地址、随机端口即可。但如果创建接收用的 MulticastSocket 对象,则该 MulticastSocket 对象必须具有指定端口,否则发送方无法确定发送数据报的目标端口。

创建一个 MulticastSocket 对象后,还需要将该 MulticastSocket 加入指定的多点广播地址,MulticastSocket 使用 joinGroup()方法来加入指定组,使用 leaveGroup()方法脱离一个组。

```
void joinGroup(InetAddress multicastAddr)
void leaveGroup(InetAddress multicastAddr)
```

MulticastSocket 用于发送、接收数据报的方法与 DatagramSocket 的完全一样。但 MulticastSocket 比 DatagramSocket 多一个 setTimeToLive(int ttl)方法,该方法用于设置在此 MulticastSocket 上发出的多播数据包的默认生存时间,以便控制多播的范围。ttl 必须在

0～255内，否则将抛出IllegalArgumentException。默认情况下，该ttl的值为1。

【例11.9】 一个简单的多点广播。

服务器端程序代码如下：

```
package code1103;
import java.net.DatagramPacket;
import java.net.InetAddress;
import java.net.MulticastSocket;
public class UDPMulticastServer {
    final static int RECEIVE_LENGTH = 1024;
    static String multicastHost = "224.0.0.1";
    static int localPort = 9998;
    public static void main(String[] args) throws Exception {
        InetAddress receiveAddress = InetAddress.getByName(multicastHost);
        if (!receiveAddress.isMulticastAddress()) {      //测试是否为多播地址
            throw new Exception("请使用多播地址");
        }
        int port = localPort;
        MulticastSocket receiveMulticast = new MulticastSocket(port);
        receiveMulticast.joinGroup(receiveAddress);
        DatagramPacket dp = new DatagramPacket(new byte[RECEIVE_LENGTH],
                RECEIVE_LENGTH);
        receiveMulticast.receive(dp);
        System.out.println(new String(dp.getData()).trim());
        receiveMulticast.close();
    }
}
```

客户端程序代码如下：

```
package code1103;
import java.net.DatagramPacket;
import java.net.InetAddress;
import java.net.MulticastSocket;
public class UDPMulticastClient {
    static String destAddressStr = "224.0.0.1";
    static int destPortInt = 9998;
    static int TTLTime = 4;
    public static void main(String[] args) throws Exception {
        InetAddress destAddress = InetAddress.getByName(destAddressStr);
        if (!destAddress.isMulticastAddress()) {        //检测该地址是否是多播地址
            throw new Exception("地址不是多播地址");
        }
        int destPort = destPortInt;
        int TTL = TTLTime;
        MulticastSocket multiSocket = new MulticastSocket();
        multiSocket.setTimeToLive(TTL);
        byte[] sendMSG = "Hello".getBytes();
        DatagramPacket dp = new DatagramPacket(sendMSG, sendMSG.length,
                destAddress, destPort);
        multiSocket.send(dp);
        multiSocket.close();
    }
}
```

11.4 网络通信实训任务

【任务描述】

使用 Java 访问某些 URL 时,如果该 URL 需要身份验证(即登录),那么就不能够直接访问,请编码实现网页模拟登录功能。

【任务分析】

使用 Java 模拟登录的基本思想是:首先用合法的用户名和密码向 Web 服务器提出登录请求,如果请求获得授权,则记录下返回的 Cookie 信息,在下次发起请求时将该 Cookie 发送过去用以表明身份,这样就能够访问带有权限的 URL 了。

【任务解决】

【例 11.10】 模拟身份验证登录。

```
package code1104;
import java.io.BufferedReader;
import java.io.InputStreamReader;
import java.net.HttpURLConnection;
import java.net.URL;
public class Login {
    public static void main(String[] args) {
        try {
            //验证登录的网页 URL
            URL url = new URL("http://localhost:8088/webapp/action.jsp");
            //打开该网页的 URL 连接
            HttpURLConnection conn = (HttpURLConnection) url.openConnection();
            //设置该网页连接可以向服务器发送信息
            conn.setDoOutput(true);
            //设置请求的提交方式
            conn.setRequestMethod("POST");
            //向服务器发送包含用户名称和密码的登录认证查询串
            conn.getOutputStream().write("username=a&password=a".getBytes());
            conn.getOutputStream().flush();
            conn.getOutputStream().close();
            //获取网页的标头信息中的 Cookie 字符串
            String cookieVal = conn.getHeaderField("Set-Cookie");
            //连接服务器
            conn.connect();
            conn.disconnect();
            //主网页的 URL
            url = new URL("http://localhost:8088/webapp/index.jsp");
            //打开该网页的 URL 连接
            conn = (HttpURLConnection) url.openConnection();
            //设置当前连接的 Cookie 信息与上次连接的 Cookie 信息一致,维持同一会话状态
            conn.setRequestProperty("Cookie", cookieVal);
            //获取服务器的主页返回结果
             BufferedReader rd = new BufferedReader(new InputStreamReader(conn.
getInputStream()));
            //输出主页结果
            String line;
            while ((line = rd.readLine()) != null) {
                System.out.println(line);
```

```
            }
            rd.close();
        } catch (Exception e) {
            System.out.println(e.getMessage());
        }
    }
}
```

习题与思考

1. 基于 TCP 与 UDP 的通信有什么区别？

2. 创建一个 URL 对象，并通过它读取指定的 WWW 资源。

3. 给定服务器 templates 目录中的文件 FileServer.java，编程实现：使服务器接收来自客户的文件名字符串，试图打开该文件并将文件内容通过 Socket 传回到客户。

4. 如何找出连接到服务器的客户端的 IP 地址？

5. 下列哪种说法是错误的？

A. TCP 是面向连接的协议，而 UDP 是无连接的协议

B. 数据报传输是可靠的，可以保证包按顺序到达

C. URL 代表统一资源定位符，通过它可以确定资源的位置

D. Socket 和 ServerSocket 分别表示连接的 Client 端和 Server 端

第12章 JSP 与 Servlet 技术

本章学习目标

基于 Java 的 Web 服务器端编程主要涉及 Servlet 和 JSP 技术，Servlet 与 JSP 之间的交互为动态网页的开发提供了优秀的解决方案。本章主要介绍 JSP 与 Servlet 技术的基本内容和使用方法。通过本章学习应该重点掌握以下主要内容：

(1) JSP 开发技术。

(2) Servlet 开发技术。

(3) JSP 与 Servlet 的混合使用。

12.1 为什么使用 JSP

网页是网络应用中最重要的一种形式，但是一般的网页是静态的。通过 JSP 技术可以实现网页的交互性、自动更新以及因时因人而变的动态网页特性。JSP 页面由 HTML 代码和嵌入其中的 Java 代码所组成。服务器在接到客户端请求页面后，对这些 Java 代码进行处理，然后将生成的 HTML 页面返回给客户端的浏览器。

【例 12.1】 编写一个 JSP 页面，使之能够显示一个数据表格，表格中的数据会根据行数进行变化。

```
<%@ page language="java" contentType="text/html; charset=gb2312" pageEncoding
= "gb2312"%>
<!DOCTYPE html PUBLIC "-//W3C//DTD HTML 4.01 Transitional//EN" "http://www.w3.
org/TR/html4/loose.dtd">
<html>
<head>
<meta http-equiv="Content-Type" content="text/html; charset=gb2312">
<title>第一个 JSP 页面</title>
</head>
<body>
<table>
    <tr>
        <td>id</td>
        <td>姓名</td>
    </tr>
    <%
```

```
        String color1 = "99ccff";
        String color2 = "88cc33";
        for (int i = 1; i <= 10; i++) {
            String color = "";
            if (i % 2 == 0) {
                color = color1;
            } else {
                color = color2;
            }
            out.println("<tr bgcolor=" + color + ">");
            out.println("<td>" + i + "</td>");
            out.println("<td>姓名" + i + "</td>");
            out.println("</tr>");
        }
    %>
</table>
</body>
</html>
```

程序可能的运行结果如图 12-1 所示。

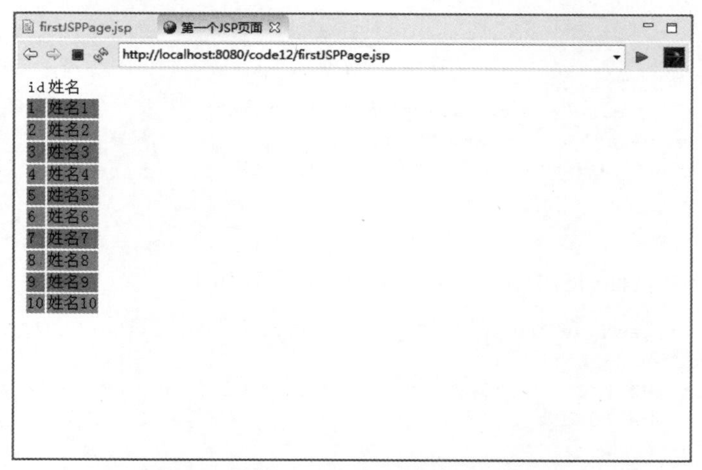

图 12-1　例 12.1 程序可能的运行结果

编写一个 JSP 页面涉及 3 个关键内容：①了解 JSP 页面的构成；②定义 JSP 页面的指令；③使用各种 JSP 内建对象。

12.2　JSP 技 术

12.2.1　JSP 工作原理

JSP 源文件是由安装在 Web 服务器上的 JSP 引擎编译执行的。例如，Tomcat 就是一种 JSP 引擎。JSP 引擎把来自客户端的请求传递给 JSP 源文件，然后 JSP 引擎再把对它的响应从 JSP 源文件传递给客户端。客户端请求和响应的过程如图 12-2 所示。

【例 12.2】　编写一个简单的 HTML 网页 JSPParameterDemo.html，该网页中包含一个表单，这个表单可以接收用户输入的数字，然后把这个数字发送到 Web 服务器，服务器根据接收的参数来决定显示内容重复的次数。

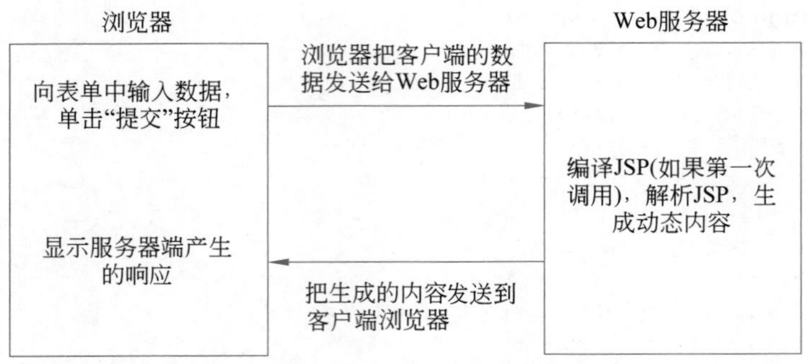

图 12-2　客户端请求和响应的过程

JSPParameterDemo.html 网页的代码如下：

```
<!DOCTYPE html PUBLIC "-//W3C//DTD HTML 4.01 Transitional//EN" "http://www.w3.
org/TR/html4/loose.dtd">
<html>
<head>
<meta http-equiv="Content-Type" content="text/html; charset=gb2312">
<title>理解 JSP 工作原理</title>
</head>
<body>
<p>请输入显示的次数:</p>
<form action="JSPParameterDemo.jsp" method="get">
    <input type="text"     name="times" />
    <input type="submit" value="提交" /></form>
</body>
</html>
```

处理请求的 JSP 页面(JSPParameterDemo.jsp)代码如下：

```
<%@ page language="java" contentType="text/html; charset=gb2312"
    pageEncoding="gb2312"%>
<!DOCTYPE html PUBLIC "-//W3C//DTD HTML 4.01 Transitional//EN" "http://www.w3.
org/TR/html4/loose.dtd">
<html>
<head>
<meta http-equiv="Content-Type" content="text/html; charset=gb2312">
<title>理解 JSP 工作原理</title>
</head>
<body>
<h1>
<%
    int times = Integer.parseInt(request.getParameter("times"));
    for (int i = 0; i < times; i++) {
        out.println("Hello, World!");
        out.println("<br/>");
    }
%>
</h1>
</body>
</html>
```

程序可能的运行结果如图 12-3 所示。

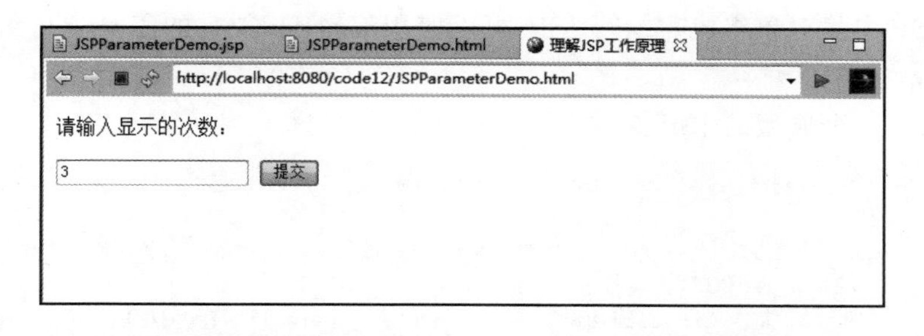

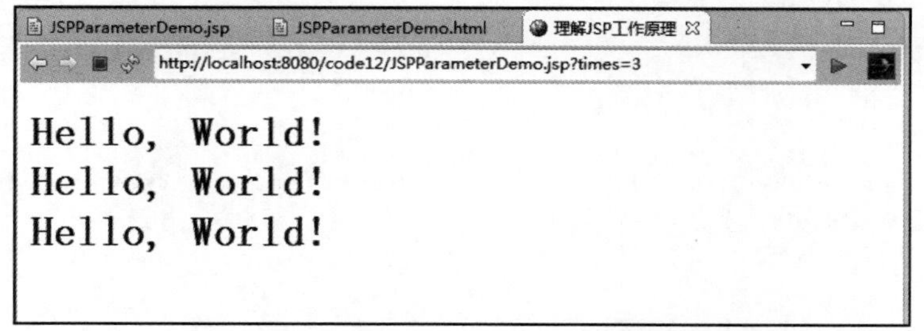

图 12-3　例 12.2 程序可能的运行结果

分析：在 JSPParameterDemo.html 网页中，当用户输入数字后单击“提交”按钮，浏览器就会把用户的数据和请求发送给 JSPParameterDemo.jsp 页面，该 JSP 页面会被执行，其生成的内容是 HTML 代码，并传给客户端浏览器进行显示。

JSP 页面生成 HTML 代码如下：

```
<!DOCTYPE html PUBLIC "-//W3C//DTD HTML 4.01 Transitional//EN" "http://www.w3.
org/TR/html4/loose.dtd">
<html>
<head>
<meta http-equiv="Content-Type" content="text/html; charset=gb2312">
<title>理解 JSP 工作原理</title>
</head>
<body>
<h1>
Hello, World!
<br/>
Hello, World!
<br/>
Hello, World!
<br/>
</h1>
</body>
</html>
```

将上述 HTML 代码与 JSP 页面代码进行比较，可以看到 JSP 页面中的 Scriptlets 代码被替换成了 HTML 网页中的内容。JSP 引擎在第一次调用 JSP 页面时，会自动编译 JSP 页面，然后这个 JSP 就驻留在内存内。因此，第一次调用 JSP 时总会有一定的延时，在接下来的调用中就不会再有延时了。

12.2.2　JSP 的构成

在 JSP 文件里，主要由模板元素、注释、脚本元素（声明、表达式、Scriptlets）、指令元素、动作元素以及 JSP 内建对象组成。编写 JSP,可以使用编辑 HTML 的工具进行编辑,编辑完成后保存成 *.jsp 文件即可。

【例 12.3】 一个典型的 JSP 文件。

```
<%@ page language="java" contentType="text/html; charset=gb2312" pageEncoding=
"gb2312"%>
<!DOCTYPE html PUBLIC "-//W3C//DTD HTML 4.01 Transitional//EN" "http://www.w3.
org/TR/html4/loose.dtd">
<!-- 这是一个典型的 JSP 文件,当前访问时间是<%=(new java.util.Date()).toString() %> -->
<%!String getDate() {
        return new java.util.Date().toString();
    }%>
<html>
<head>
<meta http-equiv="Content-Type" content="text/html; charset=gb2312">
<title>一个典型的 JSP</title>
</head>
<body>
<%@include file="header.jsp"%>
<hr />
当前时间是:<%=getDate()%>
<jsp:include page="copy.jsp" flush="true" />
</body>
</html>
```

从上述 JSP 文件中,可以总结出 5 类元素:①模板元素;②注释;③脚本元素;④指令元素;⑤动作元素。下面分别简要介绍这些元素。

1. 模板元素

模板元素是指 JSP 的静态 HTML 内容。这些模板元素是网页的框架,它影响页面的结构和美观程度。当客户端请求 JSP 页面时,它会把这些模板元素一字不变地发送到客户端。

2. 注释

JSP 中的注释有 HTML 注释、隐藏注释和 Scriptlets 中的注释 3 种。

(1) HTML 注释:在客户端显示一个注释。其语法如下:

```
<!-- commentInfo [<%=expression %>] -->
```

这种注释和 HTML 中的很像,唯一不同之处是,可以在这个注释中使用表达式,表达式的结果是不定的,由页面来决定。如例 12.3 中的注释:

```
<!--这一个典型的 JSP,当前访问时间是:<%=(new java.util.Date()).toString() %> -->
```

在客户端的 HTML 源代码中显示为

```
<!--这是一个典型的 JSP 文件,当前访问时间是:Sun Jul 04 11:03:40 CST 2010-->
```

(2) 隐藏注释:写在 JSP 文件中,但不发送给客户端。其语法如下:

```
<%-- commentInfo --%>
```

用隐藏注释标记的字符会在 JSP 编译时被忽略。

(3) Scriptlets 中的注释:由于 Scriptlets 包含的是 Java 代码,所以 Java 中的注释规范在 Scriptlets 中也能使用,常用的 Java 注释包括使用//表示单行注释,使用/*　　*/来表示多行注释。

3. 脚本元素

JSP 脚本元素是 JSP 代码中使用最频繁的元素，通常是用 Java 写的脚本代码。脚本元素主要包括：①声明(declaration)；②表达式(expression)；③Scriptlets。

(1) 声明：在 JSP 程序中声明合法的变量和方法。如例 12.3 中的声明：

```
<%!
    String getDate() {
        return new java.util.Date().toString();
}%>
```

(2) 表达式：位于<%=和%>之间的代码，在 JSP 请求处理阶段计算其值，所得的结果转换成字符串并与模板数据组合在一起。表达式在页面的位置，也就是该表达式计算结果所在的位置。如例 12.3 中的表达式：

```
<%=getDate()%>
```

(3) Scriptlets：位于<%和%>之间，是一段可以在处理请求时执行的 Java 代码。它可以产生输入，并将输出结果发到客户端的输出流里，也可以是一些流程控制语句。

4. 指令元素

指令元素用于从 JSP 发送一个消息到容器上。指令元素主要用来设置全局变量，声明类、方法和输出内容的类型等。指令并不向客户端产生任何输出，所有的指令都在 JSP 整个文件范围内有效。指令元素使用的格式如下：

```
<%@ directivename attribute="value", attribute="value" %>
```

JSP 中常见的指令元素有两种：页面指令和 include 指令。

(1) 页面(page)指令：用来定义 JSP 文件中的全局属性。如例 12.3 中的页面指令：

```
<%@ page language="java" contentType="text/html; charset=gb2312"pageEncoding=
"gb2312"%>
```

定义了 JSP 文件所使用的脚本语言是“Java”，响应的内容类型是“text/html”，使用的字符编码集是“GB2312”，JSP 页面的字符编码集是“GB2312”。

(2) include 指令：通知容器将指定位置上的资源内容包含到当前 JSP 页面中。被包含的文件内容可以被 JSP 解析，并且一经编译，内容不可变，如果要改变被包含文件的内容，必须重新编译 JSP 文件。如例 12.3 中的 include 指令：

```
<%@include file="header.jsp"%>
```

5. 动作元素

与指令元素不同，动作元素在请求处理阶段起作用。JSP 动作元素使用 XML 语法写成，它采用以下两种格式中的一种：

```
<prefix:tag attribute=value attribute-list…/>
```

或者

```
<prefix:tag attribute=value attribute-list…>
…
</prefix:tag>
```

容器在处理 JSP 时,每遇到动作元素都会根据它的标记进行特殊的处理。JSP 规范定义了一系列的标准动作,它们用 jsp 作为前缀。常见的动作元素有<jsp:include>、<jsp:forward>、<jsp:useBean>、<jsp:setProperty>、<jsp:getProperty>等。

(1) <jsp:include>:该操作允许在请求时间内在现有 JSP 页面中包含静态或者动态的资源。其语法如下:

```
<jsp:include page="fileName" flush="true"/>
```

(2) <jsp:forward>:该操作允许将请求转发给另一个 JSP、Servlet 或静态资源文件。每当遇到此操作时,就停止执行当前的 JSP,转而执行被转发的资源。其语法如下:

```
<jsp:forward page="url"/>
```

(3) <jsp:useBean>:用来在 JSP 页面中创建一个 JavaBean 实例,并指定它的名字和作用范围。其语法如下:

```
< jsp: useBean id =" id" scope =" page | request | session | application" class ="
className"/>
```

其中,id 表示实例的名字;scope 表示此对象可以使用的范围;class 指定对象的类型。

(4) <jsp:setProperty>:与 useBean 一起协作,用来设置 JavaBean 的属性。其语法如下:

```
<jsp:setProperty name="beanName" property="propertyName" value=
"propertyValue"/>
```

其中,name 指定要赋值的 JavaBean 的名字;property 指定属性名;value 指定属性值。

(5) <jsp:getProperty>:用来访问一个 JavaBean 的属性值。它访问的属性值将转换成一个 String,然后发送到输出流中。如果属性是一个对象,将调用 toString()方法。其语法如下:

```
<jsp:setProperty name="beanName" property="propertyName" />
```

12.2.3 JSP 内建对象

JSP 为了简化页面的开发提供了一些内部对象。这些内部对象不需要由 JSP 的编写者实例化,它们由容器实现和管理,在所有的 JSP 页面中都能使用内部对象。常见的 JSP 内部对象有 out 对象、request 对象、response 对象、session 对象和 application 对象。

1. out 对象

out 对象主要用来向客户端输出数据。out 对象表示为客户打开的输出流,可以使用 out 对象的 print()、println()方法向客户端发送数据。

2. request 对象

request 对象代表请求对象,通过 getParameter()方法可以得到 request 的参数。来自客户端的请求经 Servlet 容器处理后,由 request 对象进行封装。它作为 jspService()方法的一个参数由容器传给 JSP 页面。request 对象被包装成 HttpServletRequest 接口。

3. response 对象

response 对象封装了 JSP 产生的响应。和 request 对象一样,它由容器生成,作为 jspService()方法的一个参数传入 JSP 页面。因为输出流是缓冲的,所以可以设置 HTTP 状态码和 response 头。response 对象被包装成 HttpServletResponse 接口。

4. session 对象

session 对象用来保存每个用户信息，以便跟踪每个用户的操作状态。其中，session 信息保存在服务器端，session 的 ID 保存在客户端的 Cookie 中。一般情况下，用户首次登录系统时容器会给此用户分配一个唯一标识的 session ID，这个 ID 用于区分其他用户，当用户退出系统时，这个 session 就会自动消失。通过 session.setAtribute()方法把相关信息保存在 session 中，通过 session.getAtribute()方法从 session 中获取信息。

5. application 对象

application 对象为多个应用程序保存信息。对于一个容器而言，每个用户都共同使用一个 application 对象，这和 session 对象是不一样的。服务器启动后就会自动创建 application 对象，这个对象一直会保持，直到服务器关闭为止。通过 application.setAtribute()方法把相关信息保存在 application 中，通过 application.getAtribute()方法从 application 中获取信息。

【例 12.4】 使用 request 对象的例子。

数据输入的静态网页(input.html)如下：

```
<!DOCTYPE html PUBLIC "-//W3C//DTD HTML 4.01 Transitional//EN" "http://www.w3.
org/TR/html4/loose.dtd">
<html>
<head>
<meta http-equiv="Content-Type" content="text/html; charset=gb2312">
<title>数据输入</title>
</head>
<body>
<form method="post action"="requestDemo.jsp">
<table>
    <tr>
        <td>请输入登录名:</td>
        <td><input type="text" name="name"></td>
    </tr>
    <tr>
        <td>请输入密码:</td>
        <td><input type="password" name="password"></td>
    </tr>
    <tr>
        <td><input type="submit" value="登录"></td>
    </tr>
</table>
</form>
</body>
</html>
```

requestDemo.jsp 页面代码如下：

```
<%@ page language="java" contentType="text/html; charset=gb2312"
    pageEncoding="gb2312" import="java.io.*"%>
<!DOCTYPE html PUBLIC "-//W3C//DTD HTML 4.01 Transitional//EN" "http://www.w3.
org/TR/html4/loose.dtd">
<html>
<head>
<meta http-equiv="Content-Type" content="text/html; charset=gb2312">
```

```
<title>request 对象的使用</title>
</head>
<body>
Request 对象的信息:
<hr>
<%
    out.println("<br> getMethod:");
    out.println(request.getMethod());
    out.println("<br>getParameter:");
    out.println(request.getParameter("name"));
    out.println("<br>getAttributeNames:");
    java.util.Enumeration e = request.getAttributeNames();
    while (e.hasMoreElements())
        out.println(e.nextElement());
    out.println("<br>getCharacterEncoding:");
    out.println(request.getCharacterEncoding());
    out.println("<br>getContentLength: ");
    out.println(request.getContentLength());
    out.println("<br>getContentType:");
    out.println(request.getContentType());
    out.println("<br>getLocale:");
    out.println(request.getLocale());
    out.println("<br>getProtocol:");
    out.println(request.getProtocol());
    out.println("<br>getRemoteAddr:");
    out.println(request.getRemoteAddr());
    out.println("<br>getRemoteHost:");
    out.println(request.getRemoteHost());
    out.println("<br>getRemoteUser:");
    out.println(request.getRemoteUser());
    out.println("<br>getServerName:");
    out.println(request.getServerName());
    out.println("<br>getServerPort:");
    out.println(request.getServerPort());
    out.println("<br>getSession:");
    out.println(request.getSession(true));
    out.println("<br>getHeader('User-Agent')");
    out.println(request.getHeader("User-Agent"));
%>
</body>
</html>
```

程序可能的运行结果如图 12-4 所示。

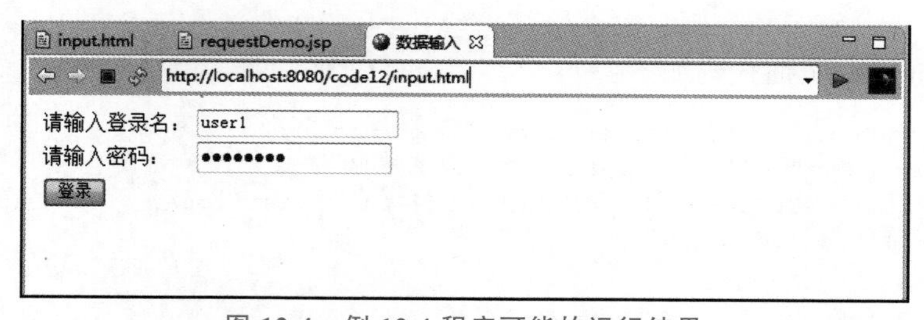

图 12-4 例 12.4 程序可能的运行结果

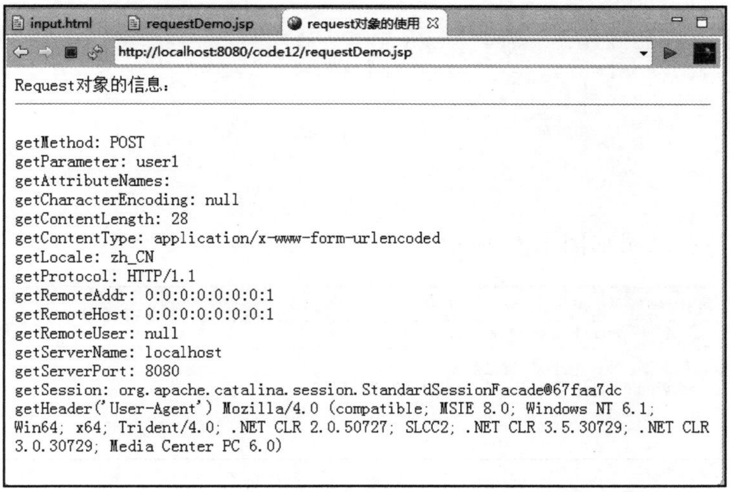

```
input.html  requestDemo.jsp  request对象的使用
http://localhost:8080/code12/requestDemo.jsp
Request对象的信息：

getMethod: POST
getParameter: user1
getAttributeNames:
getCharacterEncoding: null
getContentLength: 28
getContentType: application/x-www-form-urlencoded
getLocale: zh_CN
getProtocol: HTTP/1.1
getRemoteAddr: 0:0:0:0:0:0:0:1
getRemoteHost: 0:0:0:0:0:0:0:1
getRemoteUser: null
getServerName: localhost
getServerPort: 8080
getSession: org.apache.catalina.session.StandardSessionFacade@67faa7dc
getHeader('User-Agent') Mozilla/4.0 (compatible; MSIE 8.0; Windows NT 6.1;
Win64; x64; Trident/4.0; .NET CLR 2.0.50727; SLCC2; .NET CLR 3.5.30729; .NET CLR
3.0.30729; Media Center PC 6.0)
```

图 12-4 （续）

分析：在 input.html 中通过表单将用户输入的数据发送给 requestDemo.jsp 页面，这些数据被封装在 request 对象中。调用 request 对象的相关方法可以获取相关数据和属性值。

【例 12.5】 在 response 中使用 Cookie。

```
<%@ page language="java" contentType="text/html; charset=gb2312"
    pageEncoding="gb2312" import="java.util.*"%>
<!DOCTYPE html PUBLIC "-//W3C//DTD HTML 4.01 Transitional//EN" "http://www.w3.
org/TR/html4/loose.dtd">
<html>
<head>
<meta http-equiv="Content-Type" content="text/html; charset=gb2312">
<title>在 response 中使用 Cookie</title>
</head>
<body>
<%
    String userName = "zhangsan";
    Cookie[] cookie = request.getCookies();
    Cookie cookie_response = null;
    List list = Arrays.asList(cookie);
    Iterator it = list.iterator();
    while (it.hasNext()) {
        Cookie temp = (Cookie) it.next();
        if (temp.getName().equals(userName + "_access_time")) {
            cookie_response = temp;
            break;
        }
    }
    out.println("当前的时间:" + new Date() + "<br>");
    if (cookie_response != null) {
        out.println("上一次访问的时间:" + cookie_response.getValue());
        cookie_response.setValue(new Date().toString());
    } else {
        cookie_response = new Cookie(userName + "_access_time",
                new Date().toString());
```

```
    }
    response.addCookie(cookie_response);
    response.setContentType("text/html");
    response.flushBuffer();
%>
</body>
</html>
```

程序可能的运行结果如图 12-5 所示。

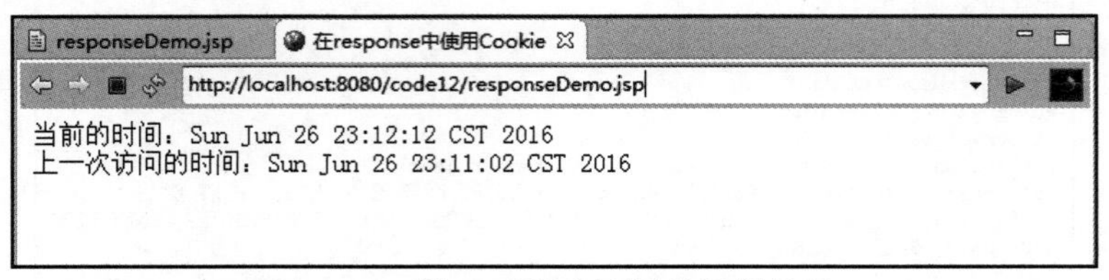

图 12-5 例 12.5 程序可能的运行结果

分析：上述程序中使用了一个 Cookie，该 Cookie 记录了用户上一次访问网页的时间。每次用户登录时，JSP 页面就通过 request 对象获取客户端的所有 Cookie，然后读取 Cookie 的值，如果有用户曾经访问过该网页，那么 request 对象中应该包含这个 Cookie，此时读取该 Cookie 的值，并且在网页中显示出来。最后更新 Cookie 的值并发送给客户端。

12.2.4 实用案例 12.1：商品信息展示

下面是商品信息展示的实用案例。

JavaBean 代码如下：

```
package code1202;
public class CommodityInfoBean {
    private String name;
    private Float price;
    private String status;

    public String getName() {
        return name;
    }

    public void setName(String name) {
        this.name = name;
    }

    public Float getPrice() {
        return price;
    }

    public void setPrice(Float price) {
        this.price = price;
    }

    public String getStatus() {
        return status;
    }

    public void setStatus(String status) {
        this.status = status;
```

```
    }
    public CommodityInfoBean() {
        this.name = "惠普笔记本电脑 CQ515";
        this.price = 2950.0f;
        this.status = "在售";
    }
}
```

showCommodityList.jsp 页面代码如下：

```
<%@ page language="java" contentType="text/html; charset=gb2312"
    pageEncoding="gb2312"%>
<!DOCTYPE html PUBLIC "-//W3C//DTD HTML 4.01 Transitional//EN" "http://www.w3.
org/TR/html4/loose.dtd">
<html>
<head>
<meta http-equiv="Content-Type" content="text/html; charset=gb2312">
<title>显示商品信息</title>
</head>
<body>
<jsp:useBean id="commodity" scope="page" class="code1202.CommodityInfoBean"/>
<h1>商品信息如下:</h1>
名称:<%=commodity.getName() %>
<br/>
价格:<%=commodity.getPrice() %>
<br/>
状态:<jsp:getProperty property="status" name="commodity"/>
<br/>
</body>
</html>
```

程序运行结果如图 12-6 所示。

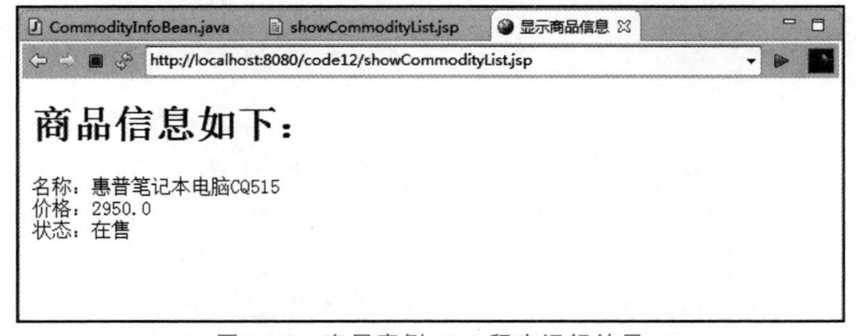

图 12-6　实用案例 12.1 程序运行结果

分析：商品信息的展示用一个 JSP 页面来实现，其中 CommodityInfoBean 类作为一个 JavaBean 存储商品信息。在 showCommodityList.jsp 页面中通过 jsp：useBean 在页面中创建 CommodityInfoBean 类的一个实例，并通过两种方式读取 JavaBean 中的信息：一是通过 JSP 表达式，即<%=commodity.getName() %>；二是通过 jsp：getProperty 来读取 JavaBean 的属性。

12.3 Servlet 技术

12.3.1 Servlet 介绍

Servlet 是用 Java 编写的服务器端程序,是由服务器端调用和执行的、按照 Servlet 自身规范编写的 Java 类。Servlet 可以看作用 Java 编写的 CGI,但是它的功能和性能比 CGI 更加强大。

Servlet 最大的好处是它可以处理客户端传来的 HTTP 请求,并返回一个响应。Servlet 是一个 Java 类,Java 语言能够实现的功能 Servlet 基本都能实现(除了图形界面外)。

Servlet 部署在容器里,它的生命周期由容器管理。Servlet 的生命周期概括为 5 个阶段:

(1) 装载 Servlet。这项操作一般是动态执行的。有些服务器提供了相应的管理功能。可以在启动的时候就装载 Servlet,并能够初始化特定的 Servlet。

(2) 创建一个 Servlet 实例。

(3) 调用 Servlet 的 init()方法。

(4) 服务。如果容器接收到对此 Servlet 的请求,那么它调用 Servlet 的 service()方法。

(5) 销毁。实例被销毁,通过调用 Servlet 的 destroy()方法来销毁 Servlet。

在 5 个阶段中,对外提供服务是最重要的阶段,service()方法是我们最关心的方法,因为它才是真正处理业务的方法。Servlet 如何为客户端提供服务,其过程如图 12-7 所示。

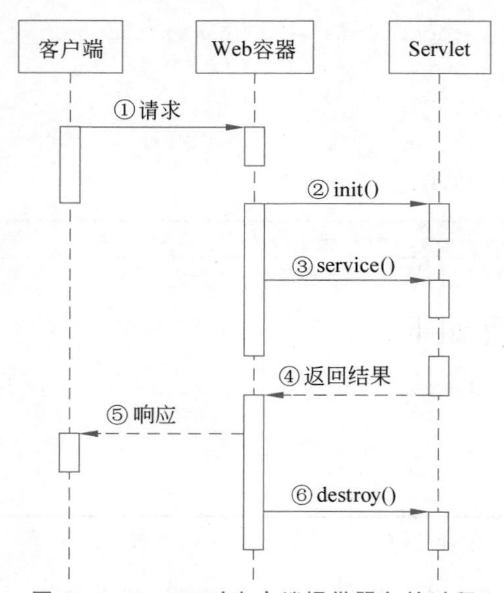

图 12-7 Servlet 对客户端提供服务的过程

将 Servlet 部署在容器中时,可以在部署描述符 web.xml 中进行 Servlet 配置。典型的 Servlet 配置包含 Servlet 的名字、Servlet 的类、初始化参数、Servlet 的映射等,其语法如下:

```
<servlet>
    <description>
    </description>
    <display-name>servletName</display-name>
```

```
        <servlet-name>servletName</servlet-name>
        <servlet-class>packageName.servletClassName</servlet-class>
    </servlet>
```

在配置 Servlet 时，必须先在<servlet-name>标记中指定 Servlet 的名字，并在<servlet-class>标记中指定 Servlet 的类名。还可以在<description>标记中添加关于 Servlet 的描述信息。

可以给一个 Servlet 做多个映射，这样就可以通过不同的方式来访问这个 Servlet。例如，可以对名为 servletName 的 Servlet 的映射配置如下：

```
<servlet-mapping>
    <servlet-name>servletName</servlet-name>
    <url-pattern>/servletName</url-pattern>
</servlet-mapping>
<servlet-mapping>
    <servlet-name>servletName</servlet-name>
    <url-pattern>/servletName/*</url-pattern>
</servlet-mapping>
<servlet-mapping>
    <servlet-name>servletName</servlet-name>
    <url-pattern>/servletName.html</url-pattern>
</servlet-mapping>
```

通过这些配置，可以使用不同的方式来访问这个 Servlet。有趣的是对于第二种映射方式，以下的访问都是有效的。

```
http://localhost:8080/ch12/servletName/afasfas?dafd
http://localhost:8080/ch12/servletName/servletName
http://localhost:8080/ch12/servletName/servletName.jsp
http://localhost:8080/ch12/servletName/servletName.jsp?name=abc
```

也就是说，它只要以/servletName/开头，都能访问到这个 Servlet。

12.3.2 Servlet 常用接口的使用

Servlet 相关类和接口主要包括以下几种类型。

（1）Servlet 实现相关：定义了用于实现 Servlet 相关的类和接口。

（2）Servlet 配置相关：主要包括 ServletConfig 接口。

（3）Servlet 异常相关：Servlet API 定义了两个异常，分别是 ServletException 和 UnavailableException。

（4）请求和响应相关：用于接收客户端的请求，并做出相应的响应。

（5）会话跟踪：用于跟踪与客户端的会话。

（6）Servlet 上下文：通过这个接口，可以在多个 Web 应用程序中共享数据。

（7）Servlet 协作：主要是 RequestDispatcher 接口，用于进行视图派发。

下面分别介绍 7 种 Servlet 常用接口的使用。

1. Servlet 实现相关

下面介绍和 Servlet 实现相关的类和接口。

（1）Servlet：其声明为

```
public interface Servlet
```

该接口是所有 Servlet 必须直接或者间接实现的接口,定义了以下方法。

void init(ServletConfig config):用于初始化 Servlet。

void destroy():销毁 Servlet。

java.lang.String getServletInfo():获得 Servlet 信息。

ServletConfig getServletConfig():获得 Servlet 配置相关信息。

void service(ServletRequest req, ServletResponse res):运行应用程序逻辑的入口点,它接收两个参数,ServletRequest 表示客户端请求的信息,ServletResponse 表示对客户端的响应。

(2) GenericServlet:其声明为

```
public abstract class GenericServlet extends java.lang.Object implements Servlet,
ServletConfig, java.io.Serializable
```

该抽象类提供了对 Servlet 接口的基本实现。它的 service()方法是一个抽象方法,GenericServlet 的派生类必须直接或者间接实现这个方法。

(3) HttpServlet:其声明为

```
public abstract class HttpServlet extends GenericServlet implements java.
io.Serializable
```

HttpServlet 类是针对使用 HTTP 的 Web 服务器的 Servlet 类,能够提供 HTTP 的功能。

HttpServlet 的子类必须实现以下方法中的一个。

void doGet(HttpServletRequest req, HttpServletResponse resp):支持 HTTP Get 请求。

void doPost(HttpServletRequest req, HttpServletResponse resp):支持 HTTP Post 请求。

void doPut(HttpServletRequest req, HttpServletResponse resp):支持 HTTP Put 请求。

void doDelete(HttpServletRequest req, HttpServletResponse resp):支持 HTTP Delete 请求。

2. Servlet 配置相关

javax.servlet.ServletConfig 接口代表了 Servlet 的配置。其声明为

```
public interface ServletConfig
```

这个接口的主要方法有以下几个。

java.lang.String getInitParameter(java.lang.String name):返回特定名字的初始化参数。

java.util.Enumeration getInitParameterNames():返回所有的初始化参数的名字。

ServletContext getServletContext():返回 Servlet 的上下文对象的引用。

3. Servlet 异常相关

(1) ServletException:其声明为

```
public class UnavailableException extends ServletException
```

它包含几个构造方法和一个获得异常原因的方法,这个方法如下。

java.lang.Throwable getRootCause():返回造成这个 ServletException 的原因。

（2）UnavailableException：其声明为

```
public class UnavailableException extends ServletException
```

当 Servlet 暂时或者永久不能使用时，就会抛出这个异常。

4. 请求和相应相关

和请求响应相关的类和接口非常多，这里重点介绍 HttpServletRequest 和 HttpServletResponse 两个接口。

（1）HttpServletRequest：其声明为

```
public interface HttpServletRequest extends ServletRequest
```

这个接口中最常用的方法就是获取请求中的参数，这个请求中的参数就是客户端表单中填写的数据。HttpServletRequest 接口可以获取由客户端传送的参数名称，也可以获取客户端正在使用的通信协议，可以获取产生请求和接收请求的服务器远端主机名及其 IP 地址等信息。

JSP 中的内建对象 request 实质上是一个 HttpServletRequest 实例。下面是 HttpServletRequest 接口的一些重要方法。

Cookie[] getCookies()：获得客户端发送的 Cookie。

HttpSession getSession()：返回和客户端关联的 Session，如果没有客户端分配 Session，则返回 null。

java.lang.String getParameter(java.lang.String name)：获得请求中名为 name 的参数的值，如果请求中没有这个参数，则返回 null。

java.lang.String[] getParameterValues(java.lang.String name)：返回请求中名为 name 的参数值，这个参数值往往是 checkbox 或者 select 控件提交的，获得的值是一个 String 数组。

java.lang.Object getAttribute(java.lang.String name)：返回由 name 指定的属性值，如果指定的属性值不存在，则会返回一个 null 值。

java.util.Enumeration getAttributeNames()：返回 request 对象的所有属性的名字集合，其结果是一个枚举的实例。

java.lang.String getCharacterEncoding()：返回请求中的字符编码方式。

int getContentLength()：返回请求的 Body 的长度，如果不确定长度，则返回－1。

java.lang.String getContentType()：返回请求的 MIME 类型，如果类型不确定，则返回－1。

java.util.Locale getLocale()：返回客户端所在的地域信息。

java.lang.String getProtocol()：获取客户端向服务器端传送数据所依据的协议名称。

java.lang.String getRemoteAddr()：获取客户端的 IP 地址。

java.lang.String getRemoteHost()：获取客户端的名字。

int getRemotePort()：获取客户端的端口号。

java.lang.String getServerName()：获取服务器的名字。

int getServerPort()：获取服务器的端口号。

void removeAttribute(java.lang.String name)：删除请求中的一个属性。

void setAttribute(java.lang.String name, java.lang.Object o)：设置名字为 name 的 request 参数值，该值由 Object 类型的参数 o 指定。

(2) HttpServletResponse：其声明为

```
public interface HttpServletResponse  extends ServletResponse
```

HttpServletResponse 代表了对客户端的 HTTP 响应。该接口允许 Servlet 设置内容长度和响应的 MIME 类型,并且提供输出流 ServletOutputStream。常用的方法有以下几种。

void addCookie(Cookie cookie)：在响应中增加一个 Cookie。

java.lang.String encodeURL(java.lang.String url)：使用 URL 和一个 sessionId 重写这个 URL。

void sendRedirect(java.lang.String location)：把响应发送到另一页面或者 Servlet 进行处理。

void setContentType(java.lang.String type)：设置响应的 MIME 类型。

void setCharacterEncoding(java.lang.String charset)：设置响应的字符编码类型。

void flushBuffer()：强制把当前缓冲区的内容发送到客户端。

ServletOutputStream getOutputStream()：返回到客户端的输出流对象。

void sendError(int sc)：向客户端发送错误的信息。例如,404 是指网页不存在或者请求的页面无效。

void setHeader(java.lang.String name, java.lang.String value)：设置指定名字的 HTTP 文件头的值,如果该值已经存在,则新值会覆盖原有的旧值。

5. 会话跟踪

和会话跟踪相关的类和接口是 HttpSession。其声明为

```
public interface HttpSession
```

这个接口被 Servlet 引擎用来实现 HTTP 客户端和 HTTP 会话之间的关联。这种关联可能在多处连续请求中持续一段给定的时间。Session 用来在无状态的 HTTP 下越过多个请求页面来维持状态和识别用户。

一个 Session 可以通过 Cookie 或者重写 URL 来维持,它的常用方法有以下几种。

long getCreationTime()：返回创建 Session 的时间。

java.lang.String getId()：返回分配给这个 Session 的标识符。一个 Http Session 的标识符是一个由服务器来建立和维护的唯一的字符串。

long getLastAccessedTime()：返回客户端最后一次发出与这个 Session 有关的请求的时间。如果这个 Session 是新建立的,则返回－1。

int getMaxInactiveInterval()：返回一个秒数,这个秒数表示客户端在不发出请求时,Session 被 Servlet 引擎维持的最长时间。在这个时间之后,Session 可能被 Servlet 引擎终止。如果这个 Session 不会被终止,则返回－1。

java.lang.Object getAttribute(java.lang.String name)：返回一个以给定的名字绑定到 Session 上的对象。如果不存在这样的绑定,则返回空值。

java.util.Enumeration getAttributeNames()：返回绑定到 Session 上的所有对象的名称集合。

void invalidate()：终止 Session,所有绑定到该 Session 上的数据都会被清除。

boolean isNew()：判断 Session 是不是新的。如果一个 Session 已经被服务器建立但是还没有收到相应客户端的请求,则这个 Session 被认为是新的。

void setAttribute(java.lang.String name, java.lang.Object value)：以给定的名字绑定给定的对象到 Session 中。已存在的同名数据会被新的数据覆盖。

void removeAttribute(java.lang.String name)：取消给定名字的对象在 Session 上的绑定。如果未找到给定名字的绑定对象,则该方法什么都不做。

void setMaxInactiveInterval(int interval)：设置一个秒数,这个秒数表示客户端在不发出请求时,Session 被 Servlet 引擎维持的最长时间。

6. Servlet 上下文

和 Servlet 上下文相关的接口有 ServletContext。其声明为

```
public interface ServletContext
```

ServletContext 对象表示一组 Servlet 共享的资源。以下是常用的方法。

java.lang.Object getAttribute(java.lang.String name)：获得 ServletContext 中名称为 name 的属性。

ServletContext getContext(java.lang.String uripath)：返回给定的 uripath 的应用的 Servlet 上下文。

void removeAttribute(java.lang.String name)：删除名称为 name 的属性。

void setAttribute(java.lang.String name, java.lang.Object object)：在 ServletContext 中设置一个属性,这个属性的名称为 name,值为 object 对象。

7. Servlet 协作

Servlet 协作主要是 RequestDispatcher 接口,它可以把一个请求转发给另一个 Servlet。其声明为

```
public interface RequestDispatcher
```

Servlet 协作包含以下方法。

void forward(ServletRequest request, ServletResponse response)：把请求转发给服务器上的另一个资源(Servlet、JSP 或者 HTML)。

void include(ServletRequest request, ServletResponse response)：把服务器上的另一个资源(Servlet、JSP 或者 HTML)包含到响应中。

12.3.3 使用 HttpServlet 处理客户端请求

Servlet 被设计成请求驱动。Servlet 的请求可能包含多个数据,当 Web 容器接收到某个对 Servlet 的请求时,它把它封装成一个 HttpServletRequest 对象,然后把此对象传给 Servlet 的对应的服务方法,服务方法通常是 doGet()和 doPost()方法。

1. doGet()方法

doGet()方法又称 GET 调用,用于获取服务器信息,并将其作为响应返回客户端。当经由 Web 浏览器,或者通过 HTML、JSP 直接访问 Servlet 的 URL 时,一般使用 Get 调用。Get 调用在 URL 里显示正在传送给 Servlet 的数据。

【例 12.6】 通过 Get 调用实现 Servlet 与网页中表单的交互。

网页代码如下：

```
<!DOCTYPE html PUBLIC "-//W3C//DTD HTML 4.01 Transitional//EN" "http://www.w3.
org/TR/html4/loose.dtd">
```

```
<html>
<head>
<meta http-equiv="Content-Type" content="text/html; charset=gb2312">
<title>使用 Get 调用来传递参数</title>
</head>
<body>
<form action="doGetDemo" method="get">
请输入参数:
<input type="text"    name="name" />
<input type="submit" value="提交" />
</form>
</body>
</html>
```

Servlet 代码如下:

```
package code1203;

import java.io.IOException;
import java.io.PrintWriter;

import javax.servlet.ServletException;
import javax.servlet.annotation.WebServlet;
import javax.servlet.http.HttpServlet;
import javax.servlet.http.HttpServletRequest;
import javax.servlet.http.HttpServletResponse;

@WebServlet(name="doGetDemo", urlPatterns="/doGetDemo")
public class DoGetDemo extends HttpServlet {
    private static final long serialVersionUID = 1L;

    protected void doGet(HttpServletRequest request,
            HttpServletResponse response) throws ServletException, IOException {
        request.setCharacterEncoding("gb2312");
        response.setContentType("text/html;charset=gb2312");
        PrintWriter out = response.getWriter();
        out.println("获得了以下参数值:name=<br>" + request.getParameter("name"));
        out.flush();
    }
}
```

程序运行结果如图 12-8 所示。

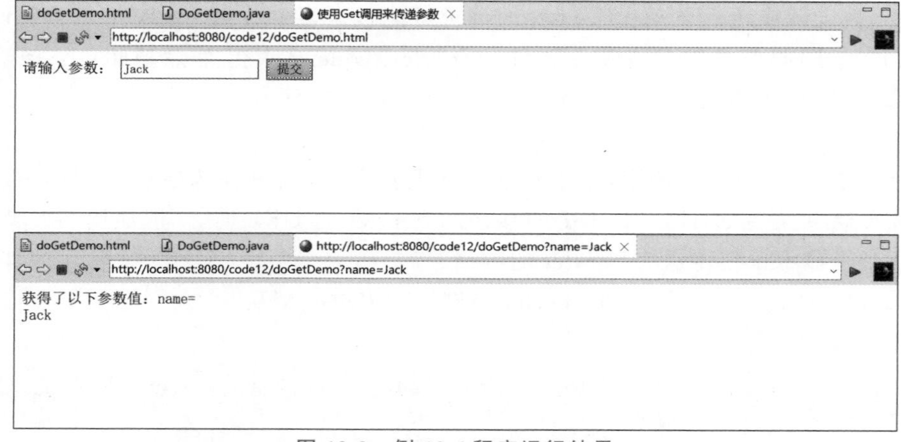

图 12-8 例 12.6 程序运行结果

分析：@WebServlet 注解用于配置 Servlet 类，可以取代在 web.xml 文件中的配置内容，其中 name 指定 Servlet 的名字，urlPatterns 指定 Servlet 映射的 URL。通过 Get 调用传递参数时，在客户端的 form 中必须指定调用的类型是 get。通过 Servlet 的 doGet()方法来处理请求，并通过 request.getParameter()方法来获得请求中的参数。注意在程序的运行结果中，请求的参数自动添加到了浏览器的地址栏中，这样可能会带来安全性方面的问题。

2. doPost()方法

doPost()方法又称 Post 调用，也用于客户端把数据传送给服务器端。使用它的好处是可以隐藏发送给服务器端的任何数据。Post 调用适合于发送大量的数据。

【例 12.7】 通过 Post 调用实现 Servlet 与网页中表单的交互。

网页代码如下：

```
<!DOCTYPE html PUBLIC "-//W3C//DTD HTML 4.01 Transitional//EN" "http://www.w3.
org/TR/html4/loose.dtd">
<html>
<head>
<meta http-equiv="Content-Type" content="text/html; charset=gb2312">
<title>使用 Post 调用来传递参数</title>
</head>
<body>
<form action="doPostDemo" method="post">
请输入参数:<br/>
<textarea rows="10" cols="50" name="data"></textarea><br/>
<input type="submit" value="提交" />
</form>
</body>
</html>
```

Servlet 代码如下：

```
package code1203;
import java.io.IOException;
import java.io.PrintWriter;
import javax.servlet.ServletException;
import javax.servlet.annotation.WebServlet;
import javax.servlet.http.HttpServlet;
import javax.servlet.http.HttpServletRequest;
import javax.servlet.http.HttpServletResponse;
@WebServlet("/doPostDemo")
public class doPostDemo extends HttpServlet {
    private static final long serialVersionUID = 1L;

    protected void doPost(HttpServletRequest request,
            HttpServletResponse response) throws ServletException, IOException {
        request.setCharacterEncoding("gb2312");
        response.setContentType("text/html;charset=gb2312");
        PrintWriter out = response.getWriter();
        out.println("获得了以下参数值:data=<br>" + request.getParameter("data"));
        out.flush();
    }
}
```

程序运行结果如图 12-9 所示。

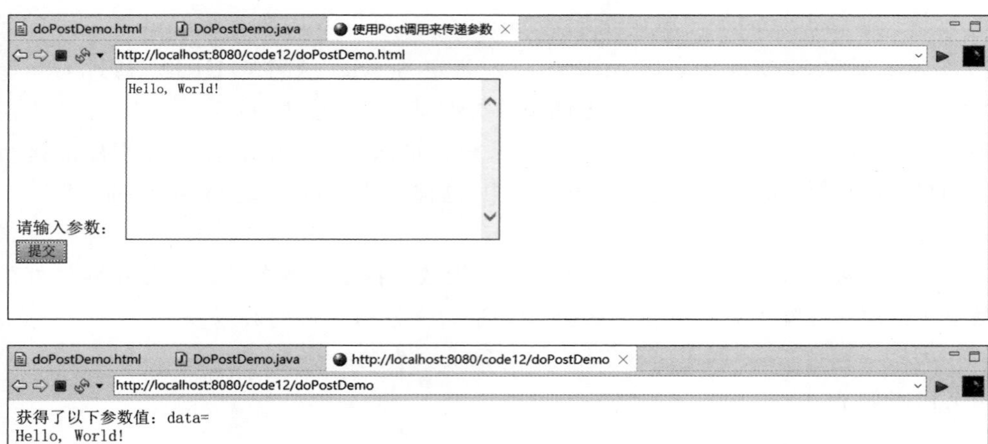

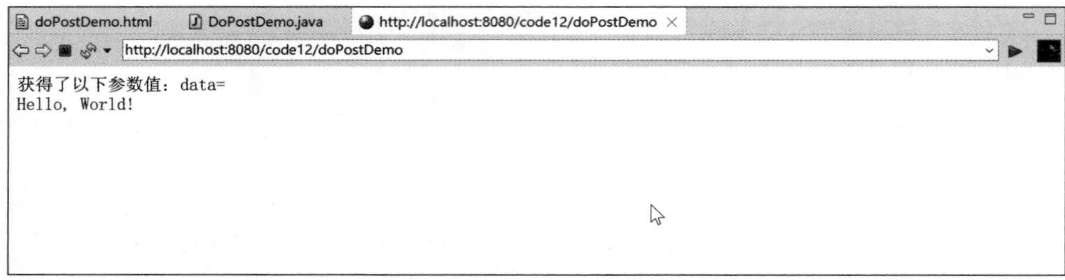

图 12-9　例 12.7 程序运行结果

分析：通过 Post 调用传递参数时，在客户端的 form 中必须指定调用的类型是 post。通过 Servlet 的 doPost()方法来处理请求，并通过 request.getParameter()方法来获得请求中的参数。可以看到，这里提交的数据不会在浏览器的地址栏中显示。

12.3.4　获得 Servlet 初始化参数

Servlet 可以配置一些初始化参数，可以在 Servlet 中获得这些初始的参数。下面看一个具体的例子。

【例 12.8】　假如 Servlet 需要知道应用程序的名字，可以把应用程序名 appName 配置为初始化参数(initParams)，然后在 Servlet 执行时来获取初始化参数 appName 的值。

Servlet 代码如下：

```
package ch1203;

import java.io.IOException;
import java.io.PrintWriter;

import javax.servlet.ServletException;
import javax.servlet.annotation.WebInitParam;
import javax.servlet.annotation.WebServlet;
import javax.servlet.http.HttpServlet;
import javax.servlet.http.HttpServletRequest;
import javax.servlet.http.HttpServletResponse;

@WebServlet(
        urlPatterns = { "/initParameterDemo" },
        initParams = {
                @WebInitParam(name = "appName", value = "第 12 章示例程序")
        })
public class InitParameterDemo extends HttpServlet {
    private static final long serialVersionUID = 1L;
```

```
    private String appName;
    public void init() throws ServletException {
        appName = getInitParameter("appName");
    }
    protected void doGet(HttpServletRequest request,
            HttpServletResponse response) throws ServletException, IOException {
        response.setCharacterEncoding("gb2312");
        response.setContentType("text/html;charset=gb2312");
        PrintWriter out = response.getWriter();
        out.println("应用程序名是:");
        out.println(appName);
    }
    protected void doPost(HttpServletRequest request,
            HttpServletResponse response) throws ServletException, IOException {
        doGet(request, response);
    }
}
```

程序运行结果如图 12-10 所示。

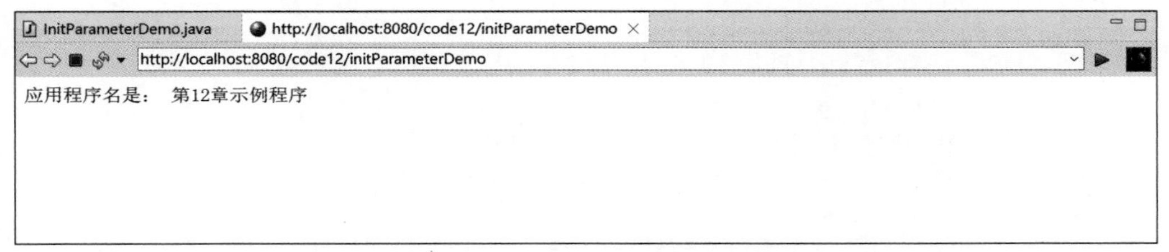

图 12-10 例 12.8 程序运行结果

分析：在 Servlet 类上使用@WebServlet 注解，对其<init-param>属性设置了 Servlet 的初始化参数，而在 Servlet 中通过 getInitParameter()方法来获取初始化参数值。

12.3.5 实用案例 12.2：基于 Session 实现简单的用户问好功能

Servlet 代码如下：

```
//GreetingDemo.java
    package ch1203;

import java.io.IOException;
import javax.servlet.RequestDispatcher;
import javax.servlet.ServletException;
import javax.servlet.annotation.WebServlet;
import javax.servlet.http.HttpServlet;
import javax.servlet.http.HttpServletRequest;
import javax.servlet.http.HttpServletResponse;
@WebServlet("/greetingDemo")
public class GreetingDemo extends HttpServlet {
    private static final long serialVersionUID = 1L;
      protected  void  doGet  ( HttpServletRequest  request,  HttpServletResponse
response) throws ServletException, IOException {
```

```
        doPost(request,response);
    }

    protected void doPost(HttpServletRequest request, HttpServletResponse
response) throws ServletException, IOException {
        String userName = "Jack";
        request.getSession().setAttribute("userName", userName);
        RequestDispatcher rd = getServletContext().getRequestDispatcher("/
showGreetingInfo");
        rd.forward(request, response);
    }
}
//ShowGreetingInfo.java
package ch1203;
import java.io.IOException;
import java.io.PrintWriter;
import javax.servlet.ServletException;
import javax.servlet.annotation.WebServlet;
import javax.servlet.http.HttpServlet;
import javax.servlet.http.HttpServletRequest;
import javax.servlet.http.HttpServletResponse;
@WebServlet("/showGreetingInfo")
public class ShowGreetingInfo extends HttpServlet {
    private static final long serialVersionUID = 1L;

    protected void doGet(HttpServletRequest request, HttpServletResponse
response) throws ServletException, IOException {
        doPost(request, response);
    }

    protected void doPost(HttpServletRequest request, HttpServletResponse
response) throws ServletException, IOException {
        String userName = (String)request.getSession().getAttribute("userName");
        response.setContentType("text/html;charset=gb2312");
        PrintWriter out = response.getWriter();
        out.println("你好," + userName);
        out.flush();
    }
}
```

程序运行结果如图 12-11 所示。

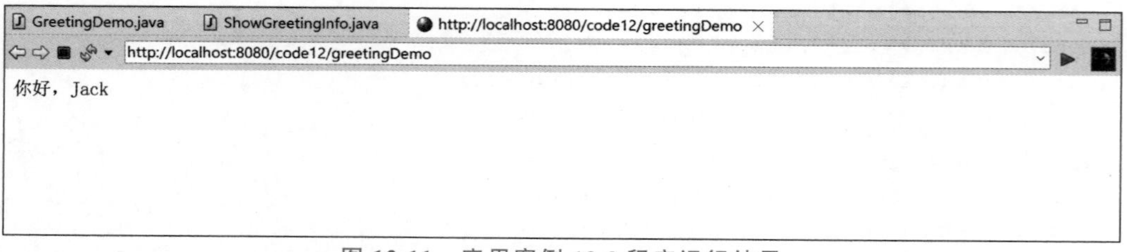

图 12-11　实用案例 12.2 程序运行结果

分析：案例12.2中请求的处理分成两步来完成。第一步由GreetingDemo.java来实现，主要是将用户名“Jack”存入session中；第二步由ShowGreetingInfo.java来实现，它从session中取出用户名变量，并将最后的显示结果传回客户端。这里两个Servlet之间的相互协作是通过RequestDispatcher和session来完成的，其中RequestDispatcher实现Servlet之间控制流的跳转，session用来作为传递数据的存储空间。

12.4 JSP和Servlet结合的方法

JSP网站开发技术标准给出了两种使用JSP的方法：模式一和模式二。模式一是JSP＋JavaBean的结合；模式二是JSP＋JavaBean＋Servlet的结合。在当今的开发中，比较偏向于使用模式二，但是模式一在小型应用开发中比较占优势。JSP模式一和模式二分别如图12-12和图12-13所示。

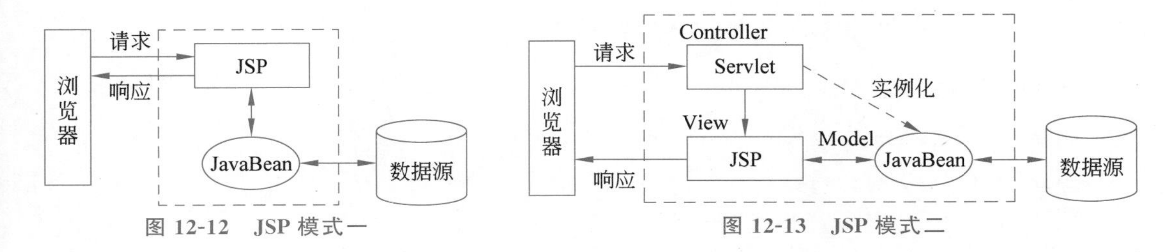

图12-12　JSP模式一　　图12-13　JSP模式二

12.4.1 模式一：JSP＋JavaBean

在模式一中，JSP页面独自响应请求，并且将处理结果返回客户。所有的数据通过JavaBean来处理，JSP实现页面的展现。模式一技术实现了页面展现与页面商业逻辑的分离。但是，大量使用此模式时，可能带来一个副作用，就是会导致在页面里嵌入大量的Java控制代码。当要处理的业务逻辑复杂时，这种情况会变得非常糟糕。所以在大型的项目里，这种方式将会导致页面维护困难。一般在小型的应用中可以考虑此模式。

12.4.2 模式二：JSP＋Servlet＋JavaBean

在模式二中，结合了JSP和Servlet技术。模式二充分利用了JSP和Servlet两种技术的优点。此模式遵循视图控制器模式(Model-View-Controler, MVC)，它的主要思想是使用一个或多个Servlet作为控制器。请求由前沿的Servlet接收并处理后，会重新定向到JSP。在Servlet作为控制器时，每个Servlet通常只实现很少一部分功能，多个Servlet控制器就可以结合起来完成复杂的任务，这样的好处是Servlet的重用性好。而JavaBean作为模型的角色，充当JSP和Servlet通信的中间工具。Servlet处理完后设置JavaBean的属性，JSP读取此JavaBean的属性，然后进行显示。在实际的项目开发中，页面设计者可以方便地使用HTML工具来开发JSP页面，而程序开发人员则可以用Java集成开发环境来开发Servlet。模式二更加明显地将显示与逻辑处理分离开，适合于大型项目的开发。

12.4.3 JSP和Servlet的选择

我们知道，所有的JSP都必须编译成Servlet，然后在Servlet容器中执行。从技术角度来看,JSP和Servlet是一样的。但二者还是有区别的。

JSP在某些方面要胜过Servlet:

(1) JSP以显示为中心,它为Web页面开发人员提供了更加方便的开发模式。

(2) JSP借助JavaBean可以把显示与内容分离。

(3) JSP由容器自动编译。

Servlet则在以下方面发挥作用:

(1) 协调输出。

(2) 处理非常简单的业务逻辑。

(3) 处理JSP不好处理的后台服务或者其他有特殊要求的问题。

在特定的软件系统环境中,选择使用JSP还是Servlet通常不是绝对的。最常见的情况是把两者结合起来使用,比如在JSP模式二中,把Servlet作为视图控制器,让它来处理请求,当Servlet处理完请求后,把处理结果转发给JSP,由JSP处理显示的问题。

12.4.4 实用案例12.3:网站计数器功能

Servlet代码如下:

```java
package code1204;

import java.io.IOException;
import javax.servlet.RequestDispatcher;
import javax.servlet.ServletException;
import javax.servlet.annotation.WebServlet;
import javax.servlet.http.HttpServlet;
import javax.servlet.http.HttpServletRequest;
import javax.servlet.http.HttpServletResponse;

@WebServlet("/websiteCounter")
public class WebsiteCounter extends HttpServlet {
    private static final long serialVersionUID = 1L;
    private static int count;

    protected void doGet(HttpServletRequest request,
            HttpServletResponse response) throws ServletException, IOException {
        count++;
        //将访问次数写入request中,以便传给JSP页面
        request.setAttribute("hitCount", count);
        String url = "/websiteCounterInfo.jsp";
        RequestDispatcher rd = getServletContext().getRequestDispatcher(url);
        rd.forward(request, response);
    }
}
```

JSP页面代码如下:

```
<%@ page language="java" contentType="text/html; charset=gb2312"
    pageEncoding="gb2312"%>
<!DOCTYPE html PUBLIC "-//W3C//DTD HTML 4.01 Transitional//EN" "http://www.w3.
org/TR/html4/loose.dtd">
```

```
<html>
<head>
<meta http-equiv="Content-Type" content="text/html; charset=gb2312">
<title>网站计数器</title>
</head>
<body>
欢迎你,你是第
<%=request.getAttribute("hitCount")%>
位访问本网站的用户!
</body>
</html>
```

程序可能的运行结果如图 12-14 所示。

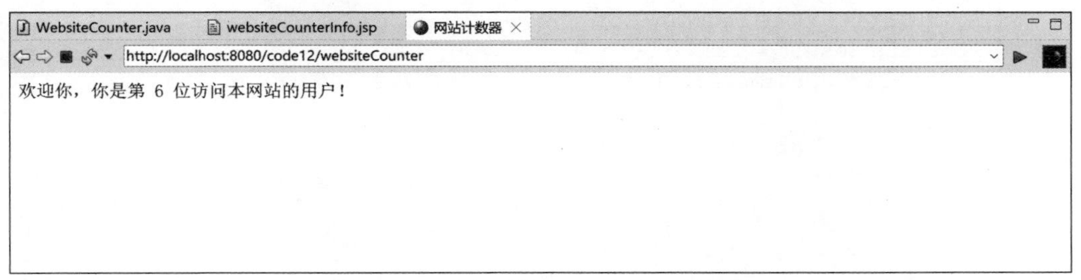

图 12-14 实用案例 12.3 程序运行结果

分析：通过一个静态变量 count 来记录网站的访问次数，每当网站被访问一次时，count 自增 1。在程序中，利用 Servlet 来处理用户的请求，并管理 count 的值；而利用 JSP 页面来显示相关信息。通过使用 RequestDispatcher 对象来实现从 Servlet 到 JSP 页面的跳转，把相应的 count 值作为 request 对象中的参数传递给 JSP 页面，以便在 JSP 页面中进行显示。

12.5 JSP 与 Servlet 开发实训任务

【任务描述】

编写一个简单的 Web 应用，实现用户登录的功能。

【任务分析】

用户登录应用程序主要包括两步：①用户在网页中输入用户名和密码；②对用户输入的用户名和密码进行验证，判断是否是合法用户，最后将登录成功与否的结果返回给用户。这里采用 JSP 模式二的方法来实现用户登录功能，即编写一个 JavaBean(UserBean 类)作为模型，其中封装了验证合法用户的代码；编写一个 Servlet(UserLoginServlet 类)作为控制器，处理用户的请求；编写一个 JSP 页面(login.jsp)作为视图，显示用户的输入界面和登录结果界面。假定合法用户的用户名是 admin，密码是 admin。

【任务解决】

UserBean 类代码如下：

```
package code1205;

public class UserBean {
    private String validUserName;
    private String validUserPassword;
```

```
    public String getValidUserName() {
        return validUserName;
    }

    public void setValidUserName(String validUserName) {
        this.validUserName = validUserName;
    }

    public String getValidUserPassword() {
        return validUserPassword;
    }

    public void setValidUserPassword(String validUserPassword) {
        this.validUserPassword = validUserPassword;
    }

    public boolean isValidUser(String name, String password) {
        boolean result = false;
        if (validUserName.equals(name) && validUserPassword.equals(password)) {
            result = true;
        }
        return result;
    }
}
```

Servlet 类代码如下：

```
package code1205;
import java.io.IOException;
import java.io.PrintWriter;
import javax.servlet.ServletException;
import javax.servlet.annotation.WebServlet;
import javax.servlet.http.HttpServlet;
import javax.servlet.http.HttpServletRequest;
import javax.servlet.http.HttpServletResponse;
@WebServlet("/userLogin")
public class UserLogin extends HttpServlet {
    private static final long serialVersionUID = 1L;
      protected void doGet (HttpServletRequest request, HttpServletResponse
response) throws ServletException, IOException {
        String name = request.getParameter("name");
        String password = request.getParameter("password");
        UserBean user = new UserBean();
        user.setValidUserName("admin");          //设置合法用户的信息
        user.setValidUserPassword("admin");
        response.setCharacterEncoding("gb2312");
        response.setContentType("text/html;charset=gb2312");
        PrintWriter out = response.getWriter();
        if(user.isValidUser(name, password)){    //调用 JavaBean 的方法来验证用户
            out.println("登录成功!");
        }else{
            out.println("用户名或者密码错误,<a href='login.jsp'>请重试。</a>");
        }
    }
```

```
    protected void doPost(HttpServletRequest request, HttpServletResponse
response) throws ServletException, IOException {
        doGet(request,response);
    }
}
```

JSP 页面代码如下：

```
<%@ page language="java" contentType="text/html; charset=gb2312" pageEncoding="
gb2312"%>
<!DOCTYPE html PUBLIC "-//W3C//DTD HTML 4.01 Transitional//EN" "http://www.w3.
org/TR/html4/loose.dtd">
<html>
<head>
<meta http-equiv="Content-Type" content="text/html; charset=gb2312">
<title>用户登录</title>
</head>
<body>
<form action="userLogin" method="post">
<table>
    <tr>
        <td>用户名:</td>
        <td><input type="text" name="name" /></td>
    </tr>
    <tr>
        <td>密码:</td>
        <td><input type="password" name="password" /></td>
    </tr>
    <tr>
        <td colspan="2"><input type="submit" value="登录" /></td>
    </tr>
</table>
</form>
</body>
</html>
```

程序可能的运行结果如图 12-15 所示。

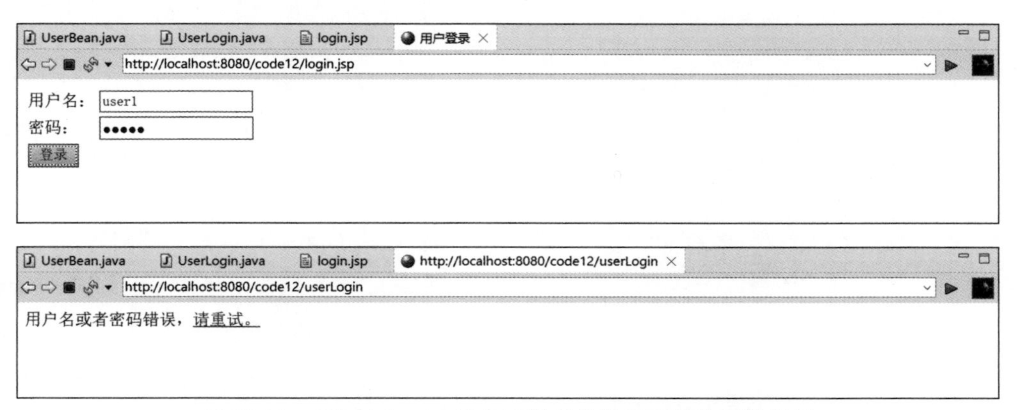

图 12-15 JSP 与 Servlet 开发实训任务程序可能的运行结果

图 12-15 (续)

1. JSP 的工作原理是什么?
2. JSP 由哪些元素构成?
3. JSP 有哪些内置对象?
4. 如何将 JSP 与 Servlet 结合起来进行 Web 应用的开发?

第13章 用 Tomcat 构建 Web 站点

本章学习目标

Web 程序开发是 Java 最擅长的一个领域，也是目前工具最多、技术最复杂的一个领域。开发 Web 程序，首先要掌握如何使用 Web 服务器。所谓 Web 服务器，就是一种能够运行 Web 程序的软件系统。本章将以 Java 领域最流行的 Tomcat 服务器为基础，结合 Eclipse IDE for Java EE Developers 主要介绍以下内容：

(1) Tomcat 和 Eclipse IDE for Java EE Developers 的安装使用。

(2) 如何在 Eclipse IDE for Java EE Developers 中结合 Tomcat 编写 Web 站点。

(3) 如何在 Eclipse IDE for Java EE Developers 中用 Tomcat 运行 Web 站点。

(4) 如何将 Web 站点发布到 Tomcat。

13.1 Tomcat 简介

Tomcat 是 Apache 软件基金会（Apache Software Foundation）的 Jakarta 项目中的一个核心项目，是目前 Java 领域应用最广泛的轻量级 Web 应用服务器。在中小型系统和并发访问用户数量不大的场合被普遍使用，并且是学习、开发和调试 Web 程序的首选。Tomcat 主要包括一个 JSP 和 Servlet 的容器。JSP 和 Servlet 是 Java 最主要的 Web 开发技术。

最初 Tomcat 由 Sun 公司的软件架构师詹姆斯・邓肯・戴维森开发。早期是为了给 Sun 公司正在开发的 JSP 和 Servlet 技术提供一个运行环境。后来他将其变为开源项目，并由 Sun 公司贡献给 Apache 软件基金会。由于 O'Reilly（国外著名的计算机类图书出版商）会为大部分开源项目出版一本相关的图书，并且将其封面设计成某个动物的素描，因此当 O'Reilly 为 Tomcat 出书时，Tomcat 的作者希望将此项目以一个动物的名字命名。他希望这种动物能够自己照顾自己，最终他将其命名为 Tomcat。而 Tomcat 的 Logo 兼吉祥物也被设计成了一只公猫，如图 13-1 所示。Tomcat 一词在英文中的含义为“公猫”，它和动画片《猫和老鼠》并没有什么渊源。目前 Tomcat 最新的稳定版本为 10.x。

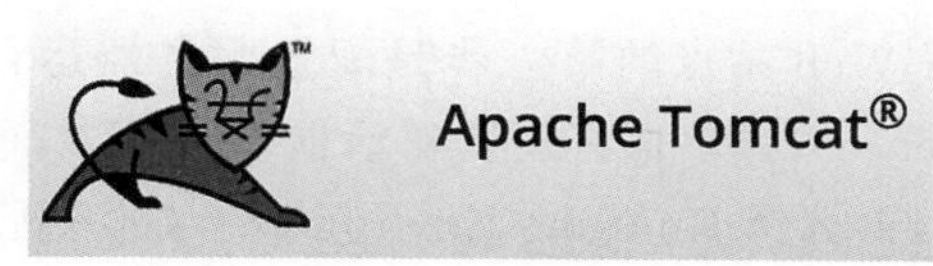

图 13-1 Tomcat 的 Logo

13.2 安装配置 Tomcat

由于 Tomcat 是基于 Java 运行的程序，因此在安装 Tomcat 之前需要首先安装好 Java SE 环境，设置好 JAVA_HOME 和 PATH 环境变量。Tomcat 软件可以在 Tomcat 官方网站(http://tomcat.apache.org)下载，如图 13-2 所示。网站提供了二进制版本(Binary Distributions)和源代码版本(Source Code Distributions)供用户下载。通常如果只是使用 Tomcat 应该下载二进制版本，而如果想为 Tomcat 这个开源项目贡献力量或者想了解 Tomcat 的工作原理就可以下载源代码版本。二进制版本分成 Core 和 Deployer 两部分。Core 是 Tomcat 程序的核心部分，要使用 Tomcat 就必须下载；而 Deployer 提供了部署发布 Web 程序的辅助功能，属于选择下载的部分。Core 之下又分为多种类型的下载格式，本书选择下载 10.0.23 版本的 zip 格式。程序下载之后，将文件解压到任意目录即可完成安装(本书选择解压到 E:\开发工具)。

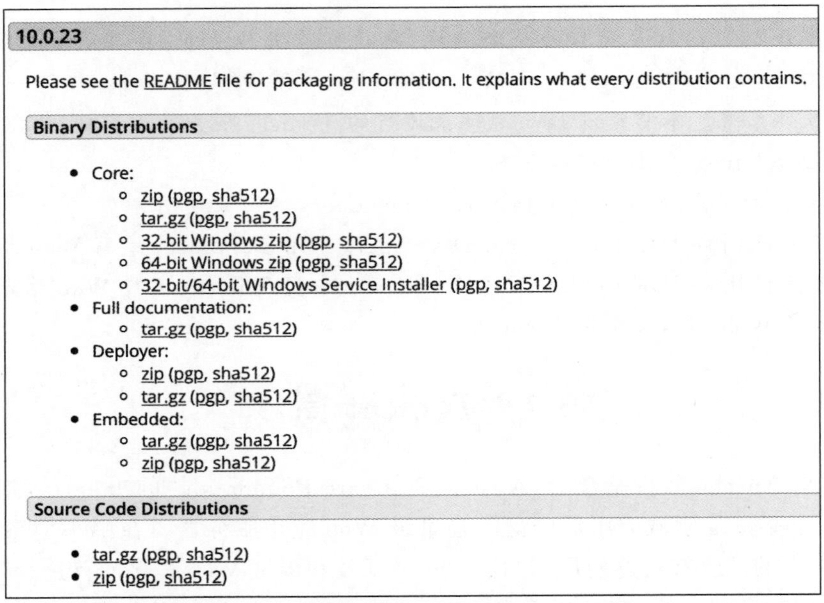

10.0.23

Please see the README file for packaging information. It explains what every distribution contains.

Binary Distributions

- Core:
  - zip (pgp, sha512)
  - tar.gz (pgp, sha512)
  - 32-bit Windows zip (pgp, sha512)
  - 64-bit Windows zip (pgp, sha512)
  - 32-bit/64-bit Windows Service Installer (pgp, sha512)
- Full documentation:
  - tar.gz (pgp, sha512)
- Deployer:
  - zip (pgp, sha512)
  - tar.gz (pgp, sha512)
- Embedded:
  - tar.gz (pgp, sha512)
  - zip (pgp, sha512)

Source Code Distributions

- tar.gz (pgp, sha512)
- zip (pgp, sha512)

图 13-2 选择下载 Tomcat

13.3 编写简单的 Web 站点

本书使用 Eclipse IDE for Java EE Developers(简称 Eclipse Java EE)进行 Web 站点的开发。Eclipse Java EE(本书使用版本为 20211202-1639)可以从 Eclipse 的官方网站(http://www.eclipse.org/downloads/)下载，下载之后解压压缩文件即可完成安装。

13.3.1 配置服务器运行环境

用 Eclipse Java EE 编写 Web 站点的第一步是配置服务器运行环境。在 Eclipse 中的菜单项 Window 中选择 Preferences，弹出偏好设置对话框，如图 13-3(a)所示。

选择对话框中的 Server，再选择 Runtime Environment，进入配置服务器运行环境的对话框，如图 13-3(b)所示。

在图 13-3(b)对话框中单击 Add 按键，弹出图 13-4(a)所示的对话框。在该对话框中，可以选择要配置环境的服务器名称。Eclipse 已经预先设置了一些常用的服务器，包括 Apache Tomcat、IBM Websphere、JBoss 等。为了配置 Tomcat 的运行环境，这里选择 Apache Tomcat 10.0(如果使用其他的 Tomcat 版本，如 9.0，则需要选择对应的版本)。

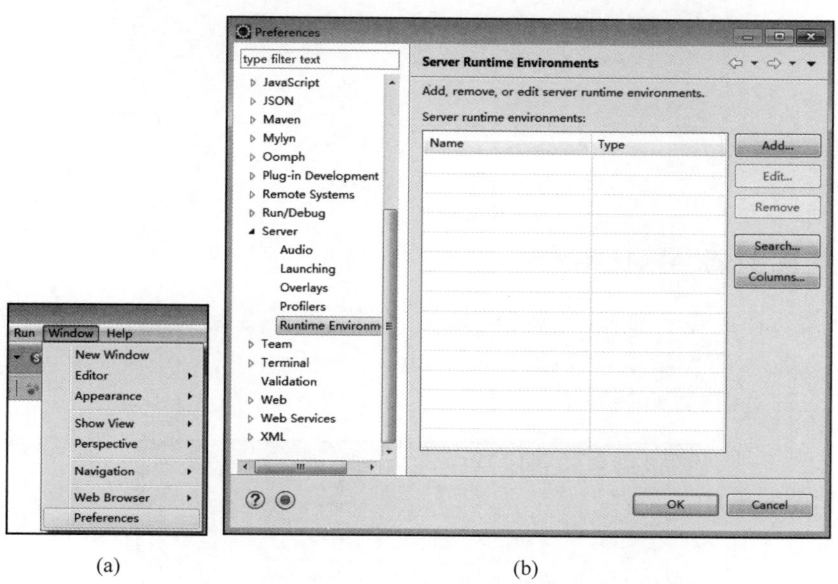

(a) (b)

图 13-3　配置服务器运行环境

在图 13-4(b)所示的对话框中输入 Tomcat 的安装位置(本书为 E:\开发工具\apache-tomcat-10.0.23)，然后单击 Finish 按钮，Tomcat 的运行环境就配置好了。

Eclipse 的设置非常人性化，如果用户事先没有安装 Tomcat，在图 13-4(b)所示对话框中可以单击 Download and Install 按钮下载 Tomcat，并完成配置。

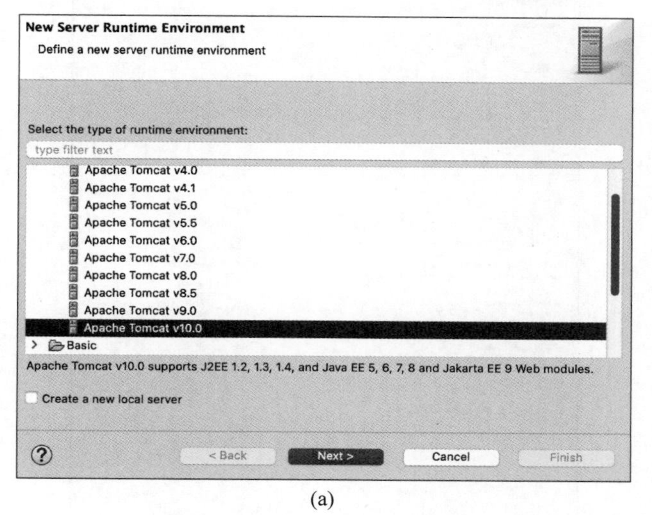

(a)

图 13-4　配置 Tomcat 运行环境

(b)

图 13-4 （续）

13.3.2 新建动态 Web 工程

Tomcat 运行环境建立之后，下一步就是建立动态 Web 工程(Dynamic Web Project)。选择菜单项 File，选择 New 选项，最后选择 Dynamic Web Project(见图 13-5(a))，就会弹出新建动态 Web 工程的对话框，如图 13-5(b)所示。

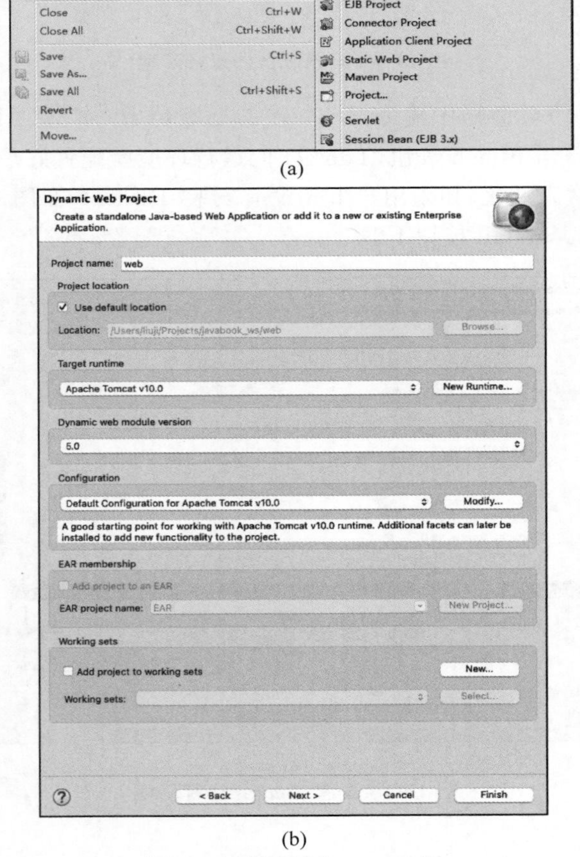

(a)

(b)

图 13-5 新建动态 Web 工程

输入工程名称(这里为 web),选择正确的 Target runtime。由于本书选择 Tomcat10.0 作为 Web 站点的运行环境,因此在 Target runtime 下拉列表框中,选择 Apache Tomcat v10.0。对话框上的其他设置保持不变,单击 Finish 按钮,一个名称为 web 的动态 Web 工程就建好了。

13.3.3 Web 工程的结构

新建的 Web 工程结构如图 13-6 所示。

图 13-6 中 src/main/java 目录放置 Java 源代码(主要是 Servlet 等 Java 类),webapp 子目录下放置 HTML、JSP、图片、视频等 Web 内容。

webapp 目录下的 WEB-INF 子目录非常重要,不能删除,也不能重命名。其下的 lib 子目录用于放置 Web 工程可能用到的外部 Java 库文件。web.xml 文件是项目的配置文件,提供了 Error Pages、Filter Mappings、Filters、Listeners、References、Servlet Mappings、Servlets、Welcome Pages 等配置。

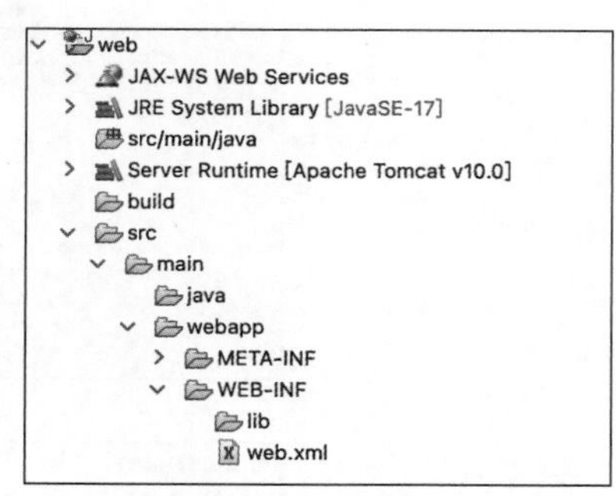

图 13-6 Web 工程的结构

13.3.4 新建 Servlet 和 JSP 程序

Servlet 和 JSP 是 Java Web 开发的主要技术。首先介绍如何创建一个 Servlet 程序,如图 13-7 所示。

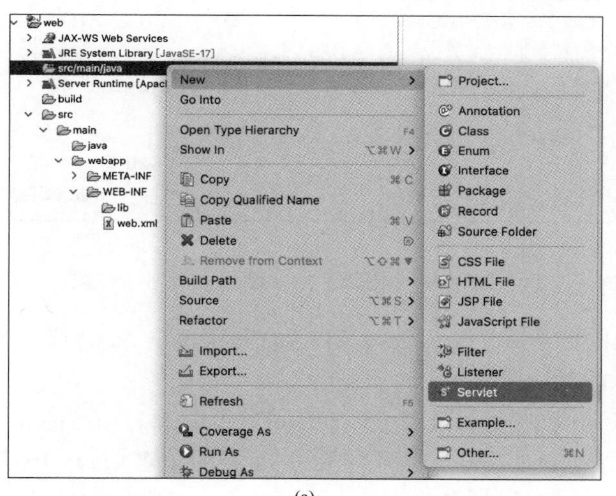

(a)

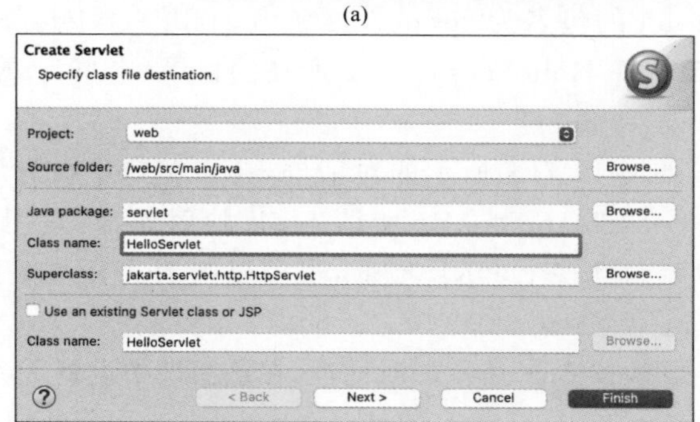

(b)

图 13-7 新建 Servlet

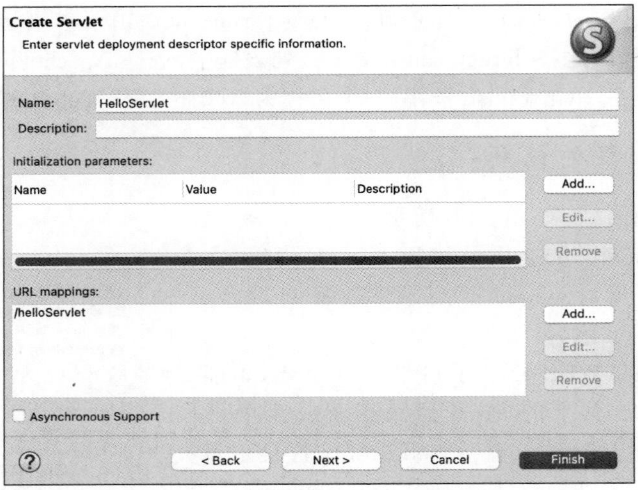

(c)

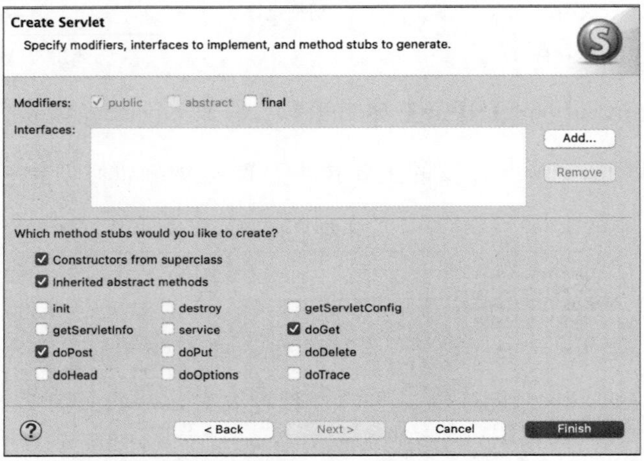

(d)

图 13-7 （续）

在 Web 工程中选中目录 src/main/java，右击，选择菜单项 New，选择 Servlet（见 13-7(a)），然后弹出图 13-7(b)所示的对话框。

所谓 Servlet，实际上是一个父类为 jakarta.servlet.http.HttpServlet 的 Java 类。因此，在图 13-7(b)对话框中，要输入 Servlet 所对应的 Java 类的包名(Java Package)以及类名(Class name)。本书新建的 Servlet 的包名为 servlet，类名为 HelloServlet。Source folder 选项指定了 HelloServlet 所对应源文件 HelloServlet.java 所在的目录（本书为/web/src/main/java），默认情况下不需要修改 Source folder。

单击 Next 按钮进入如图 13-7(c)所示的对话框，在该窗口中设置 HelloServlet 的初始化参数和 URL 映射地址(URL Mappings)。所谓 URL 映射地址，是用户访问该 Servlet 的地址。同一个 Servlet 可以映射到多个 URL 地址，也可以映射为任意名称的 URL 地址。本书将 HelloServlet 映射为/helloservlet。

单击 Next 按钮进入如图 13-7(d)所示的对话框，在该窗口中设置 HelloServlet 所要重写的父类方法。默认选择重写 doGet()和 doPost()方法。doGet()方法用于处理 HTTP 协议中 Get 调用方式的请求，doPost()方法用于处理 HTTP 协议中 Post 调用方式的请求。Get 和 Post 是浏览器通过 HTTP 协议访问 Web 站点最常见的两种调用方式。最后单击 Finish 按钮。

选择 src/main/java 目录下 servlet 包中的 HelloServlet.Java 文件，用 Eclipse 的代码编辑器，编写如例 13.1 所示的 Java 代码，保存文件。HelloServlet 的创建就完成了。注意，Servlet 3.0 开始支持用 Annotation @WebServlet 映射 Servlet。如果采用这种方式，请删除 web.xml 文件中的映射，并打开下列源代码中的@WebServlet 注解。

【例 13.1】 创建 HelloServlet.Java。

```
package servlet;
import jakarta.servlet.ServletException;
import jakarta.servlet.annotation.WebServlet;
import jakarta.servlet.http.HttpServlet;
import jakarta.servlet.http.HttpServletRequest;
import jakarta.servlet.http.HttpServletResponse;
import java.io.IOException;
import java.io.PrintWriter;
//@WebServlet("/helloServlet")
public class HelloServlet extends HttpServlet {
     protected void doGet (HttpServletRequest request, HttpServletResponse
response) throws ServletException, IOException {
         //采用 UTF-8 编码
         response.setCharacterEncoding("UTF-8");
         PrintWriter writer=response.getWriter();
         writer.println("<!DOCTYPE html>\n");
         writer.println("<html>\n");
         writer.println("<head>\n");
         writer.println("<meta charset=\"UTF-8\">\n");
         writer.println("<title>JSP</title>\n");
         writer.println("</head>\n");
         writer.println("<body>\n");
         writer.println("Hello World!<br>\n");
         writer.println("这是 Servlet 输出的信息。\n");
         writer.println("</body>");
         writer.println("</html>");
         writer.close();
     }
     protected void doPost (HttpServletRequest request, HttpServletResponse
response) throws ServletException, IOException {
         //用同样的方法处理 Get 和 Post 调用请求
         doGet(request, response);
     }
}
```

下面介绍如何创建 JSP 程序。选择 webapp 目录，右击，选择菜单项 New，选择 JSP（见图 13-8(a)），弹出如图 13-8(b)所示的对话框。在 13-8(b)对话框中输入 JSP 的文件名称（本书为 hello.jsp），单击 Finish 按钮就创建了一个名称为 hello.jsp 的 JSP 文件。

用编辑器打开 hello.jsp 文件，输入例 13.2 所示的代码，保存文件，JSP 程序的创建就完成了。

【例 13.2】 创建 hello.jsp。

```
<%@ page language="java" contentType="text/html; charset=UTF-8"
    pageEncoding="UTF-8"%>
<!DOCTYPE html>
<html>
```

```
<head>
<meta charset="UTF-8">
<title>JSP</title>
</head>
<body>
Hello World!<br>
这是 JSP 输出的信息。
</body>
</html>
```

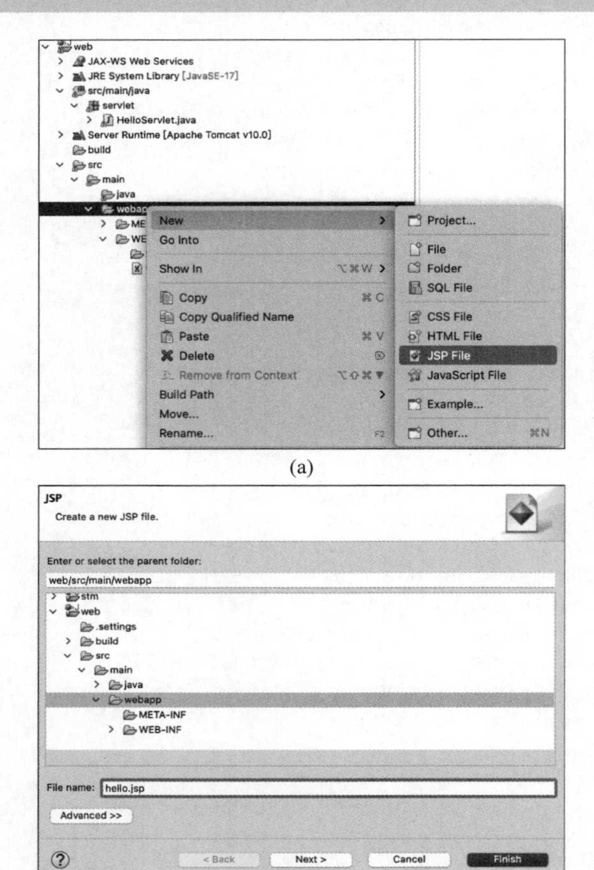

(a)

(b)

图 13-8 新建 JSP

13.4 运行 Web 站点

第一次运行 Web 站点需要一些额外的设置。首先选中 Web 工程,选择工具栏上的 Run 按钮,选择 Run As 选项,选择 Run on Server 选项(见图 13-9(a)),弹出 13-9(b)所示的对话框。

在图 13-9(b)对话框中,需要选择运行 Web 站点的服务器。由于本书使用 Tomcat 10.0.23 服务器,因此这里选择 Tomcat v10.0 Server。最后单击 Finish 按钮,Web 站点就启动了。Web 站点启动之后,可以在 Eclipse 界面下方的信息提示窗口,查看到服务器的运行状态和 Web 站点的启动情况，如图 13-9(c)所示。

Web 站点启动之后，在打开的浏览器或者操作系统的浏览器(IE 或者 Firefox 等)的地址栏中输入 HelloServlet 对应的 URL 地址(http://localhost:8080/web/helloServlet)，就可以查看 HelloServlet 的运行情况。在地址栏中输入 JSP 程序对应的 URL 地址(http://localhost:8080/web/hello.jsp)，就可以查看 JSP 程序的运行情况。hello.jsp 和 HelloServlet 的运行情况分别如图 13-9(d)和图 13-9(e)所示。

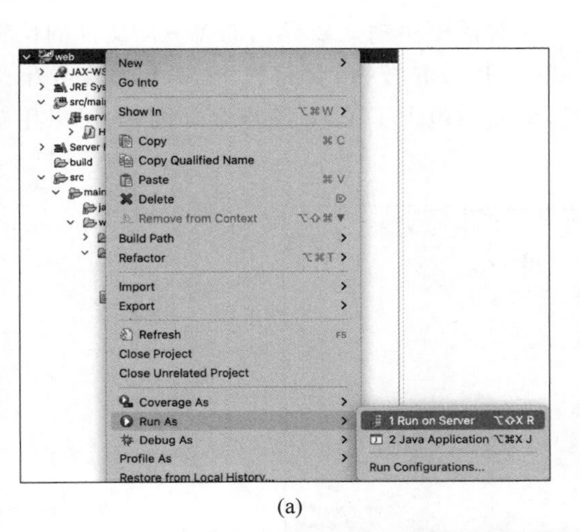

(a)

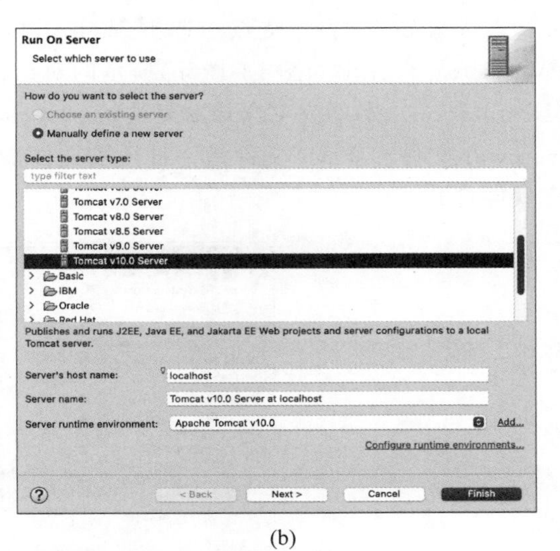

(b)

(c)

(d)

(e)

图 13-9　运行 Web 站点

13.5 发布 Web 站点

在 Eclipse Java EE 中运行的 Web 程序，只是为了便于开发调试。当一个 Web 程序完成开发之后，就需要部署到实际的服务器上运行。本节将介绍如何将 13.4 节中的 Web 程序发布到 Tomcat 中。

首先要导出 Web 站点。选中 Web 工程，右击，选择菜单项 Export，选择 WAR file(见图 13-10(a))，弹出如图 13-10(b)所示的对话框。在对话框中输入要导出的 WAR 文件的位置(Destination)，本书为 E:\开发工具\web.war。单击 Finish 按钮，就将整个 Web 站点导出为 web.war 文件。WAR 文件是一种压缩文件格式，该文件包含了 Web 站点全部内容，可以用解压软件打开。

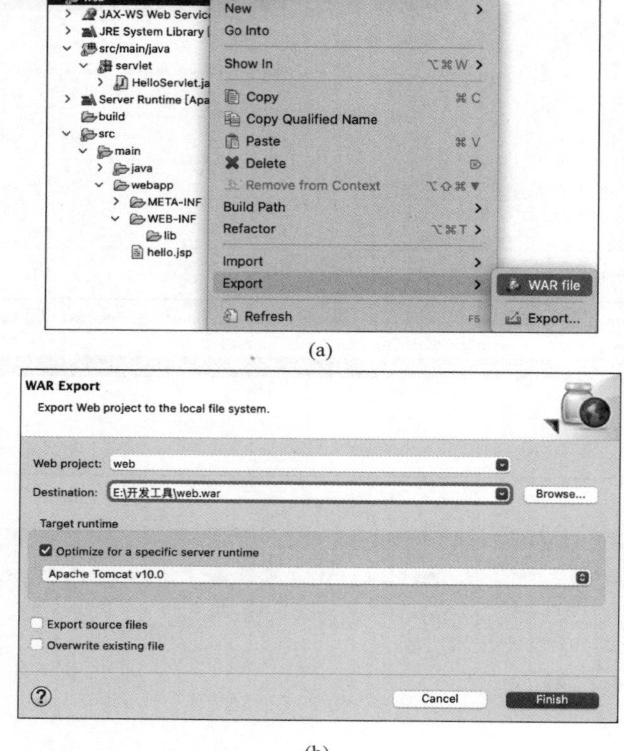

(a)

(b)

图 13-10 导出 Web 站点

下一步是部署 Web 站点。将导出的 web.war 文件复制到 tomcat 安装目录下的 webapps 文件夹(本书为 E:\开发工具\apache-tomcat-10.0.23\webapps)下，如图 13-11(a)所示。然后启动 Tomcat 服务器。即在 tomcat 安装目录下的 bin 文件夹下运行 startup.bat 程序，如图 13-11(a)所示。Tomcat 服务器启动后，在 IE 浏览器中查看 HelloServlet 和 hello.jsp 得到如图 13-12 所示的结果。

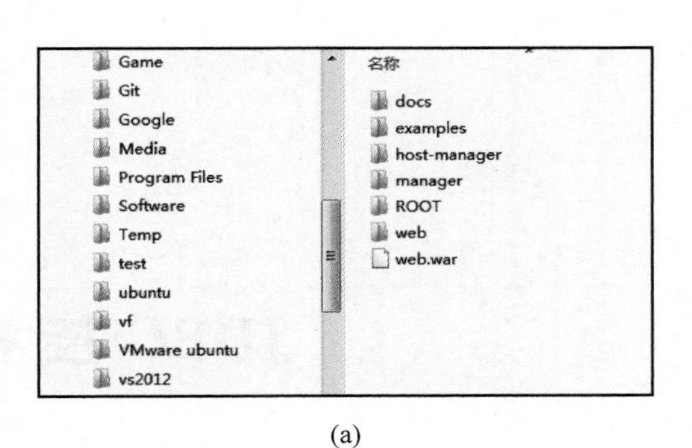

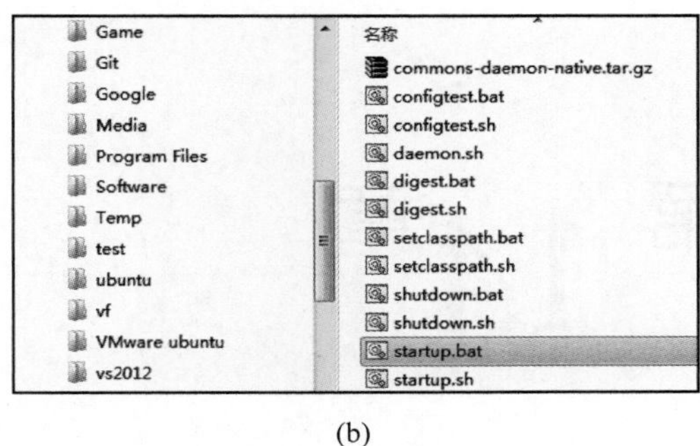

(a) (b)

图 13-11　部署 Web 站点

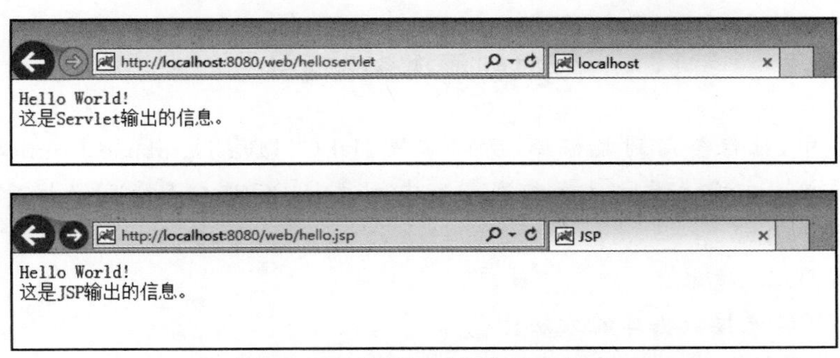

图 13-12　在 IE 中查看 Web 站点

1. 请结合网上的资料，学习 Tomcat 之外的 Web 服务器，例如 jetty、resin 等。

2. Java 的 Web 服务器，例如 Tomcat，与微软公司的 ASP.net 技术常用的服务器 IIS 有什么区别？

3. 请尝试在 Eclipse Java EE 中配置其他 Web 服务器，并运行 Servlet 和 JSP 程序。

4. 请尝试在 Tomcat 之外的服务器上发布 Web 站点。

第14章 JDBC 技术

本章学习目标

在软件开发中，往往会用到数据库，所以掌握 JDBC(Java Database Connectivity)数据库编程技术非常重要。本章主要讲述如何在应用中通过 JDBC 连接和访问数据库。通过本章学习应掌握以下主要内容：

(1) 什么是 JDBC 技术。

(2) 使用 JDBC 连接数据库的方法。

(3) JDBC 常用接口的使用。

14.1 为什么需要 JDBC

应用程序离不开对数据的处理。出于方便高效的考虑，数据往往是存储在数据库中的。这就要求应用程序能够访问和处理存储在数据库中的数据。为此，人们定义了应用程序与数据库之间进行操作的接口。JDBC 就是 Java 程序连接和存取数据库的应用程序接口。

【例 14.1】 从数据库中提取商品的基本信息。

对于一个商品管理系统，系统中的数据都存储在数据库中。这里编写系统中的一个功能模块，即通过软件从数据库中提取商品的基本信息。假设商品基本信息存储在 testDB 数据库的 commodity 表中，其中包含商品编号、商品名称、单价、状态 4 个字段信息。这里数据库使用 JavaDB。

代码实现如下：

```
package code1401;
import java.sql.*;
public class QueryCommodityList {
    public static void main(String[] args) {
        String url = "jdbc:derby:testDB;create=true";
        Connection con = null;
        Statement stmt = null;
        String query = "SELECT * FROM commodity";
        try {
            Class.forName("org.apache.derby.jdbc.EmbeddedDriver");
```

```
            con = DriverManager.getConnection(url);
            stmt = con.createStatement();
            ResultSet rs = stmt.executeQuery(query);
            System.out.println("商品信息列表:");
            System.out.println(new String(new char[63]).replace('\0', '-'));
          System.out.format("|%-4s\t|%-22s\t|%-8s\t|%-4s|\n", "编号", "名称",
"价格", "状态");
            System.out.println(new String(new char[63]).replace('\0', '-'));
            while (rs.next()) {
                int id = rs.getInt("id");
                String name = rs.getString("name");
                float price = rs.getFloat("price");
                String status = rs.getString("status");
                System.out.format("|%-4s\t|%-22s\t|%-8s\t|%-4s|\n", id, name,
price, status);
            }
            System.out.println(new String(new char[63]).replace('\0', '-'));
        } catch (ClassNotFoundException e) {
            System.err.print("类没有找到");
            System.err.println(e.getMessage());
        } catch (SQLException e) {
            System.err.println("SQL 异常:" + e.getMessage());
        } finally {
            if (stmt != null) {
                try {
                    stmt.close();
                } catch (SQLException e) {
                }
                stmt = null;
            }
            if (con != null) {
                try {
                    con.close();
                } catch (SQLException e) {
                }
                con = null;
            }
        }
    }
}
```

程序可能的运行结果如图 14-1 所示。

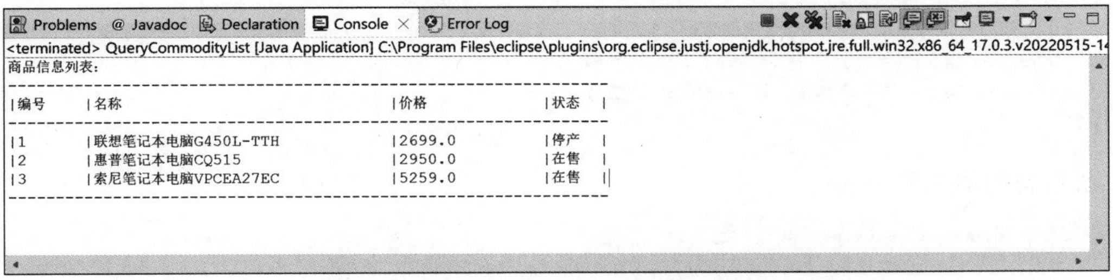

```
Problems  @ Javadoc  Declaration  Console  Error Log
<terminated> QueryCommodityList [Java Application] C:\Program Files\eclipse\plugins\org.eclipse.justj.openjdk.hotspot.jre.full.win32.x86_64_17.0.3.v20220515-1
商品信息列表:
---------------------------------------------------------------
|编号    |名称                    |价格      |状态 |
---------------------------------------------------------------
|1     |联想笔记本电脑G450L-TTH       |2699.0    |停产 |
|2     |惠普笔记本电脑CQ515          |2950.0    |在售 |
|3     |索尼笔记本电脑VPCEA27EC       |5259.0    |在售 |
---------------------------------------------------------------
```

图 14-1　例 14.1 程序可能的运行结果

本例题问题的求解涉及两个关键点：

(1) 查询数据库中的商品信息用到了 SQL 语句 SELECT。

(2) 程序通过 JDBC 访问数据库的基本操作过程。

下面详细介绍。

14.2 数据库和常用的 SQL 语句

数据库管理系统(Database Management System, DBMS)是一个软件系统,它具有存储、检索和修改数据的功能。目前主流的数据库管理系统是关系型数据库,包括 Oracle、Sybase、Microsoft SQL Server、DB2、MySQL 等产品。SQL(Structured Query Language,结构化查询语言)是目前使用的最为广泛的关系数据库查询语言。本节将对数据库技术和常用 SQL 进行简要介绍。

14.2.1 创建、删除数据库

创建数据库的语法如下:

```
CREATE DATABASE [database_name]
```

对于不同的数据库,有不同的选项,但是每个数据库都会支持以上的语句。执行上面的语句时,数据库管理系统会使用默认值创建一个名为 database_name 的数据库。

删除一个数据库的语法如下:

```
DROP DATABASE [database_name]
```

14.2.2 创建、删除表

创建表的语法如下:

```
CREATE TABLE table_name (
    column1 DATATYPE [NOT NULL] [NOT NULL PRIMARY KEY],
    column2 DATATYPE [NOT NULL],
...
)
```

创建表,必须定义表名、列名、列的类型和列的宽度。例如,创建一个商品表(commodity),包含商品编号(id)、商品名称(name)和商品价格(price)3 个字段,其建表的 SQL 语句如下:

```
CREATE TABLE commodity(
  id INT NOT NULL PRIMARY KEY,
  name VARCHAR(50),
  price FLOAT, " + "status VARCHAR(10)
)
```

删除表的语法如下:

```
DROP TABLE tableName
```

14.2.3 插入一条数据

在 SQL 语句中,INSERT 语句用来向表中添加记录。INSERT 语句的基本语法如下:

```
INSERT INTO tableName (column1, column2, …)
VALUES (value1, value2, …)
```

14.2.4 在表中删除数据

要删除表中已经存在的一条或多条记录，则应该使用 DELETE 语句。DELETE 语句可以使用 WHERE 语句来选择删除特定的记录。DELETE 语句的基本语法如下：

```
DELETE FROM tableName [WHERE …]
```

14.2.5 更新表中的数据

要修改表中已经存在的一条或多条记录，则应该使用 UPDATE 语句。UPDATE 语句可以使用 WHERE 语句来选择更新特定的记录。UPDATE 语句的基本语法如下：

```
UPDATE tableName SET column1=values1, column2=values2, … [WHERE …]
```

14.2.6 查询表中的数据

数据库查询是数据库的核心操作。使用 SELECT 语句进行数据库的查询，该语句具有灵活的使用方式和丰富的功能。其一般格式如下：

```
SELECT [ALL | DISTINCT] 目标列表达式[, 目标列表达式] …
FROM 基本表或视图[, 基本表或视图] …
[WHERE 条件表达式]
[GROUP BY 列名 1 [HAVING 内部函数表达式]]
[ORDER BY 列名 2 [ASC | DESC]]
```

其中，SELECT 子句用于列出查询结果中的属性列，FROM 子句用于列出表达式求值中需要扫描的关系；WHERE 子句实现对记录的筛选；GROUP BY 子句根据指定列名对结果集中的元组进行分组；HAVING 子句对分组进行过滤，它一般与 GROUP BY 子句配合使用；ORDER BY 子句对结果集按指定列名进行排序，ASC 是升序，DESC 是降序。

SELECT 语句的含义是：根据 WHERE 子句的条件表达式，从 FROM 子句指定的基本表或者视图中找出满足条件的记录，再按 SELECT 子句中的目标列选出记录中的属性值形成结果集。如果有 GROUP BY 子句，则将结果按“列名 1”的值进行分组。如果有 ORDER BY 子句，则结果集还要按照“列名 2”的值的升序或者降序排序。

14.2.7 条件子句

1. WHERE 语句

使用 WHERE 语句可以选择满足条件的特定的记录。在使用 WHERE 语句时，应注意：若列的数据类型为数字型，则不需要用引号；若列的数据类型为字符型，则需要用单引号把字符串括起来。

2. IN 和 NOT IN 语句

IN 语句选择列值与值列表中某一个值相等的相关行信息。NOT IN 语句选择那些不在列表中的记录。

3. BETWEEN…AND 和 NOT BETWEEN…AND 语句

BETWEEN…AND 语句选择列值在某个范围的记录。NOT BETWEEN…AND 语句选择列值不在该范围的记录。

4. LIKE 和 NOT LIKE 语句

用于查找字符串的匹配。通配符"%"匹配任意长度的字符串,而下画线"_"只匹配一个字符。

5. IS NULL 和 IS NOT NULL 语句

IS NULL 语句用于查找列值为空值的记录。IS NOT NULL 语句用于查找列值为非空值的记录。

6. 逻辑运算 AND 和 OR 语句

逻辑运算 AND 语句选择列值同时满足多个条件的记录。逻辑运算 OR 语句选择列值满足其中任意一个条件的记录。

14.3 JDBC 的结构

JDBC 是 Java 程序连接和存取数据库的应用程序接口,它为 Java 开发者使用数据库提供了统一的操作方式,它由一组 Java 类和接口组成。JDBC 使得开发人员可以使用纯 Java 的方式来连接数据库,并且进行操作。

JDBC 由上、下两层组成,上面一层是 JDBC API,下面一层是 JDBC 驱动程序 API,如图 14-2 所示。JDBC API 负责与 JDBC 管理器驱动程序 API 进行通信,将各个不同的 SQL 语句发送给它。驱动程序管理器 API(对程序员是透明的)与实际连接到数据库的各个第三方驱动程序进行通信,并且返回查询的信息,或者执行由查询规定的操作。下面分别介绍各个部分。

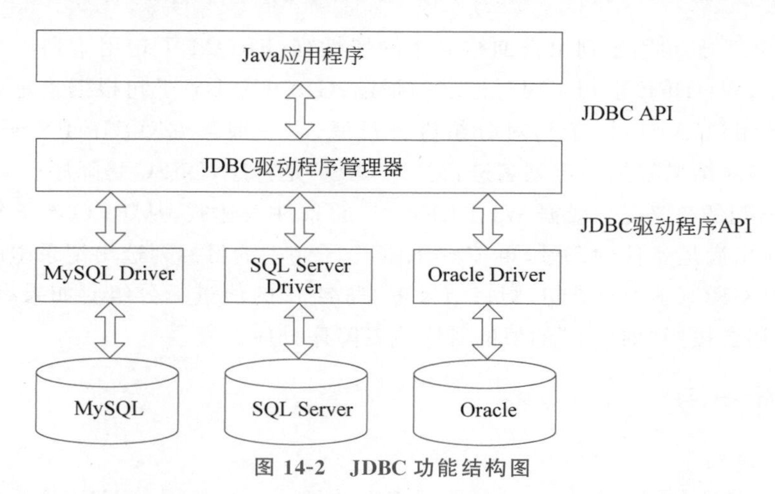

图 14-2　JDBC 功能结构图

1. Java 应用程序

Java 程序包括应用程序、Servlet 等,这些类型的程序都可以利用 JDBC 方法完成对数据库的访问和操作。

2. JDBC 驱动程序管理器

JDBC 驱动程序管理器能够动态地管理和维护数据库查询所需要的所有厂商或第三方所

提供的驱动程序对象，实现Java任务与特定驱动程序的连接，从而体现JDBC与平台无关这一特点。它完成的主要任务有为特定的数据库选择驱动程序、处理JDBC初始化调用、为每个驱动程序提供JDBC功能的入口、为JDBC调用执行参数等。

3. 驱动程序

这里的驱动程序(Driver)一般由数据库厂商或第三方提供，它由JDBC方法调用，向特定数据库发送SQL请求，并为Java程序获取结果。在必要时，驱动程序可以进行翻译或优化请求，使SQL请求符合DBMS支持的语言。

JDBC是独立于DBMS的，而每个数据库系统都有自己的协议与客户端通信，因此，JDBC利用数据库驱动程序来使用这些数据库引擎。JDBC驱动程序由数据库软件厂商和第三方提供，因此，根据使用的DBMS的不同，所需要的驱动程序也有所不同。

4. 数据库

这里的数据库指的是Java程序需要的数据库以及数据库管理系统。

14.4 通过JDBC访问数据库

通过JDBC访问数据库的过程包括加载JDBC驱动程序、建立与数据库的连接、执行SQL语句和获取数据等步骤，如图14-3所示。接下来将具体介绍每一步的详细内容。

驱动程序管理器	加载JDBC驱动程序
连接	建立与数据库的连接
SQL语句	发送SQL查询
结果集	得到结果

图14-3 一个简单的通过JDBC访问数据库的过程

14.4.1 加载JDBC驱动程序

为了要连接数据库，必须要有相应数据库的JDBC驱动程序，并将驱动程序的.jar文件加入应用程序的classpath设置中。此后再在程序中通过DriverManager类加载JDBC驱动类。

DriverManager(驱动程序管理器)类是JDBC的管理层，作用于用户和驱动程序之间。DriverManager类跟踪可用的驱动程序，并在数据库和相应驱动程序之间建立连接。另外，DriverManager类也处理诸如驱动程序登录时间限制以及登录和跟踪消息的显示等事务。

通过调用Class.forName()方法将显式地加载驱动程序类。由于这与外部设置无关，因此推荐使用这种加载驱动程序的方法。例如，以下代码加载类com.microsoft.sqlserver.jdbc.SQLServerDriver：

```
Class.forName("com.microsoft.sqlserver.jdbc.SQLServerDriver");
```

14.4.2 建立连接

当调用DriverManager类的getConnection()方法时，DriverManager类首先从它已加载的驱动程序池中找到一个可以接受该数据库URL的驱动程序，然后请求该驱动程序使用相关的数据库URL去连接到数据库中。于是，getConnection()方法将建立在JDBC URL中定义的数据库的Connection连接：

```
Connection con = DriverManager.getConnection(url, login, password);
```

JDBC 的 URL 的语法如下：

```
jdbc:<子协议>:<子名>
```

这里有 3 个部分，它们用冒号隔离。其中，jdbc 表示协议，且是唯一的，JDBC 只有这一种协议；子协议主要用于识别数据库驱动程序；子名属于专门的驱动程序，不同的专有驱动程序可以采用不同的实现。对于不同的数据库，厂商提供的驱动程序和连接的 URL 都不同，常见的数据库驱动程序和 URL 参见表 14-1 所示。

表 14-1 常见的数据库驱动程序和 URL

数据库名	驱动程序	URL
JDBC-ODBC	sun.jdbc.odbc.JdbcOdbcDriver	jdbc:odbc:[odbcsource]
Microsoft SQL Server 2008	com. microsoft. sqlserver. jdbc. SQLServerDriver	jdbc:sqlserver://[ip]:[port]; user = [user]; password=[pasword]
MySQL	org.gjt.mm.mysql.Driver	jdbc:mysql:/ip/database,user,password
JavaDB	org.apache.derby.jdbc. EmbeddedDriver	jdbc:derby:testDB;

getConnection()方法返回的 Connection 对象代表与数据库的连接。程序必须创建一个 Connection 类的实例，其中包括要连接数据库的信息。一个应用程序可与单个数据库有一个或多个连接，也可以与多个数据库有连接。

Connection 接口常用的方法有以下几种。

(1) close()方法：关闭到数据库的连接，在使用完连接后必须关闭，否则连接会保持一段比较长的时间，直到超时。

(2) createStatement()方法：创建一个 Statement 对象用于执行 SQL 语句。

(3) prepareStatement()方法：使用指定的 SQL 语句创建一个预处理语句，SQL 中的参数用“?”占位符表示。

14.4.3 执行 SQL 语句

连接一旦建立，就可用来向它所涉及的数据库传送 SQL 语句。JDBC 对可被发送的 SQL 语句类型不加任何限制。这就提供了很大的灵活性，即允许使用特定的数据库语句甚至于非 SQL 语句。然而，它要求用户自己负责确保所涉及的数据库可以处理所发送的 SQL 语句，否则将产生错误。例如，如果某个应用程序试图向不支持存储程序的 DBMS 发送存储程序调用，就会失败并将抛出异常。

JDBC 通过 Statement 接口向数据库发送 SQL 语句。Statement 提供了许多方法，常用的方法有以下几种。

(1) execute()方法：运行语句，返回是否有结果集的布尔值。

(2) executeQuery()方法：运行查询语句，返回 ResultSet 对象。

(3) executeUpdate()方法：运行更新操作，返回更新的行数。

有 3 种 Statement 对象，它们都作为在给定连接上执行 SQL 语句的包容器：Statement、PreparedStatement(从 Statement 继承而来)和 CallableStatement(从 PreparedStatement 继承而来)。Statement 对象用于执行不带参数的简单 SQL 语句；PreparedStatement 对象用于

执行带或者不带参数的预编译 SQL 语句；CallableStatement 对象用于执行对数据库已存储过程的调用。

14.4.4 检索结果

SQL 语句发送以后，返回的结果通常存放在一个 ResultSet 类的对象中，ResultSet 对象可以视为一个表，这个表中包含由 SQL 返回的列名和相应的值。

ResultSet 对象中维持了一个指向当前行的指针，通过 ResultSet.next()方法把当前的指针向下移动一行。最初它位于第一行前，因此第一次调用 next()方法将把指针置于第一行上，使它成为当前行。随着每次调用 next()方法导致指针向下移动，按照从上到下的次序获取 ResultSet 行。

ResultSet 提供了检索不同类型字段的方法，常用的方法有以下几种。

(1) getString()方法：获得在数据库中是 varchar、char 等数据类型的数据。

(2) getFloat()方法：获得在数据库中是 float 数据类型的数据。

(3) getDouble()方法：获得在数据库中是 double 数据类型的数据。

(4) getBlob()方法：获得在数据库中是 blob(二进制大型对象)数据类型的数据。

(5) getClob()方法：获得在数据库中是 clob(字符串大型对象)数据类型的数据。

一系列的 getXXX()方法提供了获取当前行中某列值的途径，在每一行内，可按任意次序获取列值。

14.4.5 关闭连接

在对象使用完毕后，应当使用 close()方法解除与数据库的连接，并关闭数据库。例如：

```
con.close();
```

图 14-3 显示了一个用简单的 JDBC 模型进行连接、执行和获取数据的过程，其中只做了一次连接。实际上 DriverManager 一次可以有多个连接，而一个 Connection 可以执行多个 SQL 语句，图 14-4 是一个复杂的通过 JDBC API 访问 SQL 数据库的过程。

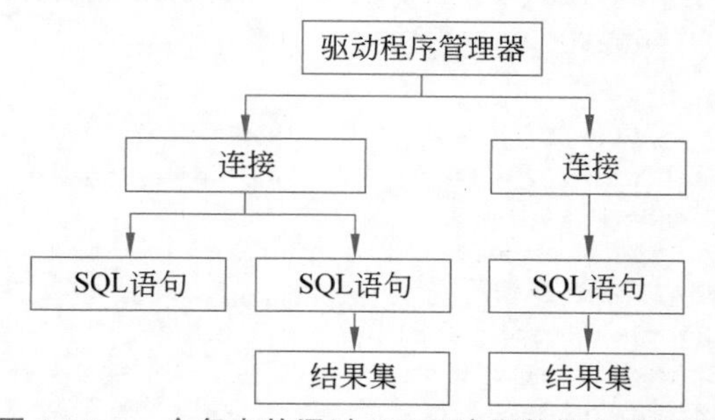

图 14-4 一个复杂的通过 JDBC 访问数据库的过程

14.4.6 通过数据库连接池获得数据库连接

数据库连接的建立和关闭是比较耗时的，频繁地建立和关闭数据库连接将降低系统性能。为了提高数据库访问的效率，可以通过数据库连接池来获取连接和释放连接。数据库连接池的核心思想是连接复用，它在内部对象池中维护一定数量的数据库连接对象，并向外提供数据库连接获取和返回的方法。外部应用需要访问数据库时,并非新建一个连接,而是从连接池中取出一个已建立的空闲连接对象;当使用完毕后,也并非将连接关闭,而是将连接重新放回连接池中,以供下一次请求访问使用。常见的Java开源数据库连接池有DBCP(Database Connection Pool)、C3P0、Druid、HikariCP等。

使用数据库连接池时,会通过配置文件对数据库连接池使用的JDBC驱动、数据库URL、初始连接数、最大连接数等参数进行设置。以DBCP为例,其配置文件dbcp.properties的常见内容如下:

```
driverClassName=org.apache.derby.jdbc.EmbeddedDriver          #JDBC驱动
url=jdbc:derby:testDB;create=true                             #数据库URL
initialSize=5                                                 #初始连接数
maxTotal=10                                                   #最大连接数
maxIdle=10                                                    #最大空闲数
```

14.4.7　实用案例14.1:查询指定商品状态的Java应用程序

应用案例14.1的代码如下:

```
package code1404;

import java.io.BufferedReader;
import java.io.File;
import java.io.FileInputStream;
import java.io.IOException;
import java.io.InputStreamReader;
import java.net.URISyntaxException;
import java.net.URL;
import java.sql.*;
import java.util.Properties;
import javax.sql.DataSource;
import org.apache.commons.dbcp2.BasicDataSourceFactory;
import code1401.QueryCommodityList2;

public class QueryCommodityStatus {
    private static Properties properties = new Properties();
    private static DataSource dataSource;

    static {
        try {
            URL fileURL = QueryCommodityList2.class.getClassLoader().
getResource("dbcp.properties");
            FileInputStream is = new FileInputStream(new File(fileURL.toURI()));
            properties.load(is);
        } catch (URISyntaxException e) {
            e.printStackTrace();
        } catch (IOException e) {
            e.printStackTrace();
        }
        try {
            dataSource = BasicDataSourceFactory.createDataSource(properties);
        } catch (Exception e) {
            e.printStackTrace();
        }
    }
```

```
    public static void main(String[] args) {
        Connection con = null;
        PreparedStatement pstm = null;
        try {
            con = dataSource.getConnection();          //获取连接
            pstm = con.prepareStatement("SELECT status FROM commodity WHERE name=?");
                                                       //创建 PreparedStatement 语句
            System.out.println("请输入要查询的商品名称:");
            String inputName = (new BufferedReader(new InputStreamReader(
                    System.in))).readLine();
            pstm.setString(1, inputName);              //对 SQL 语句中的参数进行赋值
            ResultSet rs = pstm.executeQuery();        //执行 SQL 语句
            if (rs.next()) {
                String status = rs.getString("status"); //检索结果
                System.out.println("商品\"" + inputName + "\"的状态是:" + status);
            } else {
                System.out.println("没有商品\"" + inputName + "\"的相关信息。");
            }
        } catch (SQLException e) {
            System.err.println("SQL 异常:" + e.getMessage());
        } catch (IOException e) {
            System.err.println("IO 异常:" + e.getMessage());
        } finally {
            if (pstm != null) {
                try {
                    pstm.close();                      //关闭 PreparedStatement 语句
                } catch (SQLException e) {
                }
                pstm = null;
            }
            if (con != null) {
                try {
                    con.close();                       //关闭连接
                } catch (SQLException e) {
                }
                con = null;
            }
        }
    }
}
```

程序可能的运行结果如图 14-5 所示。

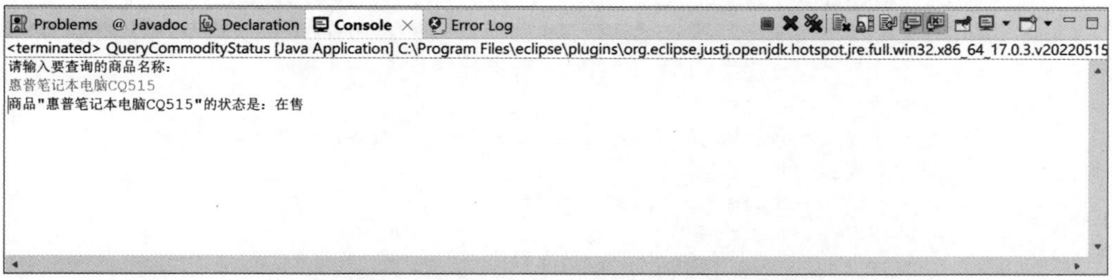

```
Problems  @ Javadoc  Declaration  Console ×  Error Log
<terminated> QueryCommodityStatus [Java Application] C:\Program Files\eclipse\plugins\org.eclipse.justj.openjdk.hotspot.jre.full.win32.x86_64_17.0.3.v2022051
请输入要查询的商品名称:
惠普笔记本电脑CQ515
商品"惠普笔记本电脑CQ515"的状态是：在售
```

图 14-5　实用案例 14.1 程序可能的运行结果

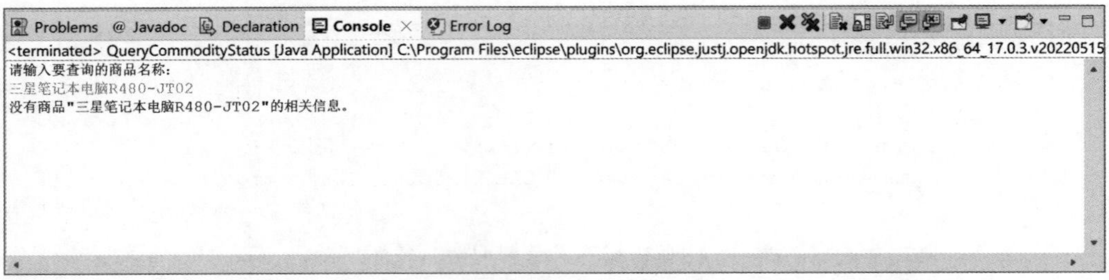

```
Problems  @ Javadoc  Declaration  Console ×  Error Log
<terminated> QueryCommodityStatus [Java Application] C:\Program Files\eclipse\plugins\org.eclipse.justj.openjdk.hotspot.jre.full.win32.x86_64_17.0.3.v20220515
请输入要查询的商品名称:
三星笔记本电脑R480-JT02
没有商品"三星笔记本电脑R480-JT02"的相关信息。
```

图 14-5 (续)

分析:按照 JDBC 连接数据库的步骤,首先按照配置文件 dbcp.properties 的内容初始化数据库连接池,再通过数据库连接池获得数据库连接,然后通过 PreparedStatement 语句,根据用户输入的商品名称查询商品的状态,如果返回的结果集不为空,则显示相应商品的状态信息;反之则显示“没有商品的相关信息”。最后关闭数据库连接。为了保证能正确释放数据库连接,将 close()语句放在 finally 语句块中。

14.4.8 实用案例 14.2:显示已有商品单价的 JSP 页面

应用案例 14.2 的代码如下:

```
<%@ page language="java" contentType="text/html; charset=gb2312"
    pageEncoding="gb2312" import="java.sql.*"%>
<!DOCTYPE html PUBLIC "-//W3C//DTD HTML 4.01 Transitional//EN" "http://www.w3.org/TR/html4/loose.dtd">
<html>
<head>
<meta http-equiv="Content-Type" content="text/html; charset=gb2312">
<title>显示商品的单价</title>
</head>
<body>
商品单价列表:
<table border="1">
    <tr>
        <td>编号</td>
        <td>名称</td>
        <td>价格</td>
    </tr>
    <%
        String url = "jdbc:derby:testDB;create=true";
        Connection con = null;
        Statement stmt = null;
        String query = "SELECT * FROM commodity";
        try {
            //加载 JDBC 驱动程序
            Class.forName("org.apache.derby.jdbc.EmbeddedDriver");
            //建立连接
            con = DriverManager.getConnection(url);
            //创建 Statement 语句
            stmt = con.createStatement();
            //执行 SQL 语句
            ResultSet rs = stmt.executeQuery(query);
```

```
            while (rs.next()) {                          //通过循环语句检索结果
                int id = rs.getInt("id");
                String name = rs.getString("name");
                float price = rs.getFloat("price");
                out.println("<tr><td>" + id + "</td><td>" + name
                        + "</td><td>" + price + "</td></tr>");
            }
        } catch (ClassNotFoundException e) {
            out.print("类没有找到异常:");
            out.println(e.getMessage());
        } catch (SQLException e) {
            out.println("SQL 异常:" + e.getMessage());
        } finally {
            if (stmt != null) {
                try {
                    stmt.close();                        //关闭 Statement 语句
                } catch (SQLException e) {
                }
                stmt = null;
            }
            if (con != null) {
                try {
                    con.close();                         //关闭连接
                } catch (SQLException e) {
                }
                con = null;
            }
        }
    %>
</table>
</body>
</html>
```

程序可能的运行结果如图 14-6 所示。

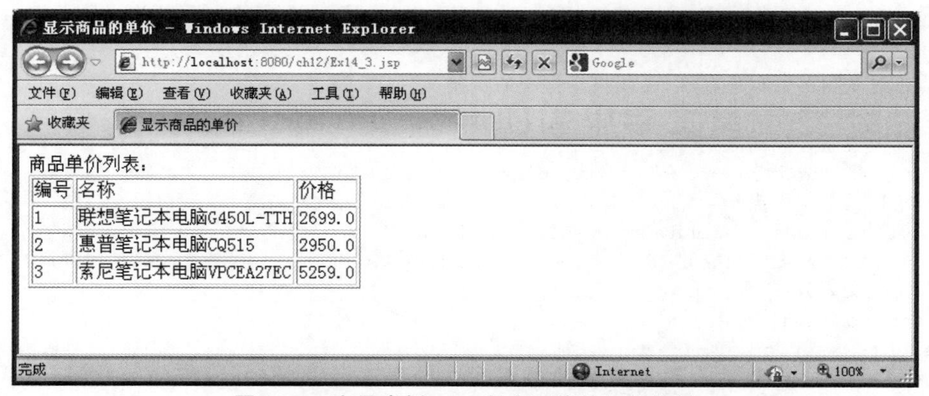

图 14-6　实用案例 14.2 程序可能的运行结果

分析：在 JSP 中访问数据库的操作步骤与在 Java 应用程序中的操作步骤是完全一致的。为了使用 JDBC 中的相关类，在 JSP 的页面指令中使用了“import="java.sql.*"”。通过 JSP 内置对象 out 将查询结果输出到网页中。

14.4.9 事务处理

数据库事务(Database Transaction)是数据库操作中的重要概念。它是指作为单个逻辑工作单元执行的一系列操作,要么完全地执行,要么完全地不执行。例如,如果用户 A 要向用户 B 转账 1000 元,此时银行管理系统至少需要执行两条数据库操作语句,一条语句是对用户 A 的账户减少 1000 元,另一条语句是对用户 B 的账户增加 1000 元。这两条操作应该是捆绑在一起的,不能只执行其中一条语句,否则会引起数据的不一致问题。事务处理可以确保除非事务性单元内的所有操作都成功完成,否则不会永久更新面向数据的资源。数据库事务需要满足 4 个特性:原子性(Atomic)、一致性(Consistency)、隔离性(Isolation)和持久性(Durability),这 4 个特性简称 ACID。

(1) 原子性:表示组成一个事务的多个数据库操作是一个不可分割的原子单元,只有所有的操作执行成功,整个事务才提交;只要事务中任何一个数据库操作失败,已经执行的任何操作都必须撤销,让数据库返回到初始状态。

(2) 一致性:事务操作成功后,数据库所处的状态和它的业务规则是一致的,即数据不会被破坏。

(3) 隔离性:在并发数据操作时,不同事务的操作之间不会相互产生干扰。用户可以通过指定连接的事务隔离级别来实现事务间的隔离性。ANSI/ISO SQL 92 标准定义了 4 个等级的事务隔离级别:READ UNCOMMITED、READ COMMITTED、REPEATABLE READ 和 SERIALIZABLE。按照这 4 个等级的排列顺序,其保证数据一致性的能力越来越强,而数据并发处理的能力则越来越弱。一般推荐使用 REPEATABLE READ 级别。在 JDBC 中对应有 4 个值来表示事务隔离级别:TRANSACTION_READ_UNCOMMITTED、TRANSACTION_READ_COMMITTED、TRANSACTION_REPEATABLE_READ 和 TRANSACTION_SERIALIZABLE。

(4) 持久性:一旦事务提交成功后,事务中所有的数据操作都必须被持久化到数据库。

典型的 JDBC 事务数据操作的代码如下:

```
Connection conn ;
try{
  //①对获得的数据库连接,关闭自动提交的机制
  conn.setAutoCommit(false);
  //②设置事务隔离级别
  con.setTransactionIsolation(Connection.TRANSACTION_REPEATABLE_READ);
  //③执行数据库操作
  Statement stmt = conn.createStatement();
  stmt.executeUpdate("INSERT INTO commodity VALUES(1,'电脑 1',2000, '停产') " );
  //④提交事务
  conn.commit();
}catch(Exception e){
  …
  //⑤回滚事务
  conn.rollback();
}finally{
  …
}
```

14.4.10 实用案例 14.3：事务操作

运用事务操作，可以同时修改指定商品的状态和单价。事务操作示例代码如下：

```
package code1404;

import java.io.BufferedReader;
import java.io.IOException;
import java.io.InputStreamReader;
import java.sql.*;

public class ModifyCommodityStatusAndPrice {

    public static void main(String[] args) throws IOException {
        String url = "jdbc:derby:testDB;create=true";
        Connection con = null;
        PreparedStatement pstm = null;

        boolean isExisted = false;

        System.out.println("请输入要修改的商品名称:");
        String inputName = (new BufferedReader(new InputStreamReader(System.
in))).readLine();

        try {
            Class.forName("org.apache.derby.jdbc.EmbeddedDriver");
            con = DriverManager.getConnection(url);
        } catch (ClassNotFoundException e) {
            System.err.print("类没有找到:");
            System.err.println(e.getMessage());
        } catch (SQLException e) {
            System.err.println("SQL 异常:" + e.getMessage());
        }

        try {
            pstm = con.prepareStatement("SELECT * FROM commodity where name=? ");
            pstm.setString(1, inputName);
            ResultSet rs = pstm.executeQuery();

            if (rs.next()) {
                isExisted = true;
                System.out.println("商品修改前的信息是:");
                int id = rs.getInt("id");
                String name = rs.getString("name");
                float price = rs.getFloat("price");
                String status = rs.getString("status");
                System.out.println(id + "\t" + name + "\t" + price + "\t" +
status);
            } else {
                System.out.println("输入的商品不存在。");
            }
        } catch (SQLException e) {
            System.err.println("SQL 异常:" + e.getMessage());
        } finally {
            if (pstm != null) {
                try {
                    pstm.close();
                } catch (SQLException e) {
                }
```

```
                pstm = null;
            }
            System.err.flush();
        }
        if (isExisted) {
            System.out.println("请输入商品新的状态:");
            String inputStatus = (new BufferedReader(new InputStreamReader
(System.in))).readLine();
            System.out.println("请输入商品新的价格:");
            Double inputPrice = Double.valueOf((new BufferedReader(new
InputStreamReader(System.in))).readLine());
            try {
                con.setAutoCommit(false);
                con.setTransactionIsolation(Connection.TRANSACTION_REPEATABLE_
READ);
                pstm = con.prepareStatement("UPDATE commodity SET price=? where
name=? ");
                pstm.setDouble(1, inputPrice.doubleValue());
                pstm.setString(2, inputName);
                pstm.executeUpdate();
                pstm.close();
                pstm = con.prepareStatement("UPDATE commodity SET status=? where
name=? ");
                pstm.setString(1, inputStatus);
                pstm.setString(2, inputName);
                pstm.executeUpdate();
                pstm.close();
                con.commit();
            } catch (SQLException e) {
                try {
                    con.rollback();
                } catch (SQLException e1) {
                    e1.printStackTrace();
                }
                System.err.println("SQL 异常:" + e.getMessage());
            } finally {
                if (pstm != null) {
                    try {
                        pstm.close();
                    } catch (SQLException e) {
                    }
                    pstm = null;
                }
                try {
                    con.setAutoCommit(true);
                } catch (SQLException e) {
                    e.printStackTrace();
                }
                System.err.flush();
            }
            try {
                pstm = con.prepareStatement("SELECT * FROM commodity where name=? ");
```

```
                pstm.setString(1, inputName);
                ResultSet rs = pstm.executeQuery();
                if (rs.next()) {
                    isExisted = true;
                    System.out.println("商品修改后的信息是:");
                    int id = rs.getInt("id");
                    String name = rs.getString("name");
                    float price = rs.getFloat("price");
                    String status = rs.getString("status");
                    System.out.println(id + "\t" + name + "\t" + price + "\t" + status);
                }
            } catch (SQLException e) {
                System.err.println("SQL 异常:" + e.getMessage());
            } finally {
                if (pstm != null) {
                    try {
                        pstm.close();
                    } catch (SQLException e) {
                    }
                    pstm = null;
                }
                System.err.flush();
            }
        }
        if (con != null) {
            try {
                con.close();
            } catch (SQLException e) {
            }
            con = null;
        }
    }
}
```

程序可能的运行结果如图 14-7 所示。

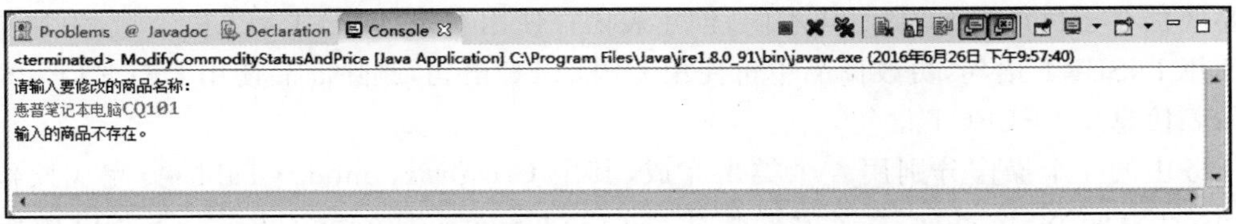

```
Problems  @ Javadoc  Declaration  Console
<terminated> ModifyCommodityStatusAndPrice [Java Application] C:\Program Files\Java\jre1.8.0_91\bin\javaw.exe (2016年6月26日 下午9:57:40)
请输入要修改的商品名称:
惠普笔记本电脑CQ101
输入的商品不存在。
```

(a)

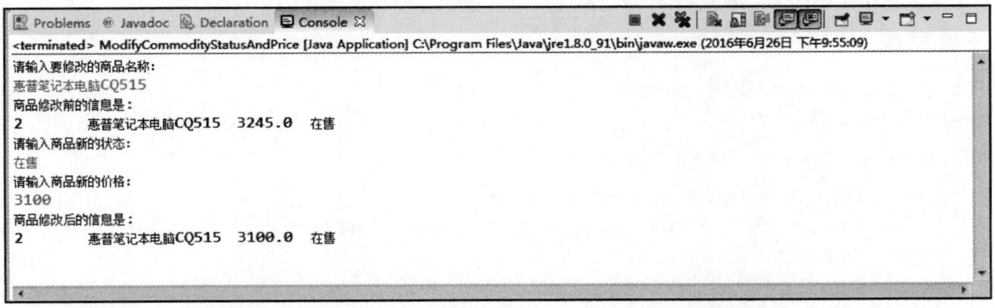

```
Problems  @ Javadoc  Declaration  Console
<terminated> ModifyCommodityStatusAndPrice [Java Application] C:\Program Files\Java\jre1.8.0_91\bin\javaw.exe (2016年6月26日 下午9:55:09)
请输入要修改的商品名称:
惠普笔记本电脑CQ515
商品修改前的信息是:
2	惠普笔记本电脑CQ515	3245.0	在售
请输入商品新的状态:
在售
请输入商品新的价格:
3100
商品修改后的信息是:
2	惠普笔记本电脑CQ515	3100.0	在售
```

(b)

图 14-7　实用案例 14.3 程序可能的执行结果

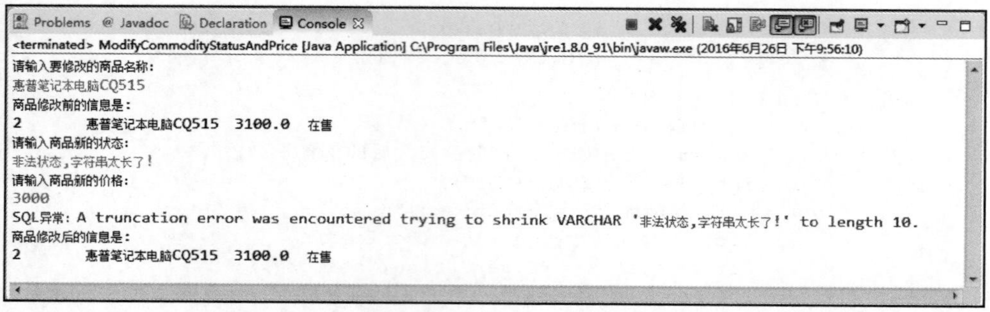

```
Problems @ Javadoc Declaration Console
<terminated> ModifyCommodityStatusAndPrice [Java Application] C:\Program Files\Java\jre1.8.0_91\bin\javaw.exe (2016年6月26日 下午9:56:10)
请输入要修改的商品名称:
惠普笔记本电脑CQ515
商品修改前的信息是:
2        惠普笔记本电脑CQ515  3100.0  在售
请输入商品新的状态:
非法状态,字符串太长了!
请输入商品新的价格:
3000
SQL异常: A truncation error was encountered trying to shrink VARCHAR '非法状态,字符串太长了!' to length 10.
商品修改后的信息是:
2        惠普笔记本电脑CQ515  3100.0  在售
```

(c)

图 14-7 (续)

分析:利用 JDBC 的事务处理操作,只有当商品价格修改操作和商品状态修改操作全部执行成功时,才会执行 commit 操作,将修改保存到数据库中;反之,只要其中任一操作失败,都会调用 rollback 操作,将数据库状态回滚到修改前的状态。

14.5 JDBC 实训任务

【任务描述】

通过程序完成以下操作:

(1) 在 testDB 数据库中创建商品表(commodity),该表有 4 个字段:商品编号(id)、商品名称(name)、单价(price)、商品状态(status)。

(2) 向商品表中添加新商品信息。

(3) 更改所有商品的单价,使之提高 10%。

(4) 删除已经停产的商品信息。

其中第(2)~(4)步在执行相关操作后,都要显示操作之后商品表的数据记录情况。

【任务分析】

按照 JDBC 访问数据库的步骤来完成任务要求的操作。执行任务中 4 步操作的程序代码的主要区别在于执行的 SQL 语句不同,创建表操作使用 CREATE TABLE 语句,添加新商品信息使用 INSERT 语句,更改商品单价使用 UPDATE 语句,删除商品使用 DELETE 语句,显示商品表信息用 SELECT 语句。

任务中的 4 个操作分别用 4 个类来完成,其中 CreateCommodityTable 类完成操作步骤(1);InsertNewCommodity 类完成操作步骤(2);UpdateCommodityPrice 类完成操作步骤(3);DeleteCommodity 类完成操作步骤(4)。数据库使用 JavaDB。

【任务解决】

```
//CreateCommodityTable.java
import java.sql.*;

public class CreateCommodityTable {

    public static void main(String[] args) {
        String url = "jdbc:derby:testDB;create=true";
        Connection con = null;
        Statement stmt = null;
        String createString;
        createString = "CREATE TABLE commodity "
```

```
                + "(id INT NOT NULL PRIMARY KEY, " + "name VARCHAR(50), "
                + "price FLOAT, " + "status VARCHAR(10))";
        try {
            Class.forName("org.apache.derby.jdbc.EmbeddedDriver");
            con = DriverManager.getConnection(url);
            stmt = con.createStatement();
            stmt.executeUpdate(createString);
            System.out.println("创建 commodity 表成功。");
        } catch (ClassNotFoundException e) {
            System.err.print("类没有找到异常:");
            System.err.println(e.getMessage());
        } catch (SQLException e) {
            System.err.println("SQL 异常:" + e.getMessage());
        } finally {
            if (stmt != null) {
                try {
                    stmt.close();
                } catch (SQLException e) {
                }
                stmt = null;
            }
            if (con != null) {
                try {
                    con.close();
                } catch (SQLException e) {
                }
                con = null;
            }
        }
    }
}
//InsertNewCommodity.java
import java.sql.*;
public class InsertNewCommodity {
    public static void main(String[] args) {
        String url = "jdbc:derby:testDB;create=true";
        Connection con = null;
        Statement stmt = null;
        String query = "SELECT * FROM commodity";
        try {
            Class.forName("org.apache.derby.jdbc.EmbeddedDriver");
            con = DriverManager.getConnection(url);
            stmt = con.createStatement();
            stmt.executeUpdate("INSERT INTO commodity "
                    + "VALUES(1,'联想笔记本计算机 G450L-TTH',2699, '停产')");
            stmt.executeUpdate("INSERT INTO commodity "
                    + "VALUES(2,'惠普笔记本计算机 CQ515',2950, '在售')");
            stmt.executeUpdate("INSERT INTO commodity "
                    + "VALUES(3,'索尼笔记本计算机 VPCEA27EC',5259, '在售')");
            ResultSet rs = stmt.executeQuery(query);
            System.out.println("商品信息列表:");
            System.out.println(new String(new char[63]).replace('\0', '-'));
```

```
            System.out.format("|%-4s\t|%-22s\t|%-8s\t|%-4s|\n", "编号", "名称", "
价格", "状态");
            System.out.println(new String(new char[63]).replace('\0', '-'));
            while (rs.next()) {
                int id = rs.getInt("id");
                String name = rs.getString("name");
                float price = rs.getFloat("price");
                String status = rs.getString("status");
               System.out.format("|%-4s\t|%-22s\t|%-8s\t|%-4s|\n", id, name,
price, status);
            }
            System.out.println(new String(new char[63]).replace('\0', '-'));
        } catch (ClassNotFoundException e) {
            System.err.print("类没有找到异常:");
            System.err.println(e.getMessage());
        } catch (SQLException e) {
            System.err.println("SQL 异常:" + e.getMessage());
        } finally {
            if (stmt != null) {
                try {
                    stmt.close();
                } catch (SQLException e) {
                }
                stmt = null;
            }
            if (con != null) {
                try {
                    con.close();
                } catch (SQLException e) {
                }
                con = null;
            }
        }
    }
}
//UpdateCommodityPrice.java
import java.sql.*;
public class UpdateCommodityPrice {
    public static void main(String[] args) {
        String url = "jdbc:derby:testDB;create=true";
        Connection con = null;
        Statement stmt = null;
        String query = "SELECT * FROM commodity";
        try {
            Class.forName("org.apache.derby.jdbc.EmbeddedDriver");
            con = DriverManager.getConnection(url);
            stmt = con.createStatement();
            stmt.executeUpdate("UPDATE commodity SET price=price*1.1");
            ResultSet rs = stmt.executeQuery(query);
            System.out.println("价格增加 10%之后,商品信息列表:");
            System.out.println(new String(new char[63]).replace('\0', '-'));
      System.out.format("|%-4s\t|%-22s\t|%-8s\t|%-4s|\n", "编号", "名称", "价格",
"状态");
            System.out.println(new String(new char[63]).replace('\0', '-'));
```

```
            while (rs.next()) {
                int id = rs.getInt("id");
                String name = rs.getString("name");
                float price = rs.getFloat("price");
                String status = rs.getString("status");
               System.out.format("|%-4s\t|%-22s\t|%-8s\t|%-4s|\n", id, name,
price, status);
            }
            System.out.println(new String(new char[63]).replace('\0', '-'));
        } catch (ClassNotFoundException e) {
            System.err.print("类没有找到异常:");
            System.err.println(e.getMessage());
        } catch (SQLException e) {
            System.err.println("SQL 异常:" + e.getMessage());
        } finally {
            if (stmt != null) {
                try {
                    stmt.close();
                } catch (SQLException e) {
                }
                stmt = null;
            }
            if (con != null) {
                try {
                    con.close();
                } catch (SQLException e) {
                }
                con = null;
            }
        }
    }
}
//DeleteCommodity.java
import java.sql.*;
public class DeleteCommodity {
    public static void main(String[] args) {
        String url = "jdbc:derby:testDB;create=true";
        Connection con = null;
        Statement stmt = null;
        String query = "SELECT * FROM commodity";
        try {
            Class.forName("org.apache.derby.jdbc.EmbeddedDriver");
            con = DriverManager.getConnection(url);
            stmt = con.createStatement();
            stmt.executeUpdate("DELETE FROM commodity WHERE status='停产'");
            ResultSet rs = stmt.executeQuery(query);
            System.out.println("删除停产商品之后的商品信息列表:");
            System.out.println(new String(new char[63]).replace('\0', '-'));
          System.out.format("|%-4s\t|%-22s\t|%-8s\t|%-4s|\n", "编号", "名称", "价格",
"状态");
            System.out.println(new String(new char[63]).replace('\0', '-'));
            while (rs.next()) {
                int id = rs.getInt("id");
                String name = rs.getString("name");
```

```
                float price = rs.getFloat("price");
                String status = rs.getString("status");
                System.out.format("|%-4s\t|%-22s\t|%-8s\t|%-4s|\n", id, name,
price, status);
            }
            System.out.println(new String(new char[63]).replace('\0', '-'));
        } catch (ClassNotFoundException e) {
            System.err.print("类没有找到异常:");
            System.err.println(e.getMessage());
        } catch (SQLException e) {
            System.err.println("SQL 异常:" + e.getMessage());
        } finally {
            if (stmt != null) {
                try {
                    stmt.close();
                } catch (SQLException e) {
                }
                stmt = null;
            }
            if (con != null) {
                try {
                    con.close();
                } catch (SQLException e) {
                }
                con = null;
            }
        }
    }
}
```

程序可能的运行结果如图 14-8～图 14-11 所示。

Problems @ Javadoc Declaration Console Error Log
<terminated> CreateCommodityTable [Java Application] C:\Program Files\eclipse\plugins\org.eclipse.justj.openjdk.hotspot.jre.full.win32.x86_64_17.0.3.v20220515
创建commodity表成功。

图 14-8　运行 CreateCommodityTable.java 的结果

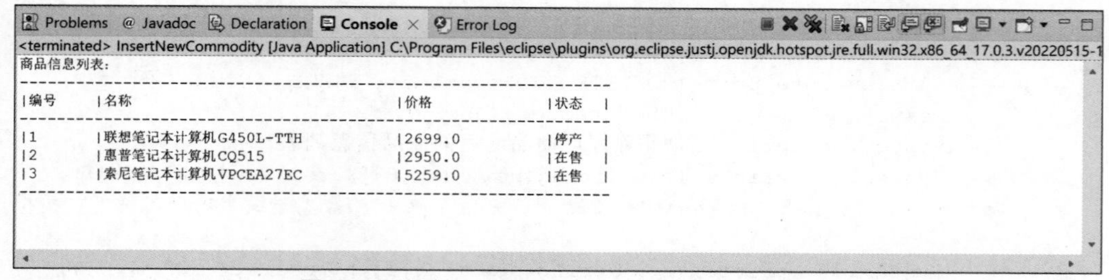

图 14-9　运行 InsertNewCommodity.java 的结果

```
<terminated> UpdateCommodityPrice [Java Application] C:\Program Files\eclipse\plugins\org.eclipse.justj.openjdk.hotspot.jre.full.win32.x86_64_17.0.3.v20220515
价格增加10%之后，商品信息列表：
------------------------------------------------------------------
|编号     |名称                                  |价格              |状态  |
------------------------------------------------------------------
|1       |联想笔记本计算机G450L-TTH                |2968.9            |停产  |
|2       |惠普笔记本计算机CQ515                    |3245.0            |在售  |
|3       |索尼笔记本计算机VPCEA27EC                |5784.9            |在售  |
------------------------------------------------------------------
```

图 14-10　运行 UpdateCommodityPrice.java 的结果

```
<terminated> DeleteCommodity [Java Application] C:\Program Files\eclipse\plugins\org.eclipse.justj.openjdk.hotspot.jre.full.win32.x86_64_17.0.3.v20220515-1416
删除停产商品之后的商品信息列表：
------------------------------------------------------------------
|编号     |名称                                  |价格              |状态  |
------------------------------------------------------------------
|2       |惠普笔记本计算机CQ515                    |3245.0            |在售  |
|3       |索尼笔记本计算机VPCEA27EC                |5784.9            |在售  |
------------------------------------------------------------------
```

图 14-11　运行 DeleteCommodity.java 的结果

1. 常用的 SQL 语句有哪些？其作用是什么？
2. JDBC 访问数据库的基本流程是什么？
3. Statement 对象和 PreparedStatement 对象的区别是什么？

第四篇　实 例 篇

第15章 Java应用开发案例

俗话“读万卷书，行万里路”。在前面的章节里已经详细地介绍了Java的各种知识，但这些知识综合在一起能够开发什么应用呢？本章给出两个应用开发案例：“基于MVC模式的简单学生信息管理”和“基于Spring Boot框架的简单学生信息管理”。两个应用开发案例都实现相同的程序功能，但采用不同的技术方案。前者采用传统的Servlet＋JSP技术，后者采用主流的软件开发框架Spring Boot。希望通过两个案例的学习，协助读者完成部分“行万里路”的过程，开启深入学习Java技术的新征程。

15.1 基于MVC模式的简单学生信息管理

15.1.1 MVC模式

本节介绍一个基于Servlet＋JSP技术对学生信息进行CRUD操作（即添加（Create）、查看（Read）、修改（Update）和删除（Delete））的简易信息管理Web程序。很多大型Web程序都具有CRUD类型的基础结构。因此，只要了解了CRUD程序的基本结构，就很容易学习更复杂的系统。

首先介绍用于Web开发的MVC模式。Java的Web开发模式，主要有3种，分别是JSP Scriptlet模式、JSP Model 1模式和JSP Model 2模式（即MVC模式）。

JSP Scriptlet模式，如图15-1所示，使用JSP实现业务逻辑，将业务代码和算法通过Scriptlet的形式（JSP页面中＜% %＞部分的Java代码）实现。当前JSP Scriptlet模式已经被淘汰，因为这种开发方法会导致业务算法和页面显示的紧密耦合，不便于程序的修改维护。

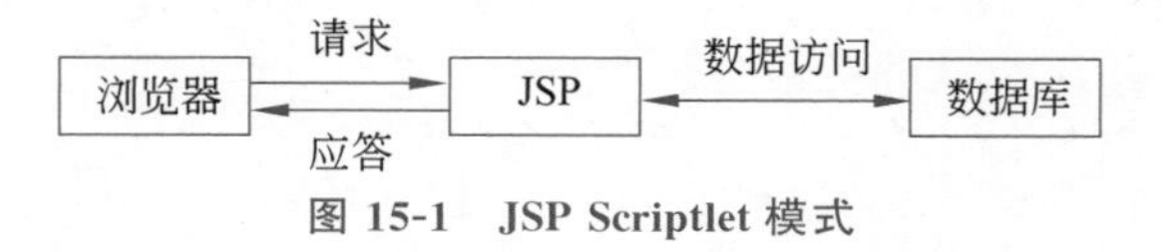

图15-1 JSP Scriptlet模式

JSP Model 1模式，如图15-2所示，是对JSP Scriptlet的改进，它将业务逻辑从JSP中抽取出来，放在Java对象（JavaBean）中，实现了业务逻辑和显示逻辑的分离。相对于JSP Scriptlet模式，JSP Model 1模式的程序耦合度较低，更容易维护。

MVC模式，即JSP Model 2模式，如图15-3所示，也就是常说的MVC Web开发模式，是

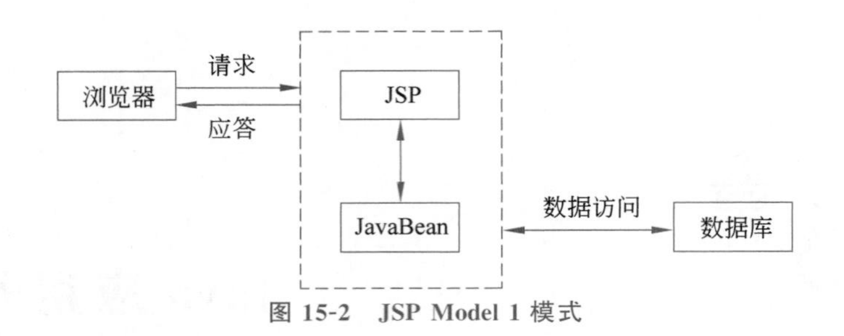

图 15-2　JSP Model 1 模式

JSP Model 1 模式的改进,它是现代 Web 开发的主流模式。在这种模式下,程序被分为 Model(模型)、View(视图)和 Controller(控制器)3 个部分,每部分负责不同的功能。Model 对应 JavaBean,封装业务逻辑和算法,实现数据库访问等具体的功能。View 对应 JSP,用于生成用户界面,实现显示逻辑。Controller 对应 Servlet,实现控制逻辑,即根据浏览器的请求,调用对应的 JavaBean 实现业务功能,然后选择对应的 View 返回给浏览器。根据不同的显示逻辑,Controller 还可能把 View 需要的数据一并发送给 View(以 JavaBean 的形式),View 使用这些 JavaBean 生成显示效果。

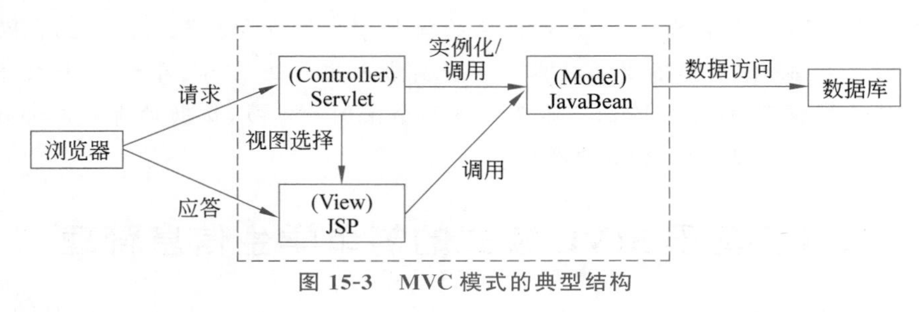

图 15-3　MVC 模式的典型结构

相对于前两种开发模式,MVC 模式各部分的逻辑更为清晰,耦合度低,因此更容易扩展和维护。目前大型的 Web 系统,几乎全部采用 MVC 模式。但 MVC 模式也有缺点,主要是结构相对复杂。

15.1.2　创建数据库

本章的案例需要使用关系型数据库 MySQL,建议在阅读下面的内容之前,首先安装数据库,可以选择 5.7 以上版本的 MySQL。由于本书不是专门讲解数据库知识的书籍,因此开发案例尽量简化了数据库的相关操作。数据库只有一个存储学生信息的表格,对应的建表 SQL 语句如下:

```
create database stm character set utf8;
use stm;
create table student(
    id varchar(10) not null ,
    name varchar(50) not null,
    age int not null,
    sex int not null,
    major varchar(50) not null,
    college varchar(50) not null,
    introduction varchar(500),primary key(id)
);
```

执行上述 SQL 语句将在 MySQL 中创建名称为 stm 的数据库，数据库采用 utf8 的字符编码以支持中文。数据库下创建名称 student 的表格，数据字段包括学号(id)、姓名(name)、年龄(age)、性别(sex)、专业(major)、学院(college)和简介(introduction)。

15.1.3 程序的基本结构

学生信息管理程序使用 Eclipse IDE for Java EE Developers(编写此书时使用 Build id 为 20211202-1639 的版本)作为开发工具，创建 Dynamic Web Project，使用 Tomcat v10.0 和 Java SE 17，学生信息管理程序的项目结构如图 15-4 所示。

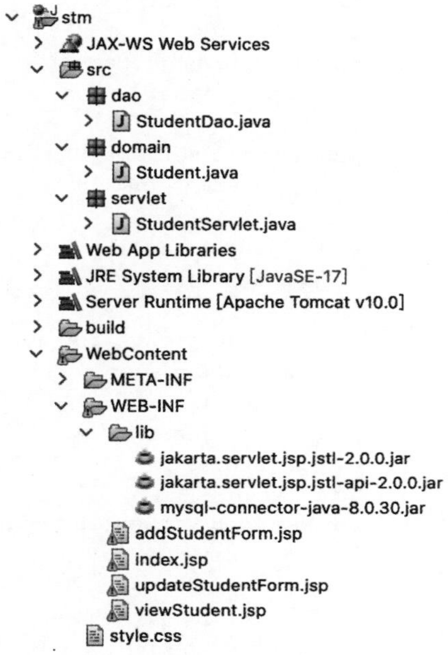

src 目录下放置 Java 源文件，各 Java 包的主要作用如下。

(1) domain 包：含有 Student.java，对应数据库表格 student，实现 MVC 模式中的 Model。

(2) dao 包：含有 StudentDao.java，用于访问数据库，实现 MVC 模式中的 Model。

(3) servlet 包：含有 StudentServlet.java，实现 MVC 模式中的 Controller。

WebContent 目录下放置 JSP 文件、CSS 文件和项目所需要的第三方库文件，其中：

(1) WebContent/WEB-INF/lib 目录包含 3 个 jar 文件，对应项目所需使用的第三方库文件。

(2) WebContent/WEB-INF/下的 jsp 文件对应项目所使用的视图(View)文件。

(3) WebContent/style.css 文件是视图显示所需要的样式文件。

```
stm
  JAX-WS Web Services
  src
    dao
      StudentDao.java
    domain
      Student.java
    servlet
      StudentServlet.java
  Web App Libraries
  JRE System Library [JavaSE-17]
  Server Runtime [Apache Tomcat v10.0]
  build
  WebContent
    META-INF
    WEB-INF
      lib
        jakarta.servlet.jsp.jstl-2.0.0.jar
        jakarta.servlet.jsp.jstl-api-2.0.0.jar
        mysql-connector-java-8.0.30.jar
      addStudentForm.jsp
      index.jsp
      updateStudentForm.jsp
      viewStudent.jsp
    style.css
```

图 15-4 学生信息管理程序的项目结构

程序使用 Servlet+JSP 技术实现 MVC 模式，下面分别对 Model、View 和 Controller 进行介绍。

15.1.4 模型

程序的模型(Model)由 domain 包和 dao 包中的 Java 类构成。本程序实现的功能是对学生信息的 CRUD 操作，这些操作由 domain.Student 和 dao.StudentDao 共同完成。

例 15.1 是 domain.Student 类的代码。Student 类是领域对象(Domain Object)，用于表示 student 表中的一行数据，它包含 id、name、age、sex、major、college 和 introduction 属性。这些属性与 student 表中的字段一一对应。由于属性的访问权限设置为 private，因此 Student 类还包含一系列的 get()和 set()函数用于提供对类属性的读取和写入操作。

【例 15.1】 Student.java。

```
package domain;
public class Student {
    private String id;                          //学号
    private String name;                        //姓名
    private int age;                            //年龄
```

```
    private int sex;                                    //性别
    private String major;                               //专业
    private String college;                             //学院
    private String introduction;                        //简介
    public String getId() {
        return id;
    }
    public void setId(String id) {
        this.id = id;
    }
    public String getName() {
        return name;
    }
    public void setName(String name) {
        this.name = name;
    }
    public int getAge() {
        return age;
    }
    public void setAge(int age) {
        this.age = age;
    }
    public int getSex() {
        return sex;
    }
    public void setSex(int sex) {
        this.sex = sex;
    }
    public String getMajor() {
        return major;
    }
    public void setMajor(String major) {
        this.major = major;
    }
    public String getCollege() {
        return college;
    }
    public void setCollege(String college) {
        this.college = college;
    }
    public String getIntroduction() {
        return introduction;
    }
    public void setIntroduction(String introduction) {
        this.introduction = introduction;
    }
}
```

dao.StudentDao 类是数据访问对象(Data Access Object,DAO),包含了对 student 表的 CRUD 操作。例 15.2 给出了对应的代码。StudentDao 类包含一系列的静态方法,其作用是:

(1) getConnection():用于建立与数据库的 JDBC 连接。本书所使用的数据库 URL 地址、用户名、密码可能与读者安装的数据库有所差异,读者注意修改对应的代码。MySQL 官方数据库驱动的名称为 com.mysql.cj.jdbc.Driver。数据库驱动需要将相应的驱动文件复制到

WebContent/WEB-INF/lib 目录下，可以在该目录下看到数据库驱动文件 mysql-connector-java-8.0.30.jar。

(2) addStudent()：添加学生信息。首先通过 getConnection()方法获取数据库连接，然后创建 PreparedStatement 的实例，将参数与 SQL 语句进行绑定，最后执行 SQL 语句并关闭数据库连接。函数如果执行成功，则返回 true；如果执行失败（如产生异常），则返回 false。

(3) updateStudent()：修改学生信息。程序逻辑与 addStudent 基本相同，区别在于执行的 SQL 语句为 update 语句。

(4) deleteStudent()：根据学号删除学生信息。程序逻辑与 addStudent 基本相同，区别在于执行的 SQL 语句为 delete 语句。

(5) getStudent()：根据学号查询学生信息。首先通过 getConnection()方法获取数据库连接，然后创建 PreparedStatement 的实例，将参数与 SQL 语句进行绑定，执行 SQL 语句，最后将获得的数据构造一个 Student 的实例，并关闭数据库连接。函数如果执行成功，则返回 Student 的实例；如果执行失败或者无法查询到学生信息，则返回 null。

(6) getAllStudents()：查询全部学生的信息。程序流程基本与 getStudent()方法相同。由于数据有多条，因此程序构造了一个类型为 ArrayList<Student> 的 students 实例。用 students 来存放返回的数据。如果学生数量为 0，则 students 的长度为 0。

【例 15.2】 StudentDao.java。

```
package dao;
import java.sql.Connection;
import java.sql.DriverManager;
import java.sql.PreparedStatement;
import java.sql.ResultSet;
import java.util.ArrayList;
import domain.Student;
public class StudentDao {
    //获取数据库连接
    private static Connection getConnection() throws Exception {
        //jdbc url 地址
        String url =
            "jdbc:mysql://localhost:3306/stm?useUnicode=true&characterEncoding=
utf8";
        //数据库用户名
        String user = "root";
        //数据库密码
        String pwd = "123456";
        //数据库驱动名
        String driver = "com.mysql.cj.jdbc.Driver";
        //加载数据库驱动
        Class.forName(driver);
        //获取数据库连接
        Connection conn = DriverManager.getConnection(url, user, pwd);
        return conn;
    }
    //添加学生信息
    public static boolean addStudent(String id, String name, int age, int sex,
            String major, String college, String introduction) {
        try {
```

```
            Connection conn = getConnection();
            PreparedStatement st = conn
                    .prepareStatement("insert into student values(?,?,?,?,?,?,?)");
            st.setString(1, id);
            st.setString(2, name);
            st.setInt(3, age);
            st.setInt(4, sex);
            st.setString(5, major);
            st.setString(6, college);
            st.setString(7, introduction);
            st.execute();
            conn.close();
            return true;
        } catch (Exception e) {
            e.printStackTrace();
            return false;
        }
    }
    //修改学生信息
    public static boolean updateStudent(String id, String name, int age,
            int sex, String major, String college, String introduction) {
        try {
            Connection conn = getConnection();
            PreparedStatement st = conn
                    .prepareStatement("update student set name=?,age=?,sex=?,
major=?,college=?,introduction=? where id=?");
            st.setString(1, name);
            st.setInt(2, age);
            st.setInt(3, sex);
            st.setString(4, major);
            st.setString(5, college);
            st.setString(6, introduction);
            st.setString(7, id);
            st.execute();
            conn.close();
            return true;
        } catch (Exception e) {
            e.printStackTrace();
            return false;
        }
    }
    //删除学生信息
    public static boolean deleteStudent(String id) {
        try {
            Connection conn = getConnection();
            PreparedStatement st = conn
                    .prepareStatement("delete from student where id=?");
            st.setString(1, id);
            st.execute();
            conn.close();
            return true;
        } catch (Exception e) {
            e.printStackTrace();
            return false;
```

```
        }
    }
    //按照学号查询学生信息
    public static Student getStudent(String id) {
        Student student = null;
        try {
            Connection conn = getConnection();
            PreparedStatement st = conn
                    .prepareStatement("select * from student where id=? ");
            st.setString(1, id);
            st.execute();
            ResultSet rs = st.getResultSet();
            if (rs.next()) {
                student = new Student();
                student.setId(rs.getString("id"));
                student.setName(rs.getString("name"));
                student.setAge(rs.getInt("age"));
                student.setSex(rs.getInt("sex"));
                student.setMajor(rs.getString("major"));
                student.setCollege(rs.getString("college"));
                student.setIntroduction(rs.getString("introduction"));
            }
            conn.close();
        } catch (Exception e) {
            e.printStackTrace();
        }
        return student;
    }
    //查询所有学生信息
    public static ArrayList<Student> getAllStudents(){
        ArrayList<Student> students = new ArrayList<Student>();
        try {
            Connection conn = getConnection();
            PreparedStatement st = conn
                    .prepareStatement("select * from student");
            st.execute();
            ResultSet rs = st.getResultSet();
            while (rs.next()) {
                Student student = new Student();
                student.setId(rs.getString("id"));
                student.setName(rs.getString("name"));
                student.setAge(rs.getInt("age"));
                student.setSex(rs.getInt("sex"));
                student.setMajor(rs.getString("major"));
                student.setCollege(rs.getString("college"));
                student.setIntroduction(rs.getString("introduction"));
                students.add(student);
            }
            conn.close();
        } catch (Exception e) {
            e.printStackTrace();
        }
        return students;
    }
}
```

15.1.5 视图

WebContent 目录下与视图(View)相关的文件包括 index.jsp、addStudentForm.jsp、updateStudentForm.jsp、viewStudent.jsp、style.css,其主要功能如下。

(1) index.jsp:学生信息管理的首页面。

(2) addStudentForm.jsp:添加学生信息的表单页面。

(3) updateStudentForm.jsp:修改学生信息的表单页面。

(4) viewStudent.jsp:查看学生信息的页面。

(5) style.css:网页的样式单,能够让网页有更好的显示效果。

JSP 视图文件在编写时使用了 JSTL(Java Server Pages Standard Tag Library)技术,该技术提供了一组自定义的标签(类似 HTML 标签),开发人员可以用 JSTL 标签替代 JSP 中的 Java 代码,从而避免 JSP Scriptlet 的使用。为了使用 JSTL,需要在 WebContent/WEB-INF/lib 目录下引入相关的第三方库,读者可以在该目录下看到 jakarta.servlet.jsp.jstl-2.0.0.jar 和 jakarta.servlet.jsp.jstl-api-2.0.0.jar 两个文件。下面分别介绍与视图相关的文件。

为了美化页面的显示效果,系统使用了 style.css 作为样式单。CSS 是 Web 页面设计常用的一类技术,但不是本书的重点,读者若想学习,可以参考其他书籍。这里只给出其代码。

【例 15.3】 style.css。

```
TD.data {
    BACKGROUND-COLOR: #eeeee0;
    BORDER-BOTTOM: #cccc99 1px solid;
    BORDER-LEFT: #ffffff 1px solid;
    BORDER-RIGHT: #cccc99 1px solid;
    BORDER-TOP: #ffffff 1px solid;
    HEIGHT: 23px
}
TD.title {
    BACKGROUND-COLOR: #3988E4;
    BORDER-BOTTOM: #000033 1px solid;
    BORDER-LEFT: #669999 1px solid;
    BORDER-RIGHT: #000033 1px solid;
    BORDER-TOP: #669999 1px solid;
    COLOR: #ffffff;
    PADDING-BOTTOM: 0px;
    PADDING-LEFT: 10px;
    PADDING-RIGHT: 10px;
    PADDING-TOP: 0px
}
TD.header {
    BACKGROUND-COLOR: #ffffff;
    BORDER-BOTTOM: #cccccc 1px solid;
    BORDER-LEFT: #ffffff 1px solid;
    BORDER-RIGHT: #cccccc 1px solid;
    BORDER-TOP: #ffffff 1px solid;
    HEIGHT: 23px
}
TD {
    FONT-FAMILY: 宋体, Verdana, Helvetica, sans-serif;
    FONT-SIZE: 12px
```

```
}
body {
    background-color: #eeeeee;
    SCROLLBAR-FACE-COLOR: #dce0e2;
    SCROLLBAR-HIGHLIGHT-COLOR: #ffffff;
    SCROLLBAR-SHADOW-COLOR: #687888;
    SCROLLBAR-3DLIGHT-COLOR: #687888;
    SCROLLBAR-ARROW-COLOR: #6e7e88;
    SCROLLBAR-TRACK-COLOR: #bcbfc0;
    SCROLLBAR-DARKSHADOW-COLOR: #dce0e2;
    FONT-SIZE: 12px
}
A:link {
    COLOR: blue;
    TEXT-DECORATION: none
}
A:visited {
    COLOR: blue;
    TEXT-DECORATION: none
}
A:hover {
    COLOR: blue;
    TEXT-DECORATION: none
}
```

style.css 通过以下代码链接到 JSP 文件，读者可以在后面 JSP 页面的代码中发现：

```
<link href="style.css"" rel="stylesheet" type="text/css">
```

如果不使用样式单，Web 页面仍然可以显示，但显示效果就会差一些。

index.jsp 是系统的首页(页面效果见图 15-5)。页面上列出了全部的学生信息。“添加学生”链接到 addStudentForm.do，“查看”链接到 viewStudent.do，“修改”链接到 updateStudentForm.do，“删除”链接到 deleteStudent.do。当单击“删除”链接时会直接由 StudentServlet 删除对应的学生信息。单击其他链接则进入相应的页面。例 15.4 给出了 index.jsp 的代码。在页面上通过使用 JSTL 的<c:forEach>标签，在 for 循环中将学生信息以表格的形式输出。

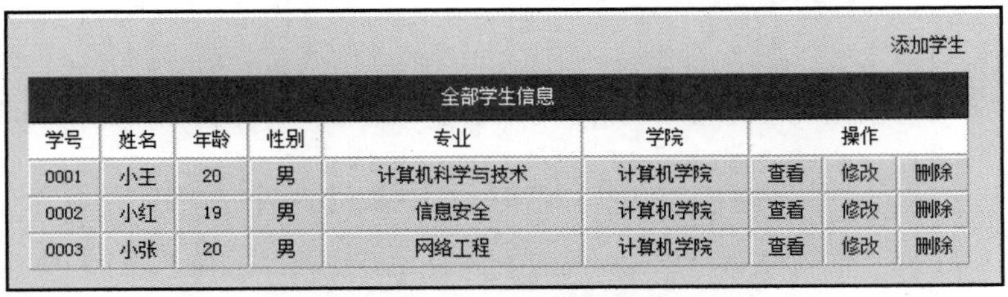

添加学生

全部学生信息

学号	姓名	年龄	性别	专业	学院	操作		
0001	小王	20	男	计算机科学与技术	计算机学院	查看	修改	删除
0002	小红	19	男	信息安全	计算机学院	查看	修改	删除
0003	小张	20	男	网络工程	计算机学院	查看	修改	删除

图 15-5　index.jsp 的显示结果

【例 15.4】　index.jsp。

```
<%@ page language="java" contentType="text/html; charset=UTF-8"
    pageEncoding="UTF-8"%>
```

```
<%@ taglib prefix="c" uri="http://java.sun.com/jsp/jstl/core"%>
<html>
<head>
<link href="style.css" rel="stylesheet" type="text/css">
<title>学生信息管理</title>
</head>
<body>
    <div align="center">
        <table width="600" cellpadding="1" cellspacing="1">
            <tr>
                <td colspan="9" align="right" height="30"><a
                    href="addStudentForm.do">添加学生</a></td>
            </tr>
        </table>
        <table width="600" cellpadding="1" cellspacing="1">
            <tr>
                <td colspan="9" align="center" class="title" height="30">
                    全部学生信息
                </td>
            </tr>
            <tr height="30">
                <td align="center" class="header">学号</td>
                <td align="center" class="header">姓名</td>
                <td align="center" class="header">年龄</td>
                <td align="center" class="header">性别</td>
                <td align="center" class="header">专业</td>
                <td align="center" class="header">学院</td>
                <td align="center" colspan="3" class="header">操作</td>
            </tr>
            <c:forEach items="${students}" var="student">
                <tr height="30">
                    <td align="center" class="data">${student.id }</td>
                    <td align="center" class="data">${student.name }</td>
                    <td align="center" class="data">${student.age }</td>
                    <td align="center" class="data">
                        <c:if test="${student.sex==1 }">
                                男
                        </c:if>
                        <c:if test="${student.sex!=1 }">
                                女
                        </c:if>
                    </td>
                    <td align="center" class="data">${student.major }</td>
                    <td align="center" class="data">${student.college }</td>
                    <td align="center" class="data"><a
                        href="viewStudent.do?id=${student.id }">查看</a></td>
                    <td align="center" class="data"><a
                        href="updateStudentForm.do?id=${student.id }">修改</a></td>
                    <td align="center" class="data"><a
                        href="deleteStudent.do?id=${student.id }">删除</a></td>
                </tr>
            </c:forEach>
        </table>
    </div>
```

```
</body>
</html>
```

为了使用 JSTL 标签，需要在 JSP 文件的头部加入标签的声明，代码如下：

```
<%@ taglib prefix="c" uri="http://java.sun.com/jsp/jstl/core"%>
```

其后就可以使用如下<c:forEach>标签：

```
<c:forEach items="${students}" var="student">
...
</c:forEach>
```

<c:forEach>标签的 items 属性指明要进行迭代的集合，var 属性指明单次迭代所使用的变量名称。${students}是 JSTL 的 EL(Expression Language)表达式，用于获取域对象中的数据。students 是保存在 JSP 内在对象(request 或 session 等)中的某个数据，使用${students}就可以访问到该数据。类似的${student.id }可以访问到 student 对象(在 var="student"声明)中的 id 属性。

<c:if>标签相当于 Java 代码的 if 语句，其代码如下：

```
<c:if test="${student.sex==1 }">
        男
</c:if>
```

上述语句判定 student 对象的 sex 属性是否为 1，如果为 1 则输出“男”。

从上面的代码可以看出，使用 JSTL 后，JSP 文件中就不再有 Java 代码。本项目只使用了极少数的 JSTL 标签，读者若想完整的学习 JSTL 的使用，可以参考其他资料。

addStudentForm.jsp 是输入学生信息的表单窗口(页面效果见图 15-6)，用户可以在该页面上输入学生信息，然后单击“提交”按键将数据提交到 addStudent.do，由 StudentServlet 进行处理。addStudentForm.jsp 的代码见例 15.5。

图 15-6　addStudentForm.jsp 的显示结果

【例 15.5】　addStudentForm.jsp。

```
<%@ page language="java" contentType="text/html; charset=UTF-8"
```

```
    pageEncoding="UTF-8"%>
<html>
<head>
<link href="style.css" rel="stylesheet" type="text/css">
<title>添加学生信息</title>
</head>
<body>
<div align="center">
<form action="addStudent.do" method="post">
<table width="500" cellpadding="1" cellspacing="1">
    <tr>
        <td colspan="9" align="center" class="title" height="30">学生信息</td>
    </tr>
    <tr height="30">
        <td align="center" class="header">学号:</td>
        <td align="center" class="data"><input name="id"></td>
    </tr>
    <tr height="30">
        <td align="center" class="header">姓名:</td>
        <td align="center" class="data"><input name="name"></td>
    </tr>
    <tr height="30">
        <td align="center" class="header">年龄:</td>
        <td align="center" class="data"><input name="age"></td>
    </tr>
    <tr height="30">
        <td align="center" class="header">性别:</td>
        <td align="center" class="data"><select name="sex">
            <option value="1">男</option>
            <option value="0">女</option>
        </select></td>
    </tr>
    <tr height="30">
        <td align="center" class="header">专业:</td>
        <td align="center" class="data"><input name="major"></td>
    </tr>
    <tr height="30">
        <td align="center" class="header">学院:</td>
        <td align="center" class="data"><input name="college"></td>
    </tr>
    <tr height="30">
        <td align="center" class="header">简介:</td>
        <td align="left" class="data"><textarea rows="10" cols="50"
            name="introduction"></textarea></td>
    </tr>
</table>
<input type="submit" value="提交"></form>
</div>
</body>
</html>
```

updateStudentForm.jsp 是修改学生信息的表单页面(页面效果见图 15-7),用户可以在该界面上修改学生信息,然后单击“提交”按键,将数据提交给 updateStudent.do,由 StudentServlet 进行处理。例 15.6 是 updateStudentForm.jsp 的代码,页面上使用 JSTL 从 ${student}中获取用户数据。

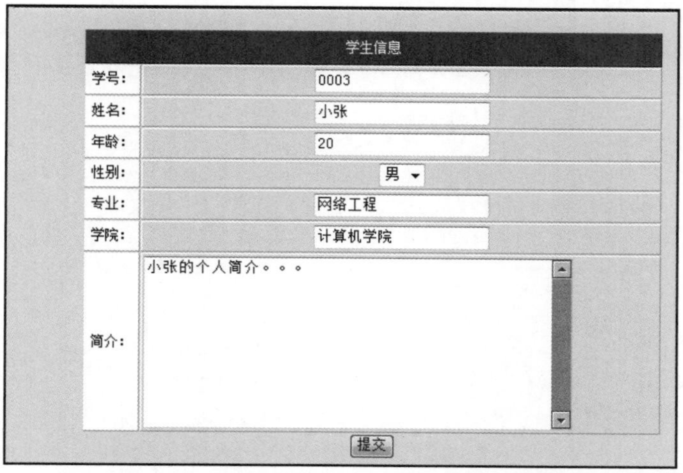

图 15-7　updateStudentForm.jsp 的显示结果

【例 15.6】　updateStudentForm.jsp。

```
<%@ page language="java" contentType="text/html; charset=UTF-8"
    pageEncoding="UTF-8"%>
<%@ taglib prefix="c" uri="http://java.sun.com/jsp/jstl/core"%>
<html>
<head>
<link href="style.css" rel="stylesheet" type="text/css">
<title>查看学生信息</title>
</head>
<body>
<div align="center">
<form action="updateStudent.do" method="post">
<table width="500" cellpadding="1" cellspacing="1">
    <tr>
        <td colspan="9" align="center" class="title" height="30">学生信息</td>
    </tr>
    <tr height="30">
        <td align="center" class="header">学号:</td>
        <td align="center" class="data"><input name="id"
            value="${student.id }" readonly></td>
    </tr>
    <tr height="30">
        <td align="center" class="header">姓名:</td>
        <td align="center" class="data"><input name="name"
            value="${student.name }"></td>
    </tr>
    <tr height="30">
        <td align="center" class="header">年龄:</td>
        <td align="center" class="data"><input name="age"
            value="${student.age }"></td>
    </tr>
    <tr height="30">
        <td align="center" class="header">性别:</td>
        <td align="center" class="data">
```

```
            <select name="sex">
                <c:if test="${student.sex==1}">
                    <option value="1" selected="selected">男</option>
                    <option value="0">女</option>
                </c:if>
                <c:if test="${student.sex!=1}">
                    <option value="1" >男</option>
                    <option value="0" selected="selected">女</option>
                </c:if>
            </select>
        </td>
    </tr>
    <tr height="30">
        <td align="center" class="header">专业:</td>
        <td align="center" class="data"><input name="major"
            value="${student.major }"></td>
    </tr>
    <tr height="30">
        <td align="center" class="header">学院:</td>
        <td align="center" class="data"><input name="college"
            value="${student.college }"></td>
    </tr>
    <tr height="30">
        <td align="center" class="header">简介:</td>
        <td align="left" class="data"><textarea rows="10" cols="50"
            name="introduction">${student.introduction }</textarea></td>
    </tr>
</table>
<input type="submit" value="提交"></form>
</div>
</body>
</html>
```

viewStudent.jsp 界面上显示了学生的详细信息(页面效果见图 15-8)。例 15.7 是 viewStudent.jsp 的代码。viewStudent.jsp 通过 JSTL 从 ${student} 中获取学生的数据,将该学生的信息以表格的形式输出。

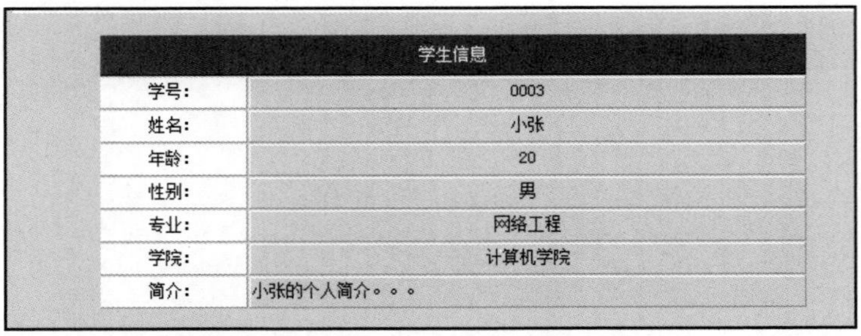

学生信息	
学号:	0003
姓名:	小张
年龄:	20
性别:	男
专业:	网络工程
学院:	计算机学院
简介:	小张的个人简介。。。

图 15-8 viewStudent.jsp 的显示效果

【例 15.7】 viewStudent.jsp。

```
<%@ page language="java" contentType="text/html; charset=UTF-8"
    pageEncoding="UTF-8"%>
<%@ taglib prefix="c" uri="http://java.sun.com/jsp/jstl/core"%>
```

```
<html>
<head>
<link href="style.css" rel="stylesheet" type="text/css">
<title>查看学生信息</title>
</head>
<body>
<div align="center">
<table width="500" cellpadding="1" cellspacing="1">
    <tr>
        <td colspan="9" align="center" class="title" height="30">学生信息</td>
    </tr>
    <tr height="30">
        <td align="center" class="header" width="100">学号:</td>
        <td align="center" class="data">${student.id }</td>
    </tr>
    <tr height="30">
        <td align="center" class="header" width="100">姓名:</td>
        <td align="center" class="data">${student.name }</td>
    </tr>
    <tr height="30">
        <td align="center" class="header" width="100">年龄:</td>
        <td align="center" class="data">${student.age }</td>
    </tr>
    <tr height="30">
        <td align="center" class="header" width="100">性别:</td>
        <td align="center" class="data">
            <c:if test="${student.sex==1 }">
            男
            </c:if>
            <c:if test="${student.sex!=1 }">
            女
            </c:if>
        </td>
    </tr>
    <tr height="30">
        <td align="center" class="header" width="100">专业:</td>
        <td align="center" class="data">${student.major }</td>
    </tr>
    <tr height="30">
        <td align="center" class="header" width="100">学院:</td>
        <td align="center" class="data">${student.college }</td>
    </tr>
    <tr height="30">
        <td align="center" class="header" width="100">简介:</td>
        <td align="center" class="data">${student.introduction }</td>
    </tr>
</table>
</div>
</body>
</html>
```

15.1.6 控制器

看过了模型和视图两部分，读者一定对模型什么时候调用、视图中的 ${students} 和

${student}等数据从哪里来,感到非常困惑。其实这一切的工作都是在控制器(Controller) servlet.StudentServlet 中完成的,例 15.8 给出了控制器的完整代码。

【例 15.8】 StudentServlet.java。

```
package servlet;
import java.io.IOException;
import java.util.ArrayList;
import dao.StudentDao;
import domain.Student;
import jakarta.servlet.ServletException;
import jakarta.servlet.annotation.WebServlet;
import jakarta.servlet.http.HttpServlet;
import jakarta.servlet.http.HttpServletRequest;
import jakarta.servlet.http.HttpServletResponse;
@WebServlet("*.do")
public class StudentServlet extends HttpServlet {
    private static final long serialVersionUID = 1L;
    //处理 GET 请求
    protected void doGet(HttpServletRequest request, HttpServletResponse response)
            throws ServletException, IOException {
        //让 GET 和 POST 请求以相同的逻辑进行处理
        doPost(request, response);
    }
    //处理 POST 请求
    protected void doPost(HttpServletRequest request, HttpServletResponse response)
            throws ServletException, IOException {
        //设置编码为 UTF-8,否则会出现中文乱码
        request.setCharacterEncoding("UTF-8");
        String requestURI=request.getRequestURI();
        if (requestURI.endsWith("/index.do")) {
            //处理对/index.do 的请求
            index(request, response);
        } else if (requestURI.endsWith("/deleteStudent.do")) {
            //处理对/deleteStudent.do 的请求
            deleteStudent(request, response);
        } else if (requestURI.endsWith("/addStudentForm.do")) {
            //处理对/addStudentForm.do 的请求
            addStudentForm(request, response);
        } else if (requestURI.endsWith("/addStudent.do")) {
            //处理对/addStudent.do 的请求
            addStudent(request, response);
        } else if (requestURI.endsWith("/updateStudentForm.do")) {
            //处理对/updateStudentForm.do 的请求
            updateStudentForm(request, response);
        } else if (requestURI.endsWith("/updateStudent.do")) {
            //处理对/updateStudent.do 的请求
            updateStudent(request, response);
        } else if (requestURI.endsWith("/viewStudent.do")) {
            //处理对/viewStudent.do 的请求
            viewStudent(request, response);
        }
    }
    //修改学生信息表单页
    private void updateStudentForm(HttpServletRequest request,
```

```
HttpServletResponse response)
            throws ServletException, IOException {
        String id = request.getParameter("id");
        Student student = StudentDao.getStudent(id);
        request.setAttribute("student", student);
        request.getRequestDispatcher("/WEB-INF/updateStudentForm.jsp").forward
(request, response);
    }
    //添加学生信息表单页
    private void addStudentForm(HttpServletRequest request, HttpServletResponse
response)
            throws ServletException, IOException {
        request.getRequestDispatcher("/WEB-INF/addStudentForm.jsp").forward
(request, response);
    }
    //首页
    private void index(HttpServletRequest request, HttpServletResponse response)
throws ServletException, IOException {
        ArrayList<Student> students = StudentDao.getAllStudents();
        request.setAttribute("students", students);
        request.getRequestDispatcher("/WEB-INF/index.jsp").forward(request,
response);
    }
    //删除学生信息
    private void deleteStudent(HttpServletRequest request, HttpServletResponse
response)
            throws ServletException, IOException {
        String id = request.getParameter("id");
        StudentDao.deleteStudent(id);
        response.sendRedirect("index.do");
    }
    //添加学生信息
    private void addStudent(HttpServletRequest request, HttpServletResponse
response) throws IOException {
        String id = request.getParameter("id");
        String name = request.getParameter("name");
        String age = request.getParameter("age");
        String sex = request.getParameter("sex");
        String major = request.getParameter("major");
        String college = request.getParameter("college");
        String introduction = request.getParameter("introduction");
        StudentDao.addStudent(id, name, Integer.parseInt(age), Integer.parseInt
(sex), major, college, introduction);
        response.sendRedirect("index.do");
    }
    //修改学生信息
    private void updateStudent(HttpServletRequest request, HttpServletResponse
response) throws IOException {
        String id = request.getParameter("id");
        String name = request.getParameter("name");
        String age = request.getParameter("age");
        String sex = request.getParameter("sex");
        String major = request.getParameter("major");
        String college = request.getParameter("college");
```

```
        String introduction = request.getParameter("introduction");
        StudentDao.updateStudent(id, name, Integer.parseInt(age), Integer.
parseInt(sex), major, college, introduction);
        response.sendRedirect("index.do");
    }
    //查看学生信息
    private void viewStudent (HttpServletRequest request, HttpServletResponse
response)
            throws ServletException, IOException {
        String id = request.getParameter("id");
        Student student = StudentDao.getStudent(id);
        request.setAttribute("student", student);
        request.getRequestDispatcher("/WEB-INF/viewStudent.jsp").forward
(request, response);
    }
}
```

StudentServlet 用@WebServlet 注解映射到“*.do”,这意味着所有发送到“.do”结尾 URL 地址的请求,都会被该 Servlet 处理。请读者重点关注 doPost()函数(doGet()函数调用了 doPost,因此 GET 和 POST 请求使用了相同的逻辑)。doPost()函数通过 RequestURI 对不同的 URL 请求进行分类处理,具体分类处理如下。

(1) /index.do: 由 index()函数处理,显示首页。

(2) /deleteStudent.do: 由 deleteStudent()函数处理,实现根据 id 删除学生的数据库操作。

(3) /addStudentForm.do: 由 addStudentForm()函数处理,显示添加学生的表单页面。

(4) /addStudent.do: 由 addStudent()函数处理,实现添加学生的数据库操作。

(5) /updateStudentForm.do: 由 updateStudentForm()函数处理,显示修改学生的表单页面。

(6) /updateStudent.do: 由 updateStudent()函数处理,实现修改学生的数据操作。

(7) /viewStudent.do: 由 viewStudent()函数处理,根据 id 显示学生信息。

上述操作按照对数据的操作方式可以分为以下两种类型。

(1) 读操作: 包括/index.do、/addStudentForm.do、/updateStudentForm.do、/viewStudent.do。

(2) 写操作: 包括/deleteStudent.do、/addStudent.do、/updateStudent.do。

对于读操作,这里选择/index.do 的处理流程详细讲解。index()函数的代码片段如下:

```
(1) ArrayList<Student> students = StudentDao.getAllStudents();
(2) request.setAttribute("students", students);
(3) request.getRequestDispatcher("/WEB-INF/index.jsp").forward(request, response);
```

其中,代码片段(1)调用 StudentDao 的 getAllStudents()函数获取全部的学生信息,得到 students。这就是控制器(Controller)对模型(Model)的调用。接着代码片段(2)将数据 students 放在 request 的属性里面。代码片段(3)的作用是视图选择,通过 forward 方式,把数据发送给视图/WEB-INF/index.jsp,并渲染该视图。读者可以将该过程与图 15-3 结合起来,可以看出 index()函数的处理流程,就是 MVC 模式处理浏览器请求的流程。

/WEB-INF/index.jsp 中的代码片段如下:

```
<c:forEach items="${students}" var="student">
    …
</c:forEach>
```

此处所引用的${students}就是通过 request.setAttribute("students", students)传递过来的。

下面再来分析写操作，这里选择/deleteStudent.do 的处理流程进行详细讲解。deleteStudent()函数的代码片段如下：

```
(1) String id = request.getParameter("id");
(2) StudentDao.deleteStudent(id);
(3) response.sendRedirect("index.do");
```

其中，代码片段(1)从 request 中获取浏览器传递过来的 id 参数。代码片段(2)调用模型(Model)StudentDao 的 deleteStudent()函数删除 id 对应的学生信息。代码片段(3)通过 sendRedirect 方式跳转视图(View)。deleteStudent 的处理流程与图 15-3 的流程一致。

受篇幅所限，其他函数的处理流程就不再一一赘述了。这些函数都简单易懂，读者可以自行分析。

有两点内容需要读者注意：

(1) 所有的 JSP 文件都放置在 WebContent/WEB-INF 目录下，例如 WebContent/WEB-INF/index.jsp。其原因在于，按照标准的 MVC 模式，任何视图之前都必须有一个控制器，即视图不允许浏览器直接访问。如果允许浏览器直接访问视图，那么访问模式就变成了"浏览器—视图—浏览器"，而不是"浏览器—控制器—视图—浏览器"。按照 Java Web 项目的规范 WEB-INF 目录下的文件不允许通过浏览器直接访问，因此本书将所有 JSP 文件都放在该目录下。

(2) forward 和 sendRedirect 是两种不同的视图跳转方式。forward 是服务器端的视图跳转。本书用 forward 方式，将处理流程跳转到 WEB-INF 下的 JSP 文件(该目录下的文件无法通过浏览器直接访问)。sendRedirect 则是浏览器的跳转。response.sendRedirect("index.do")会触发浏览器向服务端发送访问 index.do 的请求。forward 和 sendRedirect 最直观的区别在于，当使用 sendRedirect 时，浏览器的地址栏会发生变化。

15.2 基于 Spring Boot 的简单学生信息管理

15.2.1 Spring 和 Spring Boot 框架

通过 15.1 节的学习，读者们一定发现相对于 JSP Scriptlet 和 JSP Model 1，MVC 模式结构更为复杂，因此采用 MVC 模式开发 Web 程序非常烦琐。直观感受来说，实现相同的功能，需要写更多的代码。那么有没有既采用 MVC 模式，又能够减少代码量的方法呢？这样的方法是有的，那就是采用开源的 Web 开发框架。近年来应用最为广泛的框架就是 Spring。

Spring 框架(https://spring.io/projects/spring-framework/)是 2004 年 Rod Johnson 和 Juergen Hoeller 两人为核心的开发团队，以他们开发 Java EE 程序的经验为基础，所构建的一款开源软件框架。经过十多年的发展，Spring 框架已经包含了 Spring MVC、Spring Data、Spring Cloud、Spring Security 等诸多子项，被广泛应用于各类 Java EE 项目的开发。相对于其他软件框架，Spring 框架的突出特点在于：功能强大、功能多(支持所有 Java EE 应用场景),但使用简单,学习成本低。甚至没有多少编程经验的程序员,经过 1～2 周简单的培训,也能用 Spring 框架写出像模像样、没有太大错误的程序。

Spring Boot(https://spring.io/projects/spring-boot)是在 Spring 框架基础上发展起来的项目。在 Spring 框架的发展早期,一个新的 Spring 项目需要较多的初始设置。为了进一步简化初始搭建,Pivotal 团队在 Spring 框架基础上搭建了 Spring Boot 框架。Spring Boot 框架进一步降低了 Spring 框架的学习成本和开发成本,因而成了现代 Java Web 开发的主流。

接下来,本书将采用 Spring Boot 框架重写 15.1 节的学生信息管理程序。读者可以将本节的内容与 15.1 节进行对比,体会新技术如何简化了程序开发。虽然 Spring 框架功能非常多,但本书的案例并不涉及过多功能,并且 Spring Boot 框架简单易学,接下来的内容会比 15.1 节简单许多。如果说 Servlet+JSP 是手动挡汽车,那么 Spring Boot 就接近于自动驾驶。

15.2.2 Spring Boot 程序的基本结构

图 15-9 给出了 Spring Boot 程序的基本结构。这个程序结构与 15.1 节有较大差异,因为这里创建的不是 Dynamic Web Project,而是 Maven Project。在 Eclipse 创建项目时选择 Maven Project 就可以创建图 15-9 的程序结构。Maven 是一款开源的项目构建和管理工具。它是目前使用最广泛的 Java 项目构建和管理工具。限于篇幅,本书不对 Maven 进行详细介绍,读者可自行查找资料学习。本书仅仅会用到 Maven 的依赖管理功能。

Maven Project 的程序结构和 Dynamic Web Project 有很大不同,注意,不能随意更改这些目录,因为每一个目录都有约定的用途。下面给出各个目录及用途。

(1) src/main/java 目录:程序所需要的 Java 类。

(2) src/main/resources 目录:程序所使用的资源文件,例如 Spring Boot 的配置文件 application.yml。

(3) src/main/webapp 目录:程序所使用的视图资源,与 15.1 节类似,所有 JSP 文件放在 WEB-INF 目录下。

(4) src/test/java 和 src/test/resources 目录:程序测试所需要的 Java 类和资源文件,本书没有使用。

(5) pom.xml 文件:该文件是 Maven 的配置文件。

```
stm_springboot
  src/main/java
    stm
      controller
        StudentController.java
      dao
        StudentDao.java
      domain
        Student.java
      Application.java
  src/main/resources
    application.yml
  src/test/java
  src/test/resources
  JRE System Library [JavaSE-1.8]
  Maven Dependencies
  src
    main
      java
        stm
      resources
      webapp
        WEB-INF
          addStudentForm.jsp
          index.jsp
          updateStudentForm.jsp
          viewStudent.jsp
        style.css
    test
  target
  pom.xml
```

图 15-9 Spring Boot 程序的基本结构

15.2.3 pom.xml 文件

首先来介绍 pom.xml 文件,它是 Maven 的核心配置文件。本书关于 Maven 的部分也仅涉及 pom.xml 文件,其代码如例 15.9 所示。

【例 15.9】 pom.xml。

```
<project xmlns="http://maven.apache.org/POM/4.0.0" xmlns:xsi="http://www.w3.
org/2001/XMLSchema-instance" xsi:schemaLocation="http://maven.apache.org/POM/
4.0.0 https://maven.apache.org/xsd/maven-4.0.0.xsd">
    <modelVersion>4.0.0</modelVersion>
```

```
    <groupId>chapter15</groupId>
    <artifactId>stm_springboot</artifactId>
    <version>0.0.1-SNAPSHOT</version>
    <packaging>jar</packaging>
    <!-- 指定 springboot 的版本为 2.2.8.RELEASE-->
    <parent>
        <groupId>org.springframework.boot</groupId>
        <artifactId>spring-boot-starter-parent</artifactId>
        <version>2.2.8.RELEASE</version>
    </parent>
    <dependencies>
        <!-- springboot 的依赖 -->
        <dependency>
            <groupId>org.springframework.boot</groupId>
            <artifactId>spring-boot-starter</artifactId>
        </dependency>
        <!-- springboot web 的依赖 -->
        <dependency>
            <groupId>org.springframework.boot</groupId>
            <artifactId>spring-boot-starter-web</artifactId>
        </dependency>
        <!-- springboot jpa 的依赖 -->
        <dependency>
            <groupId>org.springframework.boot</groupId>
            <artifactId>spring-boot-starter-data-jpa</artifactId>
        </dependency>
        <!-- mysql 数据库连接的依赖 -->
        <dependency>
            <groupId>mysql</groupId>
            <artifactId>mysql-connector-java</artifactId>
        </dependency>
        <!-- 使用 JSP 作为视图 -->
        <dependency>
            <groupId>org.apache.tomcat.embed</groupId>
            <artifactId>tomcat-embed-jasper</artifactId>
        </dependency>
        <!-- JSP 中使用 jstl -->
        <dependency>
            <groupId>javax.servlet</groupId>
            <artifactId>jstl</artifactId>
        </dependency>
    </dependencies>
</project>
```

初看起来该文件非常复杂，读者可能认为要学习 Spring Boot 就必须先学会 Maven。其实这是误解。因为该文件是可以多次复用的，简言之，所有 Spring Boot 项目的 pom.xml 文件都与例 15.9 类似，多数项目仅仅需要对上述文件做小修改。该文件主要分成以下 3 部分。

（1）第 1 部分是与项目相关的设定。代码如下：

```
<modelVersion>4.0.0</modelVersion>
<groupId>chapter15</groupId>
<artifactId>stm_springboot</artifactId>
<version>0.0.1-SNAPSHOT</version>
<packaging>jar</packaging>
```

groupId 和 artifactId 用于唯一标识项目的名称。用 Eclipse 创建 Maven 项目时,groupId 和 artifactId 是必填项。不同的项目必须修改的部分是 groupId 和 artifactId。

version 指定打包时的软件版本,packaging 指定打包方式。当程序完成,打包发布时,根据 artifactId、version 和 packaging 就会生成一个名称为 stm_springboot-0.0.1-SNAPSHOT.jar 的发布程序。

(2) 第 2 部分涉及所使用的 Spring Boot 版本。代码如下:

```
<parent>
    <groupId>org.springframework.boot</groupId>
    <artifactId>spring-boot-starter-parent</artifactId>
    <version>2.2.8.RELEASE</version>
</parent>
```

此部分唯一可能修改的部分是 version。version 指定了 Spring Boot 的版本。本书使用 2.2.8.RELEASE。若新项目继续使用该版本,则无须修改。

(3) 第 3 部分是软件包的依赖。代码如下:

```
<dependencies>
    <!-- springboot 的依赖 -->
    <dependency>
        <groupId>org.springframework.boot</groupId>
        <artifactId>spring-boot-starter</artifactId>
    </dependency>
    …
    <!-- mysql 数据库连接的依赖 -->
    <dependency>
        <groupId>mysql</groupId>
        <artifactId>mysql-connector-java</artifactId>
    </dependency>
    …
</dependencies>
```

该部分是一系列的 dependency,每一个 dependency 指定了项目所依赖的一个软件包。此部分是 Maven 最强大的功能。当开发程序时,免不了要使用各种第三方库。例如 15.1 节的程序就使用了 3 个第三方库。这些库要从网上下载,具体到哪个网站上去下载呢? 如果一个项目需要几十个第三方库,难道需要到几十个网站上去下载吗? 使用 Maven 就可以很好地解决这个问题。开发者只要在 pom.xml 文件中声明一个 dependency,Maven 就会自动到 Maven 仓库中下载该软件包(软件包由第三方开发者提前传到 Maven 仓库)。本例中所需要的软件包有几十个,通过在 pom.xml 中的 dependency 声明,这些软件包都会自动下载。可以在项目的 Maven Dependencies 中查看到这些软件包,如图 15-10 所示。

如果要创建自己的 Spring Boot 项目,可以复用本书的 pom.xml 文件(也可复制整个项目)。复制本书的 pom.xml 文件,然后根据需要进行修改。大多数情况下,需要修改第 1 部分的 groupId 和 artifactId,第 2 部分不需要修改(除非使用其他版本的 Spring Boot),第 3 部分根据需要进行增减。第 3 部分的 dependency 可以在 Maven 提供的官方查询网站(https://mvnrepository.com)查找,也可以在搜索引擎上搜索。

15.2.4 application.yml 文件

application.yml 是 Spring Boot 的配置文件,其代码如例 15.10 所示。

Maven Dependencies
spring-boot-starter-2.2.8.RELEASE.jar - /Users/liuji/.m2/repository/org/s
spring-boot-2.2.8.RELEASE.jar - /Users/liuji/.m2/repository/org/springfra
spring-context-5.2.7.RELEASE.jar - /Users/liuji/.m2/repository/org/spring
spring-boot-autoconfigure-2.2.8.RELEASE.jar - /Users/liuji/.m2/repositor
spring-boot-starter-logging-2.2.8.RELEASE.jar - /Users/liuji/.m2/reposit
logback-classic-1.2.3.jar - /Users/liuji/.m2/repository/ch/qos/logback/log
logback-core-1.2.3.jar - /Users/liuji/.m2/repository/ch/qos/logback/logba
log4j-to-slf4j-2.12.1.jar - /Users/liuji/.m2/repository/org/apache/logging/l
log4j-api-2.12.1.jar - /Users/liuji/.m2/repository/org/apache/logging/log4j
jul-to-slf4j-1.7.30.jar - /Users/liuji/.m2/repository/org/slf4j/jul-to-slf4j/1.7.
jakarta.annotation-api-1.3.5.jar - /Users/liuji/.m2/repository/jakarta/annot
spring-core-5.2.7.RELEASE.jar - /Users/liuji/.m2/repository/org/springfra
spring-jcl-5.2.7.RELEASE.jar - /Users/liuji/.m2/repository/org/springframe
snakeyaml-1.25.jar - /Users/liuji/.m2/repository/org/yaml/snakeyaml/1.25
spring-boot-starter-web-2.2.8.RELEASE.jar - /Users/liuji/.m2/repository/
spring-boot-starter-json-2.2.8.RELEASE.jar - /Users/liuji/.m2/repository/
jackson-databind-2.10.4.jar - /Users/liuji/.m2/repository/com/fasterxml/ja
jackson-annotations-2.10.4.jar - /Users/liuji/.m2/repository/com/fasterxm
jackson-core-2.10.4.jar - /Users/liuji/.m2/repository/com/fasterxml/jacks
jackson-datatype-jdk8-2.10.4.jar - /Users/liuji/.m2/repository/com/faster
jackson-datatype-jsr310-2.10.4.jar - /Users/liuji/.m2/repository/com/fast
jackson-module-parameter-names-2.10.4.jar - /Users/liuji/.m2/repository
spring-boot-starter-tomcat-2.2.8.RELEASE.jar - /Users/liuji/.m2/reposito
tomcat-embed-websocket-9.0.36.jar - /Users/liuji/.m2/repository/org/ap
spring-boot-starter-validation-2.2.8.RELEASE.jar - /Users/liuji/.m2/repos
jakarta.validation-api-2.0.2.jar - /Users/liuji/.m2/repository/jakarta/validat
hibernate-validator-6.0.20.Final.jar - /Users/liuji/.m2/repository/org/hiber
spring-web-5.2.7.RELEASE.jar - /Users/liuji/.m2/repository/org/springfrar
spring-beans-5.2.7.RELEASE.jar - /Users/liuji/.m2/repository/org/springfr
spring-webmvc-5.2.7.RELEASE.jar - /Users/liuji/.m2/repository/org/spring
spring-aop-5.2.7.RELEASE.jar - /Users/liuji/.m2/repository/org/springfran
spring-expression-5.2.7.RELEASE.jar - /Users/liuji/.m2/repository/org/spr
spring-boot-starter-data-jpa-2.2.8.RELEASE.jar - /Users/liuji/.m2/reposit
spring-boot-starter-aop-2.2.8.RELEASE.jar - /Users/liuji/.m2/repository/
aspectjweaver-1.9.5.jar - /Users/liuji/.m2/repository/org/aspectj/aspectjw
spring-boot-starter-jdbc-2.2.8.RELEASE.jar - /Users/liuji/.m2/repository/
HikariCP-3.4.5.jar - /Users/liuji/.m2/repository/com/zaxxer/HikariCP/3.4.5
spring-jdbc-5.2.7.RELEASE.jar - /Users/liuji/.m2/repository/org/springfra
jakarta.persistence-api-2.2.3.jar - /Users/liuji/.m2/repository/jakarta/pers
jakarta.transaction-api-1.3.3.jar - /Users/liuji/.m2/repository/jakarta/trans

图 15-10 Maven 自动下载的软件包

【例 15.10】 application.yml。

```
server:
  port: 8080                                   #设置服务的端口
  servlet:
    context-path: /stm                         #设置 web 程序的 context path
spring:
  #设置数据库连接
  datasource:
    driver-class-name: com.mysql.cj.jdbc.Driver
    url: jdbc:mysql://localhost:3306/stm?useUnicode=true&characterEncoding=utf8
    username: root
    password: 123456
```

本例只进行了服务器和数据库的设置(读者根据自己数据库的情况做简单修改)。其他设置均采用 Spring Boot 的默认设置(大多数情况默认设置已经足够)。如果要修改其他设置，可以参考 Spring Boot 官方网站给出的设置参考。

15.2.5 模型

程序的模型由 domain 包和 dao 包中的 Java 类构成。

例 15.11 是 domain.Student 类的代码，代码与例 15.1 几乎相同。不同之处在于，Student 的类名前增加@Entity 注解，而在成员变量 id 之前增加了@Id 注解。这两个注解是 JPA (Java Persistence API)中的注解。JPA 是一套 Java 持久层 API。简单地说，加上了这两个注解就可以让 Student 对象与数据库中的 Student 表映射起来,不需要写一行 SQL 语句。

【例 15.11】 Student.java。

```
package stm.domain;
import javax.persistence.Entity;
import javax.persistence.Id;
@Entity
public class Student {
@Id
    private String id;                              //学号
    private String name;                            //姓名
    private int age;                                //年龄
    private int sex;                                //性别
    private String major;                           //专业
    private String college;                         //学院
    private String introduction;                    //简介
    public String getId() {
        return id;
    }
    public void setId(String id) {
        this.id = id;
    }
    public String getName() {
        return name;
    }
    public void setName(String name) {
        this.name = name;
    }
    public int getAge() {
        return age;
    }
    public void setAge(int age) {
        this.age = age;
    }
    public int getSex() {
        return sex;
    }
    public void setSex(int sex) {
        this.sex = sex;
    }
    public String getMajor() {
        return major;
    }
    public void setMajor(String major) {
        this.major = major;
    }
    public String getCollege() {
        return college;
    }
    public void setCollege(String college) {
        this.college = college;
    }
    public String getIntroduction() {
        return introduction;
```

```
    }
    public void setIntroduction(String introduction) {
        this.introduction = introduction;
    }
}
```

dao.StudentDao 类是数据访问对象，实现了对 student 表的 CRUD 操作，例 15.12 是 StudentDao 的代码，就是数据库的全部代码。该代码的核心是@Repository 注解和继承的 CrudRepository 接口。数据访问代码会动态生成。相对于例 15.2 的代码，基于 Spring Boot 的 StudentDao 有极大的简化，不懂数据库操作的程序员也可顺利地实现数据表的 CRUD 操作。

【例 15.12】 StudentDao.java。

```
package stm.dao;
import org.springframework.data.repository.CrudRepository;
import org.springframework.stereotype.Repository;
import stm.domain.Student;
@Repository
public interface StudentDao extends CrudRepository<Student, String> {}
```

15.2.6 视图

本例采用的视图 JSP 文件和 CSS 文件与 15.1.5 节完全相同，在此不再赘述。

15.2.7 控制器

控制器是联系模型和视图的核心。例 15.13 给出了控制器 controller.StudentController 的代码。StudentController 的代码，相对于例 15.8 的 StudentServlet 有极大的简化。URL 请求和处理函数的对应关系与 StudentServlet 相同。

（1）/index.do：由 index()函数处理，显示首页。

（2）/deleteStudent.do：由 deleteStudent()函数处理，实现根据 id 删除学生的数据库操作。

（3）/addStudentForm.do：由 addStudentForm()函数处理，显示添加学生的表单页面。

（4）/addStudent.do：由 addStudent()函数处理，实现添加学生的数据库操作。

（5）/updateStudentForm.do：由 updateStudentForm()函数处理，显示修改学生的表单页面。

（6）/updateStudent.do：由 updateStudent()函数处理，实现修改学生的数据操作。

（7）/viewStudent.do：由 viewStudent()函数处理，根据 id 显示学生信息。

函数名前面的@RequestMapping 注解，用于将 URL 请求地址对应到处理函数。相对于 StudentServlet，不再需要用 if 语句判断请求地址。StudentController 类名前需要用@Controller 注解，如此 StudentController 才能被 Spring Boot 框架管理。成员变量 studentDao 前面用@Autowired 注解，表明该对象的创建和销毁完全由 Spring Boot 框架自动完成(包含数据库的映射、数据库连接的获取和关闭等)。

【例 15.13】 StudentController.java。

```
package stm.controller;
import org.springframework.beans.factory.annotation.Autowired;
```

```
import org.springframework.stereotype.Controller;
import org.springframework.web.bind.annotation.RequestMapping;
import org.springframework.web.servlet.ModelAndView;
import stm.dao.StudentDao;
import stm.domain.Student;
@Controller
public class StudentController {
    @Autowired
    private StudentDao studentDao;
    //首页
    @RequestMapping("/index.do")
    public ModelAndView index() {
        ModelAndView mv = new ModelAndView();
        Iterable<Student> students = studentDao.findAll();
        mv.addObject("students", students);
        mv.setViewName("/WEB-INF/index.jsp");
        return mv;
    }
    //添加学生信息表单页
    @RequestMapping("/addStudentForm.do")
    public String addStudentForm() {
        return "/WEB-INF/addStudentForm.jsp";
    }
    //修改学生信息表单页
    @RequestMapping("/updateStudentForm.do")
    public ModelAndView updateStudentForm(String id) {
        ModelAndView mv = new ModelAndView();
        Student student = studentDao.findById(id).get();
        mv.addObject("student", student);
        mv.setViewName("/WEB-INF/updateStudentForm.jsp");
        return mv;
    }
    //查看学生信息
    @RequestMapping("/viewStudent.do")
    public ModelAndView viewStudent(String id) {
        ModelAndView mv = new ModelAndView();
        Student student = studentDao.findById(id).get();
        mv.addObject("student", student);
        mv.setViewName("/WEB-INF/viewStudent.jsp");
        return mv;
    }
    //添加学生信息
    @RequestMapping("/addStudent.do")
    public String addStudent(Student student) {
        studentDao.save(student);
        return "redirect:/index.do";
    }
    //修改学生信息
    @RequestMapping("/updateStudent.do")
    public String updateStudent(Student student) {
        studentDao.save(student);
        return "redirect:/index.do";
```

```
    }
    //删除学生信息
    @RequestMapping("/deleteStudent.do")
    public String deleteStudent(String id) {
        studentDao.deleteById(id);
        return "redirect:/index.do";
    }
}
```

与15.1.6节类似,本节选择index()和deleteStudent()两个函数来分析控制器的执行流程。对应index函数其代码如下:

```
(1)ModelAndView mv = new ModelAndView();
(2)Iterable<Student> students = studentDao.findAll();
(3)mv.addObject("students", students);
(4)mv.setViewName("/WEB-INF/index.jsp");
(5)return mv;
```

代码(1)创建一个ModelAndView对象mv,类名清楚表明了mv对象的作用。代码(2)调用studentDao的findAll()函数获取全部学生,findAll()函数由Spring Boot自动实现。代码(3)将数据放在mv对象里面。代码(4)选择视图为/WEB-INF/index.jsp。代码(5)返回模型和视图,由/WEB-INF/index.jsp渲染生成页面。

deleteStudent函数的代码如下:

```
(1)studentDao.deleteById(id);
(2)return "redirect:/index.do";
```

代码(1)调用studentDao的deleteById函数删除id对应的学生,id由Spring Boot框架自动从request中获取,deleteById由Spring Boot框架自动实现。代码(2)返回视图,redirect相当于例15.8中的sendRedirect。

从上述分析可以看出,采用Spring Boot,实现相同的逻辑,所需要的代码数量远远少于Servlet方式。由于框架自动实现了很多功能,因此开发过程被极大地简化了。

15.2.8 运行项目

最后来看看Application.java的代码,如例15.14所示。

【例15.14】 Application.java。

```
package stm;
import org.springframework.boot.SpringApplication;
import org.springframework.boot.autoconfigure.SpringBootApplication;
@SpringBootApplication
public class Application {
    public static void main(String[] args) {
        SpringApplication.run(Application.class, args);
    }
}
```

该文件是Spring Boot项目的启动类,像普通Java程序一样,执行该类的main()函数,就可以启动Web项目。通常情况下,Application类无须做任何修改。运行Application,在eclipse的控制台下看到如图15-11所示的信息,就表明项目启动成功了。此后打开浏览器,输入 http://localhost:8080/stm/index.do 就可以看到与 15.1 节相同的程序效果。

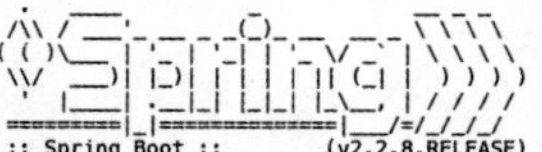


```
2022-07-31 18:19:36.668  INFO 85790 --- [           main] stm.Application                          : Starting Application on LiuJi-MBP-2018 with PID 85790 (/Users/liu
2022-07-31 18:19:36.671  INFO 85790 --- [           main] stm.Application                          : No active profile set, falling back to default profiles: default
2022-07-31 18:19:37.371  INFO 85790 --- [           main] .s.d.r.c.RepositoryConfigurationDelegate : Bootstrapping Spring Data JPA repositories in DEFAULT mode.
2022-07-31 18:19:37.456  INFO 85790 --- [           main] .s.d.r.c.RepositoryConfigurationDelegate : Finished Spring Data repository scanning in 71ms. Found 1 JPA rep
2022-07-31 18:19:38.278  INFO 85790 --- [           main] o.s.b.w.embedded.tomcat.TomcatWebServer  : Tomcat initialized with port(s): 8080 (http)
2022-07-31 18:19:38.298  INFO 85790 --- [           main] o.apache.catalina.core.StandardService   : Starting service [Tomcat]
2022-07-31 18:19:38.299  INFO 85790 --- [           main] org.apache.catalina.core.StandardEngine  : Starting Servlet engine: [Apache Tomcat/9.0.36]
2022-07-31 18:19:38.708  INFO 85790 --- [           main] org.apache.jasper.servlet.TldScanner     : At least one JAR was scanned for TLDs yet contained no TLDs. Enab
2022-07-31 18:19:38.716  INFO 85790 --- [           main] o.a.c.c.C.[Tomcat].[localhost].[/stm]    : Initializing Spring embedded WebApplicationContext
2022-07-31 18:19:38.717  INFO 85790 --- [           main] w.s.c.ServletWebServerApplicationContext : Root WebApplicationContext: initialization completed in 2005 ms
2022-07-31 18:19:38.964  INFO 85790 --- [           main] com.zaxxer.hikari.HikariDataSource       : HikariPool-1 - Starting...
2022-07-31 18:19:39.740  INFO 85790 --- [           main] com.zaxxer.hikari.HikariDataSource       : HikariPool-1 - Start completed.
2022-07-31 18:19:39.792  INFO 85790 --- [           main] o.hibernate.jpa.internal.util.LogHelper  : HHH000204: Processing PersistenceUnitInfo [name: default]
2022-07-31 18:19:39.870  INFO 85790 --- [           main] org.hibernate.Version                    : HHH000412: Hibernate ORM core version 5.4.17.Final
2022-07-31 18:19:40.117  INFO 85790 --- [           main] o.hibernate.annotations.common.Version   : HCANN000001: Hibernate Commons Annotations {5.1.0.Final}
2022-07-31 18:19:40.344  INFO 85790 --- [           main] org.hibernate.dialect.Dialect            : HHH000400: Using dialect: org.hibernate.dialect.MySQL57Dialect
2022-07-31 18:19:41.077  INFO 85790 --- [           main] o.h.e.t.j.p.i.JtaPlatformInitiator       : HHH000490: Using JtaPlatform implementation: [org.hibernate.engin
2022-07-31 18:19:41.085  INFO 85790 --- [           main] j.LocalContainerEntityManagerFactoryBean : Initialized JPA EntityManagerFactory for persistence unit 'defaul
2022-07-31 18:19:41.353  WARN 85790 --- [           main] JpaBaseConfiguration$JpaWebConfiguration : spring.jpa.open-in-view is enabled by default. Therefore, databas
2022-07-31 18:19:41.482  INFO 85790 --- [           main] o.s.s.concurrent.ThreadPoolTaskExecutor  : Initializing ExecutorService 'applicationTaskExecutor'
2022-07-31 18:19:41.829  INFO 85790 --- [           main] o.s.b.w.embedded.tomcat.TomcatWebServer  : Tomcat started on port(s): 8080 (http) with context path '/stm'
2022-07-31 18:19:41.831  INFO 85790 --- [           main] stm.Application                          : Started Application in 5.637 seconds (JVM running for 6.15)
```

图 15-11　启动 Spring Boot 项目